高水平专业群建设成果活页式系列双语教材

网络工程实践（活页式）

Practice of Network Engineering

主　编　钟文基　宁爱民　苗志锋
副主编　吴丽萍　陆　腾　石　巍
　　　　黎明明　邓丽萍　莫年发
　　　　劳　薇　陈启龙

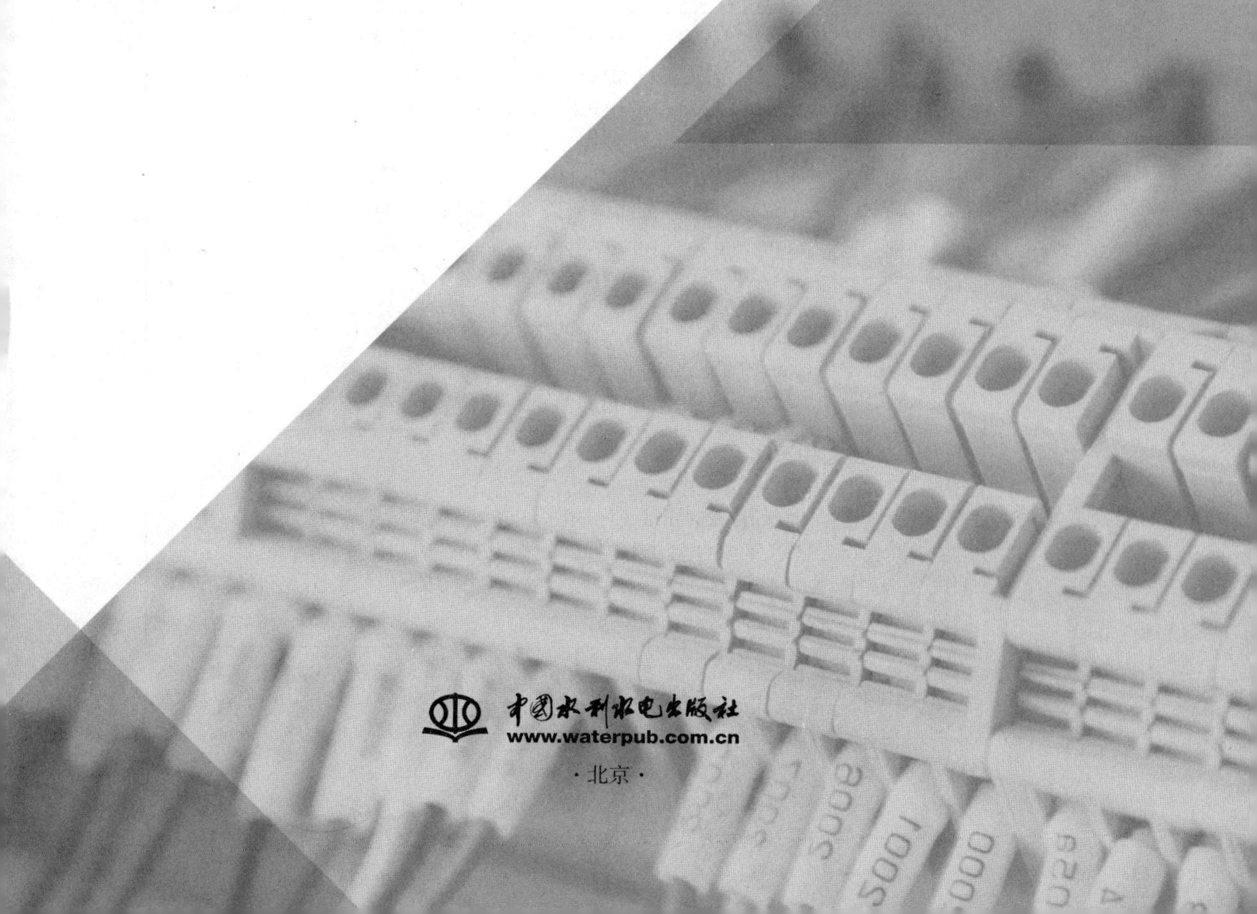

中国水利水电出版社
www.waterpub.com.cn
·北京·

内 容 提 要

本书完整叙述了数据通信与IP网络技术的基础知识及其相关内容，融合了路由技术、交换技术、防火墙技术和计算机网络的知识，全书以网络工程项目为主线，从eNSP的使用，小型局域网、中型局域网、大型园区网的构建，到网络的安全防护。由浅入深，由易到难，全面介绍了路由和交换技术的主要内容，完整地展示交换机、路由器、防火墙等网络设备在企业网中的应用。通过本书的学习，结合实验训练，读者可以对数据通信和IP网络技术有一个全面的了解，同时掌握交换机、路由器和防火墙等网络通信设备的配置与管理技能。

本书可作为高职高专计算机网络技术、现代通信技术、物联网技术等相关专业的课程教材，也适合对计算机网络技术感兴趣的相关技术人员使用。

图书在版编目（CIP）数据

网络工程实践 ： 活页式 ： 汉文、英文 / 钟文基，宁爱民，苗志锋主编. -- 北京 ： 中国水利水电出版社，2023.5

高水平专业群建设成果活页式系列双语教材

ISBN 978-7-5226-1520-2

Ⅰ. ①网… Ⅱ. ①钟… ②宁… ③苗… Ⅲ. ①计算机网络－教材－汉、英 Ⅳ. ①TP393

中国国家版本馆CIP数据核字(2023)第082156号

书　　名	高水平专业群建设成果活页式系列双语教材 **网络工程实践（活页式）** WANGLUO GONGCHENG SHIJIAN (HUOYESHI)
作　　者	主编　钟文基　宁爱民　苗志锋
出版发行	中国水利水电出版社 （北京市海淀区玉渊潭南路1号D座　100038） 网址：www.waterpub.com.cn E-mail：sales@mwr.gov.cn 电话：(010) 68545888（营销中心）
经　　售	北京科水图书销售有限公司 电话：(010) 68545874、63202643 全国各地新华书店和相关出版物销售网点
排　　版	中国水利水电出版社微机排版中心
印　　刷	北京密东印刷有限公司
规　　格	184mm×260mm　16开本　27印张　678千字
版　　次	2023年5月第1版　2023年5月第1次印刷
印　　数	0001—1500册
定　　价	**79.00**元

Summary

This book describes the basic knowledge of data communication and IP network technology and its related contents, and integrates the knowledge of routing technology, exchange technology, firewall technology and computer network. The whole book takes the network engineering project as the main line, from the use of eNSP, the construction of small local area network, medium local area network, large campus network, to the safety protection of the network. From the elementary to the profound, from the easy to the difficult, the main content of routing and exchange technology is introduced, and the application of network equipment such as switch, router and firewall in enterprise network is displayed completely. Learning through this book, combined with experimental training, readers can have a comprehensive understanding of data communication and IP network technology, while grasping the switch, router and firewall network communication equipment configuration and management skills.

This book can be used as a curriculum material for computer network technology, modern communication technology, Internet of things technology and other related specialties in higher vocational colleges, and is also suitable for relevant technicians interested in computer network technology.

前　言

随着计算机技术和网络技术的迅速发展和日益普及，计算机网络已经成为人们生活的一个重要组成部分，培养大批熟练掌握网络技术的高端技能型人才是当前社会发展的迫切需求。路由和交换技术是实践性非常强的课程，要想掌握网络设备的配置及应用技术，必须在学习一定理论知识的基础上，以实际工程项目为依托，做到理论结合实践，方能取得理想的学习效果。

本教材编写的目的是改革传统教材以理论知识传播为主的模式，采用项目式编写和教学，全文共10个工程项目，对每个项目都有具体的理论知识讲解和实践操作步骤演示，以真实的案例来阐释不同行业企业网络的组建和应用，以便能更好地满足高等职业院校以就业为导向的技能型人才培养的需求。

教材的编写从四个方面出发：一是校企“双元”合作开发。教材由具有丰富教学经验的教师和企业技术支持一线工程师共同编写，所有案例均源于企业真实项目，反映最新的ICT技术成果。二是“新形态一体化”教材建设，实现教学资源共建共享。发挥“互联网＋教材”的优势，配套精品在线开放课程、自动评分系统等数字资源，便于学生即时学习和个性化学习，有助于教师创新教学模式。三是“工作手册式”设计，助力“三教”改革落地。通过工作任务和职业能力分析，把网络工程实践中的各项能力要求落实在具体的步骤中，相当于一本详细的工作手册。四是“工匠精神”融入，推动“课程思政”进课堂。融入企业文化、团队精神、劳动教育等内容。

本教材作为广西高水平院校和专业群建设的成果，校企合作“双元”开发。教材由钟文基、宁爱民、苗志锋担任主编，由吴丽萍、陆腾、石巍、黎明明、邓丽萍、莫年发、劳薇、陈启龙担任副主编。其中，宁爱民、黎明明负责项目1、项目2的编写，苗志锋、邓丽萍负责项目3、项目4的编写，陆腾、石巍负责项目5、项目6的编写，莫年发、劳薇、陈启龙负责项目7、项目8的编写，钟文基、吴丽萍负责项目9、项目10的编写。

本书可作为高等职业院校计算机网络技术、现代通信技术、物联网技术等相关专业的课程教材，也适合对计算机网络技术感兴趣的相关技术人员使用。

由于计算机网络技术发展迅速，加之编者水平有限，教材中难免有疏漏和不妥之处，恳请广大读者和专家批评指正，编者电子邮箱：396860606@qq.com。

编者

2022 年 12 月

Preface

With the rapid development and increasing popularity of computer technology and network technology, computer networks have become an important component of people's lives. Cultivating a large number of high-end skilled talents proficient in network technology is an urgent need for current social development. Routing and switching technology is a highly practical course. To master the configuration and application technology of network equipment, it is necessary to learn certain theoretical knowledge, rely on actual engineering projects, and achieve the combination of theory and practice in order to achieve ideal learning results.

The purpose of writing this textbook is to reform the traditional mode of theoretical knowledge dissemination in textbooks, adopting a project-based writing and teaching approach. The entire text consists of 10 engineering projects, each with specific theoretical knowledge explanations and practical operation steps demonstrated. Real cases are used to explain the establishment and application of enterprise networks in different industries, in order to better meet the needs of vocational colleges for employment-oriented skilled talent cultivation.

The compilation of textbooks starts from four aspects: firstly, the "dual" cooperative development between schools and enterprises. The textbook is jointly written by teachers with rich teaching experience and frontline engineers in enterprise technical support. All cases are derived from real projects in the enterprise and reflect the latest ICT technology achievements. The second is the construction of "new form integration" textbooks, achieving the co-construction and sharing of teaching resources. Give full play to the advantages of "Internet plus textbooks", support high-quality online open courses, automatic scoring system and other digital resources, facilitate students' real-time learning and personalized learning, and help teachers innovate teaching models. The third is the "work manual style" design, which helps to implement the "three education" reform. By analyzing work tasks and professional abilities, the various ability requirements in network engineering practice are implemented in specific steps, which is equivalent to a detailed work manual. The fourth is to integrate the spirit of craftsmanship and promote the integration of "curriculum ideology and politics" into the classroom. Integrating corporate culture, team spirit, labor education, and other content.

This textbook, as a result of the construction of high-level universities and professional groups in Guangxi, is developed through the "dual" cooperation between schools and enterprises. The textbook is edited by Zhong Wenji, Ning

Aimin, and Miao Zhifeng, with Wu Liping, Lu Teng, Shi Wei, Li Mingming, Deng Liping, Mo Nianfa, Lao Wei, and Chen Qilong serving as deputy editors. Among them, Ning Aimin and Li Mingming are responsible for writing projects 1 and 2, Miao Zhifeng and Deng Liping are responsible for writing projects 3 and 4, Lu Teng and Shi Wei are responsible for writing projects 5 and 6, Mo Nianfa, Lao Wei, and Chen Qilong are responsible for writing projects 7 and 8, and Zhong Wenji and Wu Liping are responsible for writing projects 9 and 10.

This book can be used as a course material for computer network technology, modern communication technology, Internet of things technology and other related majors in vocational colleges, and is also suitable for technical personnel interested in computer network technology.

Due to the rapid development of computer network technology and the limited level of editors, there are inevitably omissions and inadequacies in the textbooks. We kindly request readers and experts to criticize and correct us. The editor's email address is: 396860606@qq.com.

Authors
December 2022

目　录

项目 1

项目 1　eNSP 使用

1.1　项目导入

近些年来，针对越来越多的 ICT 从业者的对真实网络设备模拟的需求，不同的 ICT 厂商开发出了针对自家设备的仿真平台软件。但目前行业中推出的仿真平台软件普遍存在着仿真程度不够高、仿真系统更新不够及时、软件操作不够方便等一系列问题，这些问题也困扰着广大 ICT 从业者，同时也极大地影响了模拟真实设备的操作体验，降低了用户了解相关产品进行操作和配置的兴趣。

为了避免现行仿真软件存在的这些问题，国内 ICT 企业研发出了一款界面友好、操作简单并且具备极高仿真度的数通设备模拟器——eNSP（Enterprise Network Simulation Platform）。这款仿真软件运行是物理设备的 VRP 操作系统，最大限度地模拟真实设备环境，用户可以利用 eNSP 模拟工程开局与网络测试，协助用户高效地构建企业优质的 ICT 网络。eNSP 支持对接真实设备，数据包的实时抓取，有助于深刻理解网络协议的运行原理，更好地进行网络技术的钻研和探索，很好地模拟路由交换的各种实验。本项目主要介绍计算机网络组建的基础知识及 eNSP 模拟器的使用。

1.2　相关知识

1.2.1　路由器

路由器（Router）工作在网络层，是目前连接互联网和局域网的主要网络设备，通过路由表选择最佳路径顺序发送数据。路由器是互联网的枢纽，广泛应用于各种行业服务。

路由器收到数据包后，会根据数据包中的目标 IP 地址选择一条最优的路径，并将数据包转发给下一个路由器，路径上最后的路由器负责将数据包送交目的主机。数据包在网络上的传输好像是体育运动中的接力赛一样，每一个路由器负责将数据包按照最优的路径向下一跳路由器进行转发，通过多个路由器一站一站地接力，最终将数据包通过最优路径转发到目的地。

路由器能够决定数据报文的转发路径。如果有多条路径可以到达目的地，则路由器会通过计算来决定最佳下一跳。计算的原则会随着实际使用的路由协议不同而不同。如果实施了一些特别的策略路由，数据包通过的路径可能会随策略路由的规则而转发出去。

路由器转发数据包的关键是路由表。每个路由器中都保存着一张路由表，表中每条路由项都指明了数据包要到达某网络或某主机应通过路由器的哪个物理接口发送，以及可到达该路径的是哪个路由器，或者不再经过别的路由器而直接到达目的地。

路由表中包含以下关键项：

（1）目的地址（Destination）：用来标识 IP 包的目的地址或目的网络。

（2）网络掩码（Network mask）：通过 IP 地址和网络掩码进行逻辑与计算，可以得到相应的网段信息。如目的地址为 192.168.1.0，掩码为 /24，相与后便得到一个 C 类的网段信息（192.168.1.0/8）。网络掩码的另一个作用是当路由表中有多条目的地址相同的路由信息时，路由器将选择其掩码最长的一项作为匹配项。

（3）输出接口（Output interface）：指明 IP 包所经由的下一个路由器的接口地址。

（4）下一跳地址（Next hop address）：指明 IP 包所经由的下一个路由器的接口地址。

1.2.2　交换机

交换机工作在数据链路层，用于在网络内进行数据转发。

1.2.2.1　交换机的工作原理

（1）交换机根据收到数据帧中的源 MAC 地址建立该地址同交换机端口的映射，并将其写入 MAC 地址表中。

（2）交换机将数据帧中的目的 MAC 地址同已建立的 MAC 地址表进行比较，以决定由哪个端口进行转发。

（3）如数据帧中的目的 MAC 地址不在 MAC 地址表中，则向所有端口转发。这一过程称为泛洪（Flooding）。

（4）广播帧和组播帧向所有的端口转发。

1.2.2.2　交换机的三个主要功能

（1）以太网交换机了解每一端口相连设备的 MAC 地址，并将地址与相应的端口映射起来存放在交换机缓存中的 MAC 地址表中。

（2）转发 / 过滤：当一个数据帧的目的地址在 MAC 地址表中有映射时，它被转发到连接目的节点的端口而不是所有端口（如该数据帧为广播 / 组播帧则转发至所有端口）。

（3）消除回路：当交换机包括一个冗余回路时，以太网交换机通过生成树协议避免回路的产生，同时允许存在后备路径。

1.2.2.3　交换机的工作特性

（1）交换机的每一个端口所连接的网段都是一个独立的冲突域。

（2）交换机所连接的设备仍然在同一个广播域内，也就是说，交换机不隔绝广播（唯一的例外是在配有 VLAN 的环境中）。

（3）交换机依据帧头的信息进行转发，因此说交换机是工作在数据链路层的网络设备（此处所述交换机仅指传统的二层交换设备）。

1.2.2.4　交换机的分类

依照交换机处理帧时不同的操作模式，交换机主要可分为三类：

（1）直通式（Cut Through）。直通方式的以太网交换机可以理解为在各端口间是纵横交叉的线路矩阵电话交换机。它在输入端口检测到一个数据包时，检查该包的包头，获取包的目的地址，启动内部的动态查找表转换成相应的输出端口，在输入与输出交叉处接通，把数据包直通到相应的端口，实现交换功能。由于不需要存储，延迟非常小、交换非常快，这是它的优点。

它的缺点是，因为数据包内容并没有被以太网交换机保存下来，所以无法检查所传送的数据包是否有误，不能提供错误检测能力。由于没有缓存，不能将具有不同速率的输入/输出端口直接接通，而且容易丢包。

（2）存储转发（Store &Forward）。存储转发方式是计算机网络领域应用最为广泛的方式。它把输入端口的数据包检查，在对错误包处理后才取出数据包的目的地址，通过查找表转换成输出端口送出包。正因如此，存储转发方式在数据处理时延时大，这是它的不足，但是它可以对进入交换机的数据包进行错误检测，有效地改善网络性能。尤其重要的是它可以支持不同速度的端口间的转换，保持高速端口与低速端口间的协同工作。

（3）碎片隔离（Fragment Free）。这是介于前两者之间的一种解决方案。它检查数据包的长度是否够 64 字节，如果小于 64 字节，说明是假包，则丢弃该包；如果大于 64 字节，则发送该包。这种方式也不提供数据校验。它的数据处理速度比存储转发方式快，但比直通式慢。

1.2.2.5　二层、三层、四层交换机

二层交换（也称为桥接）是基于硬件的桥接。基于每个末端站点的唯一 MAC 地址转发数据包。二层交换的高性能可以产生增加各子网主机数量的网络设计。其仍然有桥接所具有的特性和限制。

三层交换是基于硬件的路由选择。路由器和第三层交换机对数据包交换操作的主要区别在于物理上的实施。

四层交换的简单定义是数据包的传输不仅仅依据 MAC 地址（第二层交换）或源 IP 地址、目标 IP 地址（第三层路由）进行交换，还依据 TCP/UDP 端口地址（第四层地址）进行交换。也就是说第四层交换除了考虑三层的 IP 地址外还要考虑对端口地址的处理。因为不同的端口地址代表了不同的业务协议，所以第四层交换不仅仅进行了物理上的交换，还包括了业务上的交换。第四层交换的交换域是由源端和终端 IP 地址、TCP 和 UDP 端口共同决定，因此，第四层交换机是真正的“会话交换机”。

1.2.3　防火墙

防火墙（Firewall）是一种位于内部网络与外部网络之间的网络安全系统，是一项信息安全的防护系统，依照特定的规则，允许或是限制传输的数据通过。与路由器相比，防火墙提供了更丰富的安全防御策略，提高了安全策略下数据报转发效率，由于防火墙用于边界安全，因此往往兼备 NAT、VPN 功能。

1.2.3.1　防火墙的分类

防火墙分为包过滤防火墙、代理型防火墙和状态检测防火墙。

（1）包过滤防火墙（Packet Filtering）利用定义的规则过滤数据包，防火墙直接获得数据包的源 IP 地址、目标 IP 地址、TCP/UDP 的源端口和目标端口。包过滤防火墙简单，但是缺乏灵活性，对一些动态协商端口没有办法设置规则。另外，包过滤防火墙对每个数据包都要进行策略检查，策略过多容易导致性能下降。

（2）代理型防火墙（Application Gateway）使得防火墙作为一个访问的中间节点，对客户端（Client）来说防火墙是一个服务器（Server），对服务器（Server）来说防火墙是一个客户端（Client）。代理型防火墙安全性较高，但是开发代价很大。对每一种应用开发一个对应的代理服务是很难做到的，因此代理型防火墙不能支持很丰富的业务，只能针对某些应用提供代理支持。

（3）状态检测防火墙是一种高级通信过滤，它检查应用层协议信息并且监控基于连接的应用层协议状态。对于所有连接，每一个连接状态信息都将被维护并用于动态地决定数据包是否被允许通过防火墙或丢弃。

现在的防火墙主流产品都是状态检测防火墙，状态检测防火墙是一个高性能和高安全的完美结合。

1.2.3.2　防火墙的工作模式

防火墙有三种工作模式，分别是路由模式、透明模式和混合模式：

（1）工作在路由模式的防火墙就像路由器那样工作。防火墙的每个接口连接一个网络，防火墙的接口就是所连接子网的网关。报文在防火墙内首先通过入接口找到进入域信息，然后通过查找转发表，根据出接口找到出口域，再根据两个域确定域间关系，然后使用配置在这个域间关系上的安全策略进行各种操作。

（2）透明模式的防火墙简单理解可以被看作一台以太网交换机。防火墙的接口不能配 IP 地址，这个设备处于现有的子网内部，对于网络中的其他设备，防火墙是透明的。报文转发的出接口，是通过查找桥接的转发表得到的。在确定域间关系之后，安全模块的内部仍然使用报文的 IP 地址进行各种安全策略的匹配。

（3）混合模式的防火墙是指防火墙一部分接口工作在透明模式，另一部分接口工作在路由模式。提出混合模式的概念，主要是为了解决防火墙在纯粹的透明模式下无法使用双机热备份功能的问题。双机热备份所依赖的 VRRP 需要在接口上配置 IP 地址，而透明模式无法实现这一点。

1.2.3.3　防火墙的安全区域

防火墙一般保留四个安全区域：

（1）非受信区（Untrust）：较低的安全区域，其安全优先级为 5。

（2）非军事化区（DMZ）：中度级别的安全区域，其安全优先级为 50。

（3）受信区（Trust）：较高级别的安全区域，其安全优先级为 85。

（4）本地区域（Local）：最高级别的安全区域，其安全优先级为 100。

如果有必要，用户还可以自行设置新的安全区域并定义其安全优先级别，最多可以设置 16 个安全区域。

域间的数据流分两个方向：

入方向（Inbound）：数据由低级别的安全区域向高级别的安全区域传输的方向。

出方向（Outbound）：数据由高级别的安全区域向低级别的安全区域传输的方向。

1.2.4　VRP 系统

VRP (Versatile Routing Platform)，通用路由平台，拥有一致的网络界面、用户界面和管理界面，为用户提供了灵活丰富的应用解决方案。

VRP 平台以 TCP/IP 协议栈为核心，实现了数据链路层、网络层和应用层的多种协议，在操作系统中集成了路由技术、QoS 技术、VPN 技术、安全技术和 IP 语音技术等数据通信要件，并以 IP 转发引擎(TurboEngine)技术作为基础，为网络设备提供了出色的数据转发能力。

随着网络技术和应用的飞速发展，VRP 平台在处理机制、业务能力、产品支持等方面也在持续演进。到目前为止，VRP 已经开发出了 VRP1、VRP2、VRP3、VRP5 和 VRP8 五个版本。

VRP5 是一款分布式网络操作系统，具有高可靠、高性能、可扩展的架构设计。

VRP8 是新一代网络操作系统，具有分布式、多进程、组件化的架构，支持分布式应用和虚拟化技术，能够适应未来的硬件发展趋势和企业急剧膨胀的业务需求。

1.2.5　IP 地址

1.2.5.1　逻辑地址与物理地址

IP 地址是 TCP / IP 模型的网络层中，用以标识网络中主机的逻辑地址。

逻辑地址是与数据链路层的物理地址即硬件地址相对应的。

物理地址是 OSI 参考模型第二层数据链路层的地址，如 MAC 地址。它固化在网卡的硬件结构中，只要主机（或设备）的网卡不变，即使把该主机从一个网络移到另一个网络，从地球的一端移到另一端，该主机的 MAC 地址也是不变的。也就是说，MAC 地址是一种平面化的地址，其不能提供关于主机所处的网络位置信息。

OSI 参考模型第三层及以上层所用的地址都可以成为逻辑地址。逻辑地址只是一个编号，计算机的逻辑地址不是固定的，而是可以任意更改的。如 IP 地址，它是 OSI 参考模型第三层网络层的地址，所以有时又被称为网络地址，该地址是随着主机（或设备）所处网络位置不同而变化的。如果把主机（或设备）从一个网络移到另一个网络，该主机的 IP 地址也会相应地发生改变。也就是说，IP 地址是一种结构化的地址，可以提供关于主机所处的网络位置信息。

逻辑地址和物理地址的关系有点类似于人的姓名和住址的关系。

1.2.5.2　IP 地址的结构与表示

IP 地址由 32 位二进制构成，分为网络号（又称网络 ID）和主机号（又称主机 ID）两部分。其中，网络号用于标识该主机所在的网络，而主机号则表示该主机在相应网络中的序号。正是因为网络号给出的网络位置信息，才使得路由器能够在通信子网中为 IP 数据包选择一条合适的传输路径。IP 地址结构如图 1.1 所示。

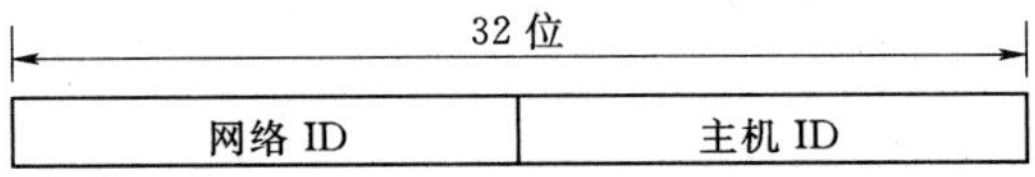

图 1.1　IP 地址结构

32 位的 IP 地址采用点分十进制表示，即 IP 地址分成四段 8 位二进制，采用四段十进制表示，各段十进制之间用点号隔开。

【例 1.1】 某 IP 地址为 11000000 10101000 00000001 00000001，使用点分十进制表示为 192.168.1.1。

1.2.5.3　IP 地址的分类

IP 地址被分为 A、B、C、D、E 五类，其中 A、B、C 类用于普通的主机地址，D 类用于提供网络组播服务或作为网络测试之用，E 类保留给未来扩充使用，如图 1.2 所示。

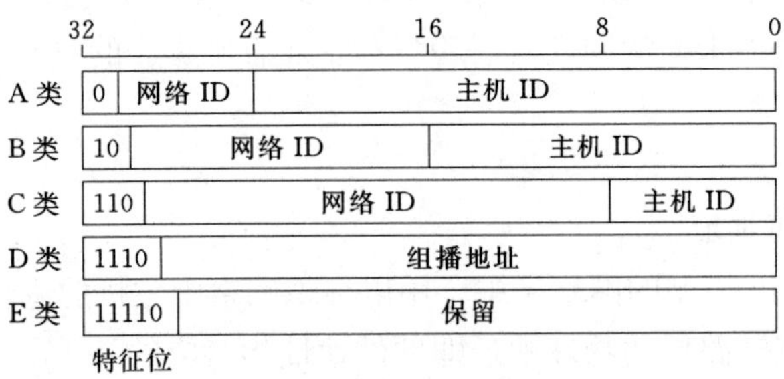

图 1.2　IP 地址的分类

1.2.5.4　默认子网掩码

子网掩码又称子网屏蔽码，用 32 位二进制表示，用来识别 IP 地址的网络号。

子网掩码是按照整个 IP 地址的位模式使用的，其中网络号部分各位全为 1，主机号部分各位全为 0。实际使用时，子网掩码也采用 4 位十进制表示。

A、B、C 三类网络的标准默认掩码见表 1.1。

表 1.1　　A、B、C 三类 IP 地址的子网掩码

类别	二进制表示的默认子网掩码	十进制表示的默认子网掩码
A	11111111.00000000.00000000.00000000	255.0.0.0
B	11111111. 11111111.00000000.00000000	255.255.0.0
C	11111111.11111111. 11111111.00000000	255.255.255.0

子网掩码在计算 IP 地址的网络号中使用。把 IP 地址跟子网掩码按位相与，结果为该 IP 地址的网络号。

【例 1.2】 有 IP 地址为 192.168.0.2，子网掩码为 255.255.255.0，求该 IP 的网络号。

解：把 IP 地址和子网掩码转换为二进制，然后按位相与得：

　　　11000000 10101000 00000000 00000010

按位相与 11111111 11111111 11111111 00000000

　　　11000000 10101000 00000000 00000000

所以该 IP 的网络号为 11000000 10101000 00000000 00000000，转换为点分十进制表示为：192.168.0.0。

1.2.5.5 特殊 IP 地址

（1）主机位全为 0 的 IP 地址表示该 IP 网络，不能分配给主机使用，如 192.168.0.0。

（2）主机位全为 1 的 IP 地址表示该 IP 网络的广播地址，不能分配给主机使用，如 192.168.0.255。

（3）网络位全为 0 的 IP 地址表示主机号。

（4）A 类 IP 网络 127.0.0.0 用于 TCP/IP 协议的环回测试，不能分配给主机使用，如 127.0.0.1。

1.2.5.6 私有 IP 地址

在 IP 地址资源中，还保留了一部分被称为私有地址（Private Address）的地址资源供局域网内部实现 IP 网络时使用。

所有以私有地址为目标地址的 IP 数据包都不能被路由到外面的因特网上，这些以私有地址作为逻辑标识的主机若要访问外面的因特网，必须采用网络地址翻译（Network Address Translation，NAT）或应用代理（Proxy）方式。

私有地址分为三类：

（1）A 类网络：10.0.0.0。

（2）B 类网络：172.16.0.0～172.31.0.0。

（3）C 类网络：192.168.0.0～192.168.255.0。

1.2.5.7　IP 地址的计算

IP 地址的计算主要应用有两种：判断 IP 地址的网络类别和判断两个 IP 地址是否在同一 IP 网络。

（1）判断 IP 地址的网络类别。方法：把 IP 地址最左边的十进制数转换为二进制数，根据最高位的特征判断。

【例 1.3】 请判断 IP 地址 138.11.22.33 属哪类 IP 地址？它的默认子网掩码是什么？

解：$(138)_{10}=(10001010)_2$

因为该 IP 地址的特征位为 10，所以该 IP 属 B 类地址，默认子网掩码为 255.255.0.0。

（2）判断两个 IP 地址是否在同一 IP 网络。只有 IP 地址在同一 IP 网络（即网络号相同）的主机才能直接通信，不同 IP 网络的主机必须经过路由器转发才能通信。用来向其他 IP 网络转发数据的路由器地址称为缺省网关。

方法：把 IP 地址跟子网掩码按位相与，结果为网络号。网络号相同的 IP 地址表示在同一网络。

【例 1.4】 有两台主机，它们的 IP 地址分别是 201.1.2.3 和 201.1.3.3，使用默认子网掩码，请问它们能直接通信吗？

解：$(201)_{10}=(11001001)_2$

因为该 IP 地址的特征位为 110，所以该 IP 属 C 类地址，默认子网掩码为 255.255.255.0。

```
         11001001 00000001 00000010 00000011
按位相与 11111111 11111111 11111111 00000000
------------------------------------------------
         11001001 00000001 00000010 00000000
```

即 IP 地址 201.1.2.3 的网络号为 201.1.2.0。

```
         11001001 00000001 00000011 00000011
按位相与 11111111 11111111 11111111 00000000
------------------------------------------------
         11001001 00000001 00000011 00000000
```

即 IP 地址 201.1.3.3 的网络号为 201.1.3.0。

因为两台主机的 IP 地址的网络号不相同，所以它们不能直接通信。

1.2.5.8　IP 地址的四要素

给主机设置 IP 地址时，必须同时设置子网掩码，以明确该主机从属的 IP 网络。所以 IP 地址和子网掩码是成对使用的。

设置好 IP 地址和子网掩码两个要素之后，只要主机 IP 地址的网络号相同，就可以保证主机在局域网内通信了。

当主机需要跟其他网络（如 Internet）通信时，必须通过路由器对数据进行转发，该路由器称为网关。这时，必须在主机的 IP 地址设置中添加上网关的地址才能实现与其他网络通信，称为缺省网关。缺省网关为 IP 地址的第三要素。

虽然在 TCP/IP 网络（如 Internet）中是使用 IP 地址进行通信的，但 IP 地址不方便记忆，所以人们引入了方便记忆的域名。域名便于记忆，但是计算机不能识别，使用 DNS 服务器可以将域名解析为计算机可以识别的 IP 地址，IP 的第四要素即为 DNS 服务器。

1.3　项目实施

1.3.1　安装 eNSP

（1）打开 eNSP 安装包→单击“安装 eNSP”→选择“中文（简体）”→“确定”，如图 1.3 所示。

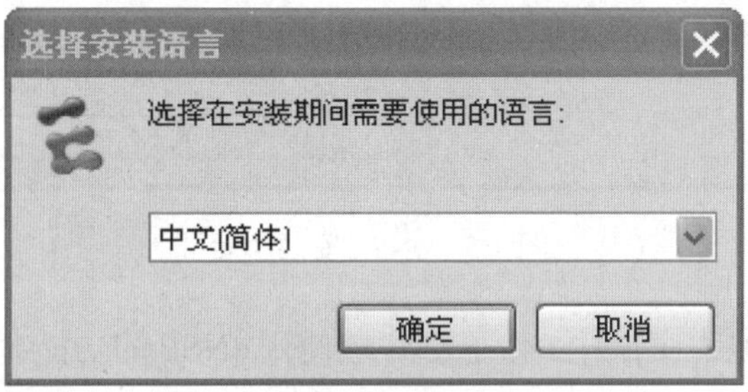

图 1.3　安装 eNSP

（2）单击“下一步”安装即可，如图 1.4 所示。

图 1.4　eNSP 过程

（3）保持默认选择，单击“下一步”，如图 1.5 所示。

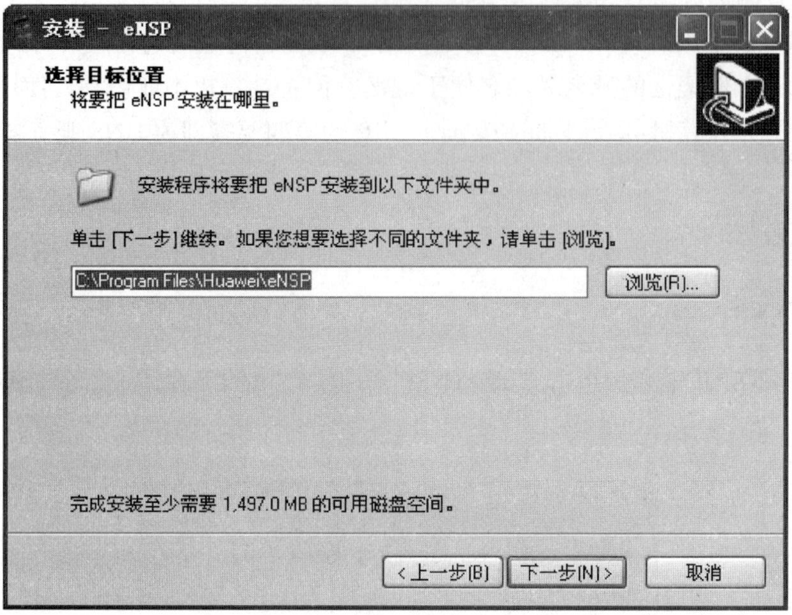

图 1.5　eNSP 安装位置

（4）在选择安装其他程序框中，为了更好地使用 eNSP 的功能，建议把所有选项均选中，如图 1.6 所示。

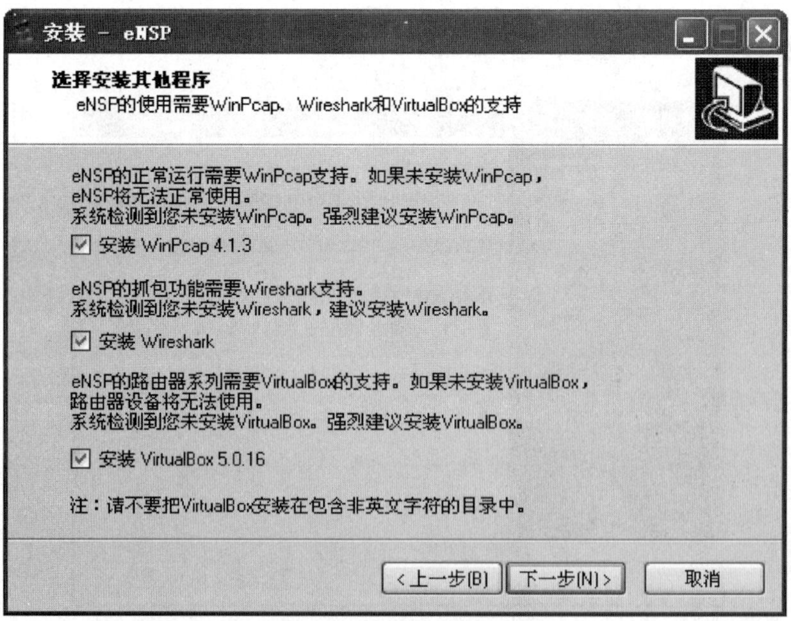

图 1.6　eNSP 组件安装

（5）保持默认单击“下一步”，安装过程进行中请耐心等待，如图 1.7 所示。

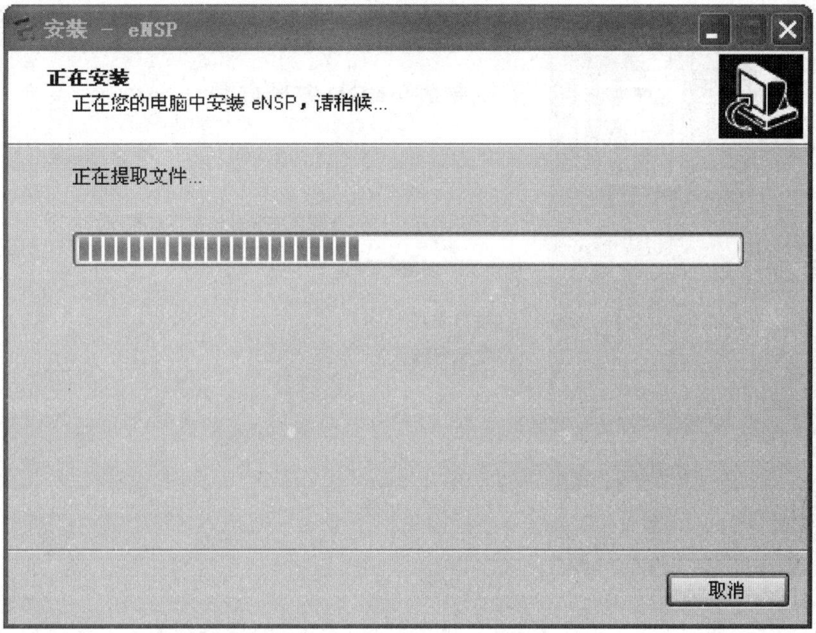

图 1.7　eNSP 安装过程

（6）在稍后弹出的组件安装中，单击“下一步”依次安装，如图 1.8 所示。

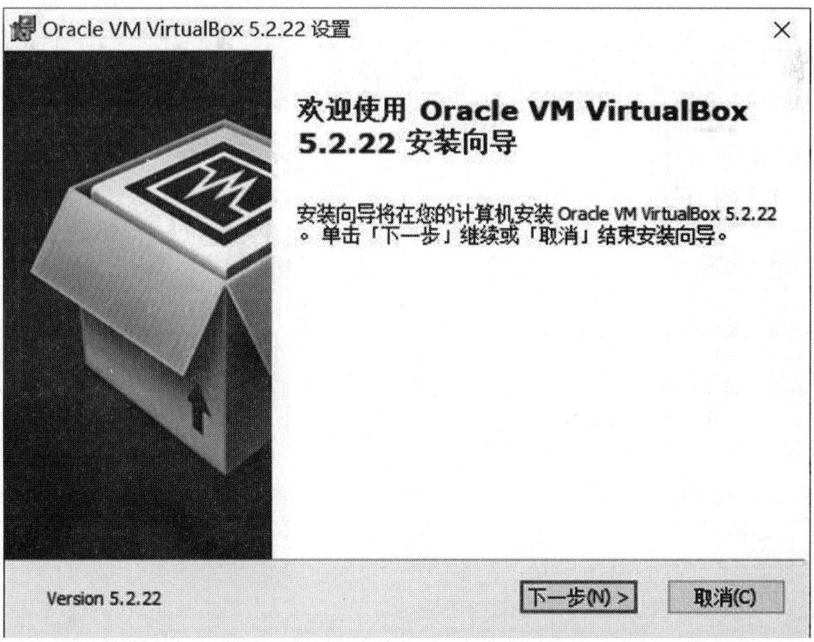

图 1.8　eNSP 组件 VirtualBox 安装

（7）保持默认单击“下一步”，直到安装结束，如图 1.9 所示。

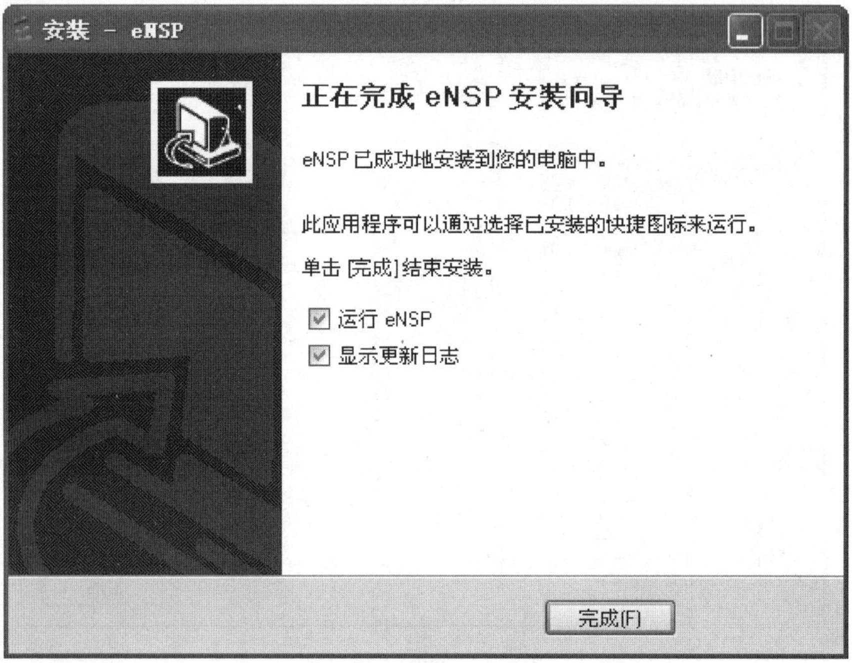

图 1.9　eNSP 安装完成

（8）安装后的界面如图 1.10 所示。

图 1.10　eNSP 界面

1.3.2　使用 eNSP 组建小型网络项目

（1）选择路由器设备后，在下方的设备窗口选择合适的路由器拖到工作区，就添加了一个路由器设备，如图 1.11 所示。

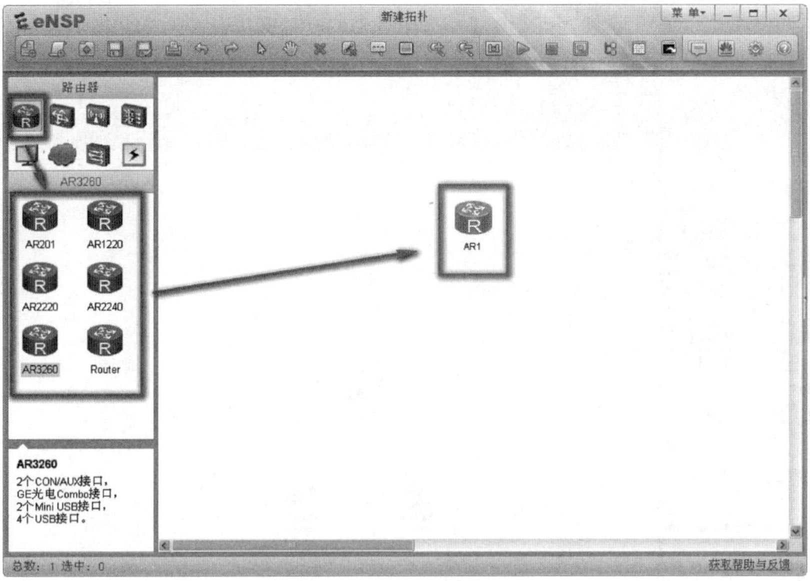

图 1.11　添加路由器

（2）同样方法，添加交换机、PC 机设备，如图 1.12 所示。

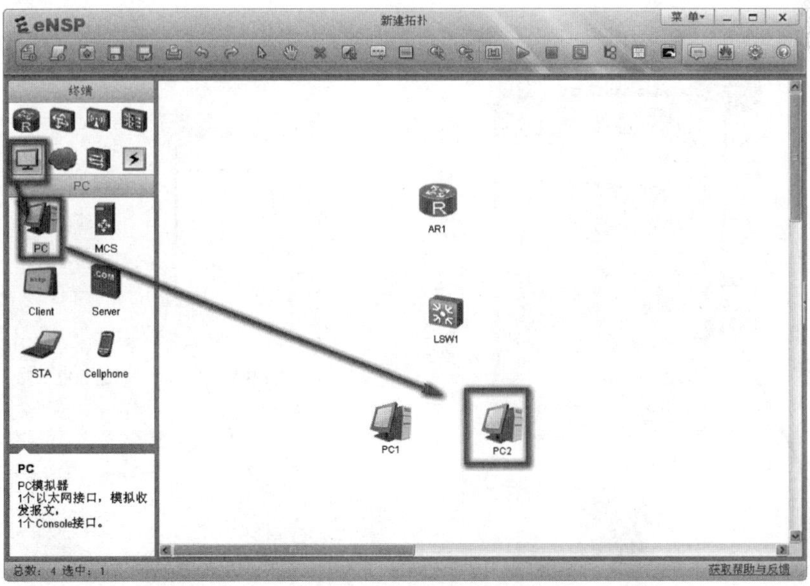

图 1.12　添加交换机、PC 机

（3）选择合适的线缆，把刚拖进工作区的路由器、交换机、PC 机连接起来，此处我们选择自动选择线缆，如图 1.13 所示。

（4）启动设备，如图 1.14 所示。

（5）设备全部启动完毕后，端口指示灯变为绿色。在需要配置的路由器或者交换机上，单击右键，选择 CLI，就可以配置设备了，如图 1.15 所示。

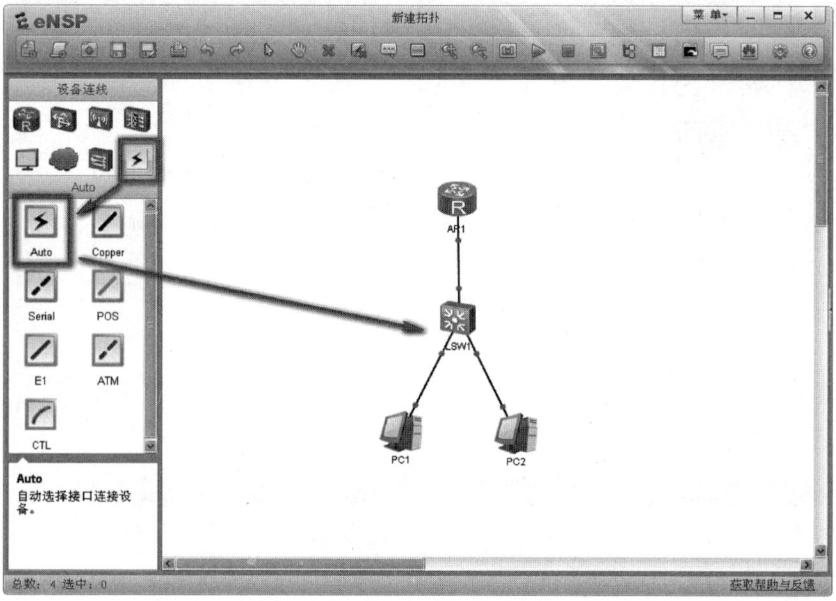

图 1.13　用线缆连接设备

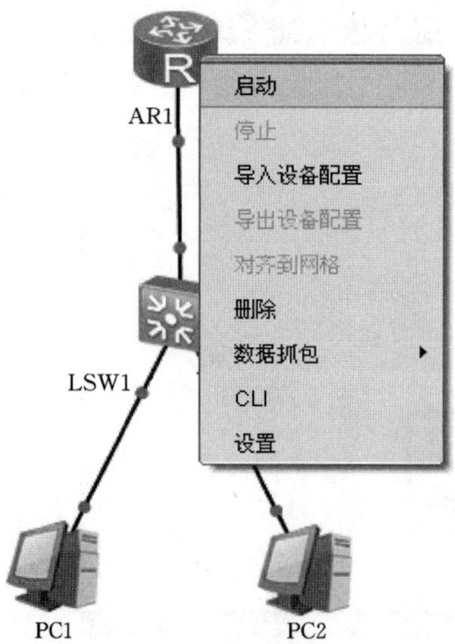

图 1.14　启动设备

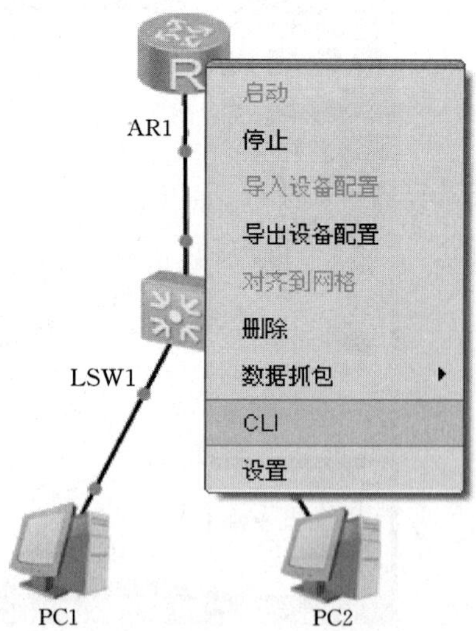

图 1.15　通过 CLI 配置路由器

（6）CLI 命令配置界面，如图 1.16 所示。

（7）单击右键，选择“设置”，在弹出的窗口中就可以进行 PC 机的设置，如图 1.17 所示。

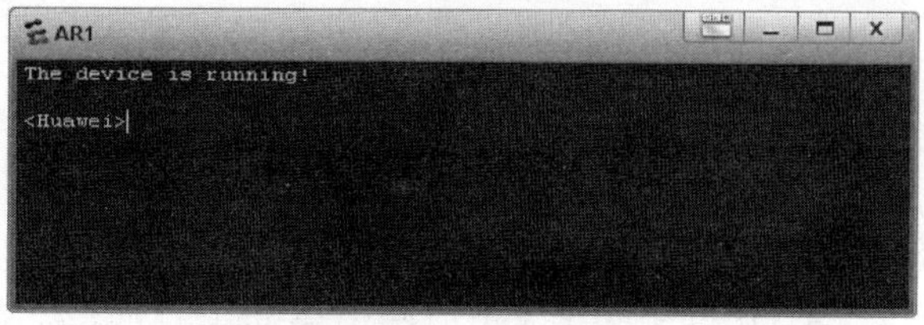

图 1.16　路由器 CLI 命令配置界面

图 1.17　PC 机设置

1.3.3　熟悉 VRP 基本操作

1.3.3.1　不同视图的切换

不同视图的切换，如图 1.18 所示。

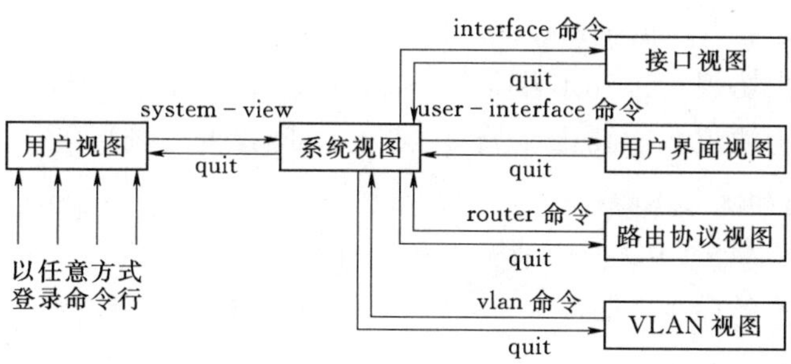

图 1.18　不同视图的切换

- 用户视图：设备启动后的缺省视图，可查看启动后基本运行状态和统计信息。
- 系统视图：配置系统全局通用参数的视图。

- 路由协议视图：配置路由协议参数的视图。
- 接口视图：配置接口参数的视图。
- 用户界面视图：配置登录设备的各个用户属性的视图。

1.3.3.2　命令级别

- 访问级（0 级）：网络诊断工具命令、从本设备出发访问外部设备的命令。
- 监控级（1 级）：用户系统维护、业务故障诊断的命令。
- 系统级（2 级）：业务配置命令。
- 管理级（3 级）：关系到系统基本运行，系统支撑模块的命令。

1.3.3.3　基本的命令操作

（1）路由器基本操作。

- 进入路由器后，是位于用户视图，该视图只能完成一些基本的命令：<Huawei>。
- 从用户视图切换到系统视图：输入 system 或者简写 sys。
- 从系统视图切换到接口视图（假设是 s0/0 接口）：输入 interface s0/0。
- 从一个视图返回到上级视图：quit。
- 配置路由器接口地址（在接口视图下）：ip address 1.1.1.1 24。
- 查看当前路由器配置：dis current。
- 查看已经保存的启动配置：dis saved-config。
- 查看路由器的版本信息：dis version。
- 更改路由器的名字：sysname。
- 清除路由器现有的配置：reset save。
- 重新启动路由器：reboot。
- 设置路由器的系统时间：clock。
- 保存配置：save。
- 查看接口状态：dis interface。
- 配置静态路由协议：ip route-static 192.168.1.0 24 172.16.1.253。
- 配置默认路由协议：ip route-static 0.0.0.0　0 172.16.2.253。
- 查看路由表：dis ip route。
- 查看路由器负载（CPU 使用情况）：dis cpu。
- 取消已经配置的命令：undo ××××，×××× 为先前配置的命令。

（2）VLAN 及相关配置。

- 显示所有交换机端口：dis cur interface。
- 进入交换机端口（假设是 E0/0/3）：Interface e0/0/3。
- 交换机端口描述：description。

- 交换机端口双工设置：duplex。
- 交换机端口速度设置：speed。
- 交换机端口流控设置：flow-control。
- 交换机端口聚合设置（将 e0/0/1 和 e0/0/3 端口配置为一个聚合组）：

  Link-aggregation e0/0/1 to e0/0/3 both。
- 查看端口 5 状态：dis interface e0/0/5。
- 新建 VLAN：（假如新建立 VLAN 3）vlan 3。
- 将端口划入 VLAN：（将 e0/0/4 划入 VLAN 3）。

```
Interface e0/0/4
Port link-type access
Port default vlan 3
```

- 删除 VLAN：undo vlan 3。
- 为 VLAN 增加描述：description。
- 设置端口 5 为 TRUNK 口，并允许通过所有 VLAN：

```
Interface e0/0/5
Port link-type trunk
Port trunk permit vlan all
```

- 为 VLAN5 设置 IP 地址：

```
Interface vlan-interface 5
Ip address 5.5.5.5 24
```

- 查看 VLAN 配置：dis vlan。
- 查看 VLAN 接口状态：dis interface vlan-interface。
- nat 配置（路由器或防火墙上）：

```
nat address-group 1 202.38.160.100 202.38.160.105    #定义公网地址池
acl number 2001                                      # 定义进行网络地址转换的私网 IP 地址段
rule permit source 10.110.10.0 0.0.0.255             # 允许 10.110.10.0/24 网段的地址转换
quit

interface serial 3/0/0
nat outbound 2001 address-group 1                    #在路由器的外网口进行 NAT 地址转换
```

如果没有多余的公网地址池，则直接在外网接口上采用：

```
Int s0/3/0
nat outbound 2001                                    #直接复用外网口的地址进行 NAT
```

- 给交换机设置管理地址（假设 VLAN 1 是管理 VLAN）。

```
Interface vlan-interface 1
```

```
Ip address 172.16.1.2 24        #配置管理 IP 地址
```

- 设置交换机远程管理用户和口令。

```
Local-user huawei               #创建账号
password simple huawei          #设置密码
service-type telnet level 3     #设置权限
user-interface vty 0 4          #配置本地或远端用户名和口令认证
authentication-mode scheme      #配置本地 TELNET 用户，用户名为“huawei”，密码为“huawei”，
                                 权限为最高级别 3(默认为级别 1)
```

1.3.4　小型办公局域网组网

1.3.4.1　项目需求

某小型办公局域网拓扑如图 1.19 所示，请对交换机及路由器进行配置，实现办公局域网内计算机能顺利访问互联网，并具备简单的安全隔离功能。

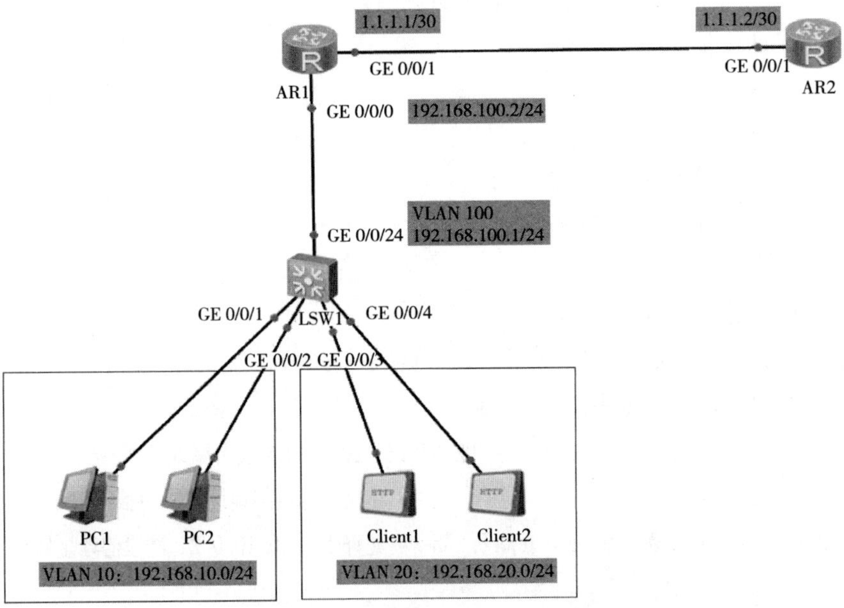

图 1.19　小型办公局域网拓扑

1.3.4.2　配置思路

（1）交换机 LSW1 实现计算机接入和简单的网络隔离功能，在交换机上进行 VLAN 的划分，把相应的 PC 放入 VLAN 中。

（2）在交换机 LSW1 上配置每个 VLAN 的 IP 地址，配置默认路由。

（3）配置路由 AR1 各个接口的 IP 地址。

（4）在路由器 AR1 配置到 AR2 的默认路由，回内网 VLAN10 和 VLAN20 的静态路由。

（5）路由器 AR1 配置 PAT，实现网络地址转换。

1.3.4.3 配置步骤

（1）交换机 LSW1 VLAN 的划分。

```
<Huawei>system-view                                #进入系统视图
[Huawei]sysname LSW1                               #把设备名称修改为 LSW1
[LSW1]undo info-center enable                      #关闭消息提示
[LSW1]vlan batch 10 20 100                         #划分 vlan10、vlan20、vlan100
[LSW1]port-group group-member g0/0/1 to g0/0/2     #把端口 G0/0/1、G0/0/2 加入到临时端口组中
[LSW1-port-group]port link-type access             #配置端口组中所有端口的链路类型为 Access
[LSW1-port-group]port default vlan 10              #把端口组中所有端口放入 vlan 10 中
[LSW1-port-group]quit                              #退出当前端口组模式

[LSW1]port-group group-member g0/0/3 to g0/0/4     #把端口 G0/0/3、G0/0/4 加入到临时端口组中
[LSW1-port-group]port link-type access             #配置端口组中所有端口的链路类型为 Access
[LSW1-port-group]port default vlan 20              #把端口组中所有端口放入 vlan 20 中
[LSW1-port-group]quit                              #退出当前端口组模式

[LSW1]int g0/0/24                                  #进入端口 G0/0/24
[LSW1-GigabitEthernet0/0/24]port link-type access  #配置端口的链路类型为 Access
[LSW1-GigabitEthernet0/0/24]port default vlan 100  #把端口放入 vlan 100 中
[LSW1-GigabitEthernet0/0/24]quit                   #退出当前端口模式
```

（2）在交换机 LSW1 各个 VLAN IP 地址及默认路由的配置。

```
[LSW1]interface vlanif 10                          #创建 VLANIF10 接口
[LSW1-Vlanif10]ip address 192.168.10.1 24          #配置 IP 地址，此 IP 地址是 VLAN10 用户的网关
                                                   地址
[LSW1-Vlanif10]quit                                #退出当前模式

[LSW1]interface vlanif 20                          #创建 VLANIF20 接口
[LSW1-Vlanif20]ip address 192.168.20.1 24          #配置 IP 地址，此 IP 地址是 VLAN20 用户的网关
                                                   地址
[LSW1-Vlanif20]quit                                #退出当前模式

[LSW1]interface vlanif 100                         #创建 VLANIF100 接口
```

```
[LSW1-Vlanif100]ip address 192.168.100.1 24          #配置 IP 地址，此 IP 地址是用于跟路由器 AR1 通信的 IP 地址
[LSW1-Vlanif100]quit                                 #退出当前模式
[LSW1]ip route-static 0.0.0.0 0 192.168.100.2        #配置默认路由，所有发往未知网络的数据将数据包的下一跳送到路由器 AR1 中
```

（3）配置路由 AR1 各个接口的 IP 地址。

```
<Huawei>system-view                                  #进入系统视图
[Huawei]sysname AR1                                  #把设备名称修改为 AR1

[AR1]interface GigabitEthernet 0/0/1                 #进入端口 G0/0/1
[AR1-GigabitEthernet0/0/1]ip address 1.1.1.1 30      #配置 IP 地址为 1.1.1.1，子网掩码为 30 位
[AR1-GigabitEthernet0/0/1]quit                       #退出当前接口视图

[AR1]interface g0/0/0                                #进入端口 G0/0/0
[AR1-GigabitEthernet0/0/0]ip address 192.168.100.2 24  #配置 IP 地址为 192.168.100.2，子网掩码为 24 位
[AR1-GigabitEthernet0/0/0]quit                       #退出当前接口视图
```

（4）在路由器 AR1 配置路由信息。

```
[AR1]ip route-static 192.168.10.0 24 192.168.100.1   #配置回内网 VLAN10 的静态路由
[AR1]ip route-static 192.168.20.0 24 192.168.100.1   #配置回内网 VLAN20 的静态路由

[AR1]ip route-static 0.0.0.0 0 1.1.1.2               #配置到互联网的默认路由
```

（5）路由器 AR1 网络地址转换配置。

```
[AR1]acl number 2000                                 #配置基本 ACL，编号为 2000
[AR1-acl-basic-2000]rule  permit source any          #创建一条规则，放行所有的源地址
[AR1-acl-basic-2000]quit                             #退出

[AR1]interface g0/0/1                                #进入外网端口 G0/0/1
[AR1-GigabitEthernet0/0/1]nat outbound 2000          #ACL 编号 2000 所匹配的地址，能通过外网口 G0/0/1 进行地址转换访问外网
[AR1-GigabitEthernet0/0/1]quit                       #退出
```

1.3.4.4 结果测试

给内网 PC 设置地址，然后使用 PING 测试访问互联网，显示网络联通正常，如图 1.20 所示。

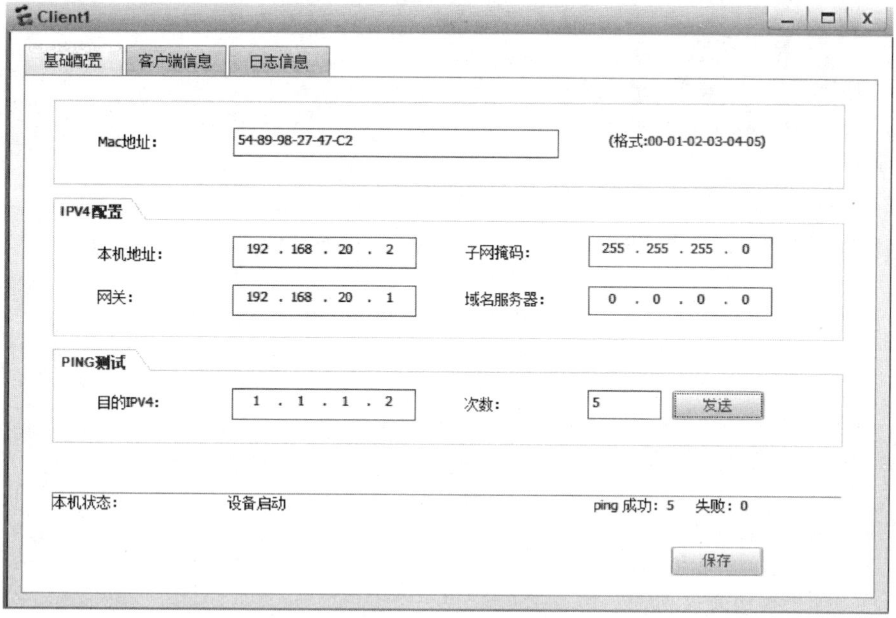

图 1.20　内网 PC 访问外网测试

1.4　思政融入

党的二十大报告对全面建成社会主义现代化强国两步走战略安排进行了宏观展望，重点部署未来 5 年的战略任务和重大举措，展现了一幅令人向往、催人奋进的壮美蓝图：展望 2035 年，我国经济实力、科技实力、综合国力大幅跃升，人均国内生产总值迈上新的大台阶，达到中等发达国家水平；实现高水平科技自立自强，进入创新型国家前列。中国 ICT 企业自主创新、攻坚克难，中国科技人员孜孜以求、执着奋斗，创造了一个又一个的科技成果，当代大学生要塑造艰苦奋斗、吃苦耐劳的灵魂，践行工匠精神，在科技创新、网络强国上贡献青春的力量。

1.5　思考与训练

1. 路由器工作在哪一层？它是如何进行工作的？
2. 路由表包括哪些关键信息？
3. 交换机的三个主要功能是什么？
4. 防火墙的工作模式有哪三种？
5. 公有 IP 地址与私有 IP 地址的区别是什么？

Project 1

Project 1　Use of eNSP

1.1　Project import

In recent years, according to the demand of more and more ICT practitioners for real network equipment simulation, different ICT manufacturers have developed simulation platform software for their own equipment. However, at present, the simulation platform software launched in the industry generally has a series of problems, such as insufficient simulation degree, insufficient timely update of simulation system and inconvenient software operation. These problems also perplex the majority of ICT practitioners. At the same time, they also greatly affect the operation experience of simulating real equipment and reduce the interest of users in understanding the operation and configuration of relevant products.

In order to avoid these problems of the current simulation software, domestic ICT enterprises have developed a friendly interface, simple operation, and has a very high degree of simulation of digital communication equipment simulator - eNSP (Enterprise Network Simulation Platform). This simulation software runs the VRP operating system of physical equipment to simulate the real equipment environment to the greatest extent. eNSP can be used to simulate the start of the project and network test, which will help you efficiently build an enterprise high-quality ICT network. eNSP supports docking with real devices and real-time capture of data packets, which can help you deeply understand the operation principle of network protocol, study and explore network technology better, and simulate various experiments of routing switching. This project mainly introduces the use of eNSP simulator.

1.2 Relevant knowledge

1.2.1　Router

Router works in the network layer. It is the main network equipment connecting the Internet and LAN. It selects the best path and sends data sequentially through the routing table. Router is the hub of the Internet, which is widely used in various industries and services.

After receiving the packet, the router will select an optimal path according to the target IP address in the packet and forward the packet to the next router. The last router on the path is responsible for sending the packet to the destination host. The transmission of data packets on the network is like a relay race in sports. Each router is responsible for forwarding the data packets to the next hop router according to the optimal path, and finally forwarding the data packets to the destination through the optimal path through the relay of multiple routers.

The router can determine the forwarding path of data packets. If there are multiple paths to the destination, the router will determine the best next hop through calculation. The calculation principle will vary with the actual routing protocol. If some special policy routing is implemented, the path through which the packet passes may be forwarded with the rules of policy routing.

The key of router forwarding packets is routing table. A routing table is saved in each router. Each routing item in the table indicates which physical interface of the router the data packet should be sent through to reach a network or a host, and which router can reach the path, or directly reach the destination without passing through other routers.

The routing table contains the following key items:

(1) Destination: used to identify the destination address or destination network of IP packets.

(2) Network mask: the corresponding network segment information can be obtained through the logic and calculation of IP address and network mask. For example, if the destination address is 192.168.1.0 and the mask is /24, a class C network segment information (192.168.1.0/8) will be obtained after matching. Another function of netmask is that when there are multiple routing informations with the same address in the routing table, the router will select the longest item of its mask as the matching item.

(3) Output interface: indicates the interface address of the next router through which the IP packet passes.

(4) Next hop address: indicates the interface address of the next router through which the IP packet passes.

1.2.2 Switch

The switch works in the data link layer and is used for data forwarding in the network.

1.2.2.1 How the switch works

(1) The switch establishes the mapping between the address and the switch port according

to the source MAC address in the received data frame, and writes it into the MAC address table.

(2) The switch compares the destination MAC address in the data frame with the established MAC address table to determine which port to forward.

(3) If the destination MAC address in the data frame is not in the MAC address table, it will be forwarded to all ports. This process is called flooding.

(4) Broadcast frames and multicast frames are forwarded to all ports.

1.2.2.2 Three main functions of the switch

(1) The Ethernet switch knows the MAC address of the equipment each port connected to, maps the address with the corresponding port and stores them in the MAC address table in the switch cache.

(2) Forwarding/Filtering: when the destination address of a data frame is mapped in the MAC address table, it is forwarded to the port connecting the destination node instead of all ports (if the data frame is a broadcast/multicast frame, it is forwarded to all ports).

(3) Eliminate loop: when the switch includes a redundant loop, the Ethernet switch avoids the generation of loop through spanning tree protocol, and allows the existence of backup path.

1.2.2.3 Operating characteristics of switch

(1) The network segment connected to each port of the switch is an independent conflict domain.

(2) The devices connected to the switch are still in the same broadcast domain, that is, the switch is not isolated from broadcasting (the only exception is in the environment with VLAN).

(3) The switch forwards according to the information of the frame header, so the switch is a network device working in the data link layer (the switch here only refers to the traditional layer2 switching equipment).

1.2.2.4 Classification of switches

According to the different operation modes when the switch uhen processing frames, it can be divided into three categories:

(1) Cut Through. The straight through Ethernet switch can be understood as a vertical and horizontal line matrix telephone switch between ports. When a data packet is detected at the input port, switch checks the packet header of the packet, obtains the destination address of the packet, starts the internal dynamic look-up table to convert it into the corresponding output port, connects it at the intersection of input and output, and connects the data packet directly to the corresponding port to realize the switching function. Because there is no need for storage, the delay is very small and the exchange is very fast, which are its advantages.

Its disadvantages are that because the packet content is not saved by the Ethernet switch, it is unable to check whether the transmitted packet is wrong and provide error detection capability. Because there is no cache, the input/output ports with different rates cannot be connected directly, and it is easy to lose packets.

(2) Store & Forward. Store and forward mode is the most widely used mode in the field of computer network. It checks the data packet of the input port, takes out the destination address of the data packet after processing the wrong packet, and converts it into the output port to send the packet through the look up table. For this reason, the store and forward mode has a large delay in data processing, which is its deficiency, but it can detect the error of data packets entering the switch and effectively improve the network performance. In particular, it can support the conversion between ports with different speeds and maintain the cooperative work between high-speed ports and low-speed ports.

(3) Fragment Free. This is a solution between the first two. It checks whether the length of the data packet is enough for 64 bytes. If it is less than 64 bytes, it indicates that it is a false packet, and the packet is discarded; If it is longer than 64 bytes, the packet will be sent. This method also does not provide data verification. Its data processing speed is faster than store and forward mode, but slower than straight through mode.

1.2.2.5 Layer 2, 3 and 4 switches

Layer 2 switching (also known as bridging) is a hardware based bridging. Forward packets based on the unique MAC address of each end site. The high performance of layer 2 switching can produce a network design that increases the number of hosts in each subnet. It still has the characteristics and limitations of bridging.

Layer 3 switching is hardware based routing. The main difference between router and layer 3 switch in packet switching operation lies in the physical implementation.

The simple definition of layer 4 switching is that the transmission of data packets is not only based on MAC address (layer 2 switching) or source IP address and destination IP address (layer 3 routing), but also based on TCP/UDP port address (layer 4 address). In other words, layer 4 switching should consider not only the IP address of layer 3, but also the processing of port address. Because different port addresses represent different service protocols, layer 4 switching not only carries out physical switching, but also includes service switching. The switching domain of layer 4 switch is determined by the source and terminal IP addresses, TCP and UDP ports. Therefore, layer 4 switch is a real “session switch”.

1.2.3 Firewall

Firewall is a network security system between internal network and external network. And it is also an information security protection system that allows or restricts the passage of transmitted data according to specific rules. Compared with routers, firewalls provide richer security defense strategies and improve the efficiency of datagram forwarding under security policies. Because firewalls are used for border security, they often have both NAT and VPN functions.

1.2.3.1 Classification of firewall

Firewall is divided into packet filtering firewall, proxy firewall and state firewall.

(1) Packet filtering firewall uses defined rules to filter data packets. The firewall directly obtains the source IP address, target IP address, source port and target port of TCP/UDP. Packet filtering firewall is simple, but it lacks flexibility. There is no way to set rules for some dynamic negotiation ports. In addition, the packet filtering firewall should check the policy of each packet. Too many policies are easy to lead to performance degradation.

(2) Application gateway makes the firewall as an intermediate node for access. For the client, the firewall is a server, and for the server, the firewall is a client. Proxy firewall has high security, but the development cost is very high. It is difficult to develop a corresponding proxy service for each application, so the proxy firewall can not support very rich services, and can only provide proxy support for some applications.

(3) State detection firewall is a kind of advanced communication filtering. It checks the application layer protocol information and monitors the connection based application layer protocol state. For all connections, each connection status information will be maintained and used to dynamically determine whether packets are allowed to pass through the firewall or discarded.

Now the mainstream products of firewall are state detection firewall. State detection firewall is a perfect combination of high performance and high security.

1.2.3.2 Working mode of firewall

Firewall has three working modes: routing mode, transparent mode and mixed mode.

(1) Firewalls working in routing mode work like routers. Each interface of the firewall is connected to a network, and the interface of the firewall is the gateway of the connected subnet. In the firewall, the message first finds the entry domain information through the entry interface, then finds the exit domain according to the exit interface by looking up the forwarding table, then determines the relationship between the domains according to the two domains, and then uses the security policy configured on the relationship between the domains to carry out various operations.

(2) The firewall in transparent mode can be regarded as an Ethernet switch. The interface of the firewall cannot be equipped with IP address. This device is inside the existing subnet. The firewall is transparent to other devices in the network. The outgoing interface of message forwarding is obtained by looking up the bridged forwarding table. After determining the inter

domain, the internal security module still uses the IP address of the message to match various security policies.

(3) Mixed mode firewall means that part of the firewall interface works in transparent mode and the other part works in routing mode. The concept of mixed mode is proposed to solve the problem that the firewall cannot use the dual computer hot backup function in the pure transparent mode. The VRRP that dual computer hot backup relies on needs to configure the IP address on the interface, which cannot be achieved in transparent mode.

1.2.3.3 Security zone of firewall

Firewall generally reserves four security zones:

(1) Untrust: a lower security area with a security priority of 5.

(2) Demilitarized zone (DMZ): a medium level security area with a security priority of 50.

(3) Trust: a high-level security zone with a security priority of 85.

(4) Local: the highest level security zone with a security priority of 100.

If necessary, users can also set new security zones and define their security priorities. Up to 16 security zones can be set.

The data flow between domains is divided into two directions:

Inbound: Data is transmitted from the high-level security area to the low-level security area.

Outbound: Data is transmitted from the low-level security area to the high-level security area.

1.2.4 VRP system

VRP (Versatile Routing Platform), namely the general routing platform, providing users with flexible and rich application solutions.

The VRP platform takes TCP/IP protocol stack as the core, implements a variety of protocols in data link layer, network layer and application layer, integrates data communication elements such as routing technology, QoS technology, VPN technology, security technology and IP voice technology in the operating system, and provides excellent data forwarding capability for network devices based on IP forwarding engine (Turbo Engine) technology.

With the rapid development of network technology and applications, the VRP platform is also evolving in processing mechanism, business capabilities, product support and so on. So far, five versions of VRP have been developed, VRP1, VRP2, VRP3, VRP5 and VRP8.

VRP5 is a distributed network operating system with high reliability, high performance and scalable architecture design.

VRP8 is a new generation network operating system. It has a distributed, multi-process and component architecture, supports distributed application and virtualization technology, and can adapt to the future hardware development trend and the rapidly expanding business needs of enterprises.

1.2.5 IP address

1.2.5.1 Logical and physical addresses

IP addresses are the logical addresses in the network layer of the TCP/IP model that identify the hosts in the network.

The logical address corresponds to the physical address, or hardware address, of the data link layer.

The physical address is the address of the second data link layer of the OSI reference model, such as the MAC address. It is solidified in the hardware structure of the network card. As long as the network card of the host (or device) is unchanged, the MAC address of the host is unchanged, even if the host is moved from one network to another or from one end of the Earth to the other. That is, a MAC address is a planar address that does not provide network location information about where the host is located.

Addresses used in the third and higher layers of the OSI reference model can be logical addresses. The logical address is only a number. The logical address of a computer is not fixed, but can be changed at will. For example, an IP address, which is the address of the third layer of the OSI reference model network layer, is sometimes referred to as a network address, which varies depending on the network location of the host (or device). If a host (or device) is moved from one network to another, the IP address of the host changes accordingly. That is, an IP address is a structured address that can provide information about the network location of the host.

The relationship between a logical address and a physical address is somewhat similar to that between a person's name and address.

1.2.5.2 Structure and representation of IP address

IP addresses are composed of 32-bit binaries and are divided into network number (also known as network ID) and host number (also known as host ID). The network number is used to identify the network where the host is located, and the host number is the serial number of the host in the corresponding network. It is the network location information given by the network number that enables the router to choose an appropriate transmission path for IP packets in the communication subnet. The IP address structure is shown in Figure 1.1.

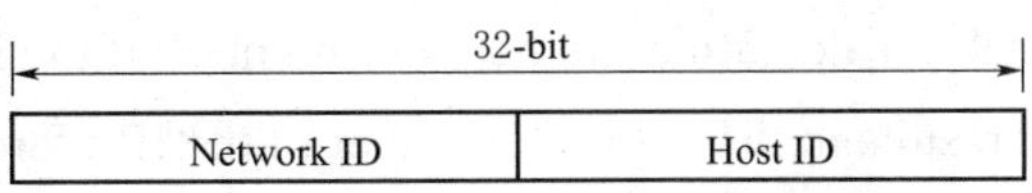

Figure 1.1 IP address structure

32-bit IP addresses are represented in dot decimal, that is, IP addresses are divided into four segments of 8-bit binary, four segments of decimal, and the decimal segments are separated by dots.

Example 1.1: An IP address is 11000000 10101000 00000001 00000001, use dot decimal representation as 192.168.1.1。

1.2.5.3 Classification of IP addresses

IP addresses are divided into five categories: A, B, C, D and E. Classes A, B, C are used as common host addresses, Class D is used to provide network multicast services or for network testing, and Class E is reserved for future expansion. As shown in Figure 1.2.

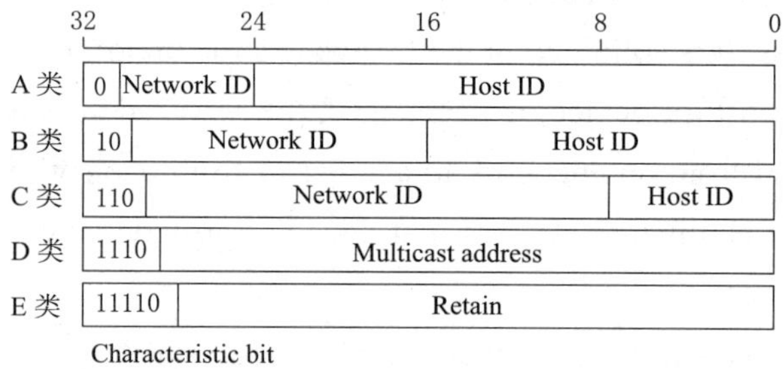

Figure 1.2 Classification of IP addresses

1.2.5.4 Default subnet mask

A subnet mask, also known as a subnet mask, is a 32-bit binary representation that identifies the network number of an IP address.

The subnet mask is used according to the bit mode of the entire IP address, where the network number part is all 1 and the host number part is all 0. In practice, the subnet mask is also represented in 4-bit decimal.

The standard default masks for Class A, Class B, and Class C networks are shown in Table 1.1.

Table 1.1 Subnet Mask of Class A, B and C IP Addresses

Category	Default subnet mask in binary representation	Default subnet mask in decimal
A	11111111.00000000.00000000.00000000	255.0.0.0
B	11111111. 11111111.00000000.00000000	255.255.0.0
C	11111111. 11111111. 11111111.00000000	255.255.255.0

The subnet mask is used in calculating the network number of the IP address. Bitwise sum of IP address and subnet mask results in the network number of the IP address.

Example 1.2: If the IP address is 192.168.0.2 and the subnet mask is 255.255.255.0, calculate the network number of the IP.

Convert the IP address and subnet mask to binary, and then sum by phase:

11000000 10101000 00000000 00000010

Sum by phase 11111111 11111111 11111111 00000000

11000000 10101000 00000000 00000000

Therefore, the network number of this IP is //000000 10101000 00000000 00000000, which is converted to Dot-decimal notation. 192.168.0.0.

1.2.5.5 Special IP address

(1) An IP address with all host bits of 0 indicates the IP network and cannot be assigned to the host. Such as 192.168.0.0.

(2) An IP address with all host bits of 1 indicates the broadcast address of the IP network and cannot be assigned to the host. Such as 192.168.0.255.

(3) The IP address with all network bits of 0 indicates the host number.

(4) Class A IP network 127.0.0.0 is used for loopback test of TCP/IP protocol and cannot be assigned to the host. Such as 127.0.0.1.

1.2.5.6 Private IP address

In the IP address resources, a part of the address resources called private address are reserved for use when implementing the IP network within the LAN.

All IP packets with private address as the target address cannot be routed to the external Internet. If these hosts with private address as the logical identification want to access the external Internet, they must adopt network address translation (NAT) or application proxy.

The private address is divided into three blocks:

(1) Class A network: 10.0.0.0

(2) Class B network: 172.16.0.0~172.31.0.0

(3) Class C network: 192.168.0.0~192.168.255.0

1.2.5.7 Calculation of IP address

There are two main applications of IP address calculation: judging the network type of IP address and judging whether two IP addresses are in the same IP network.

(1) Judge the network type of IP address. Methods: convert the decimal number on the left of the IP address into binary number, and judge according to the characteristics of the highest bit.

Example 1.3: What kind of IP address is the IP address 138.11.22.33? What is its default subnet mask?

$(138)_{10} = (10001010)_2$

Because the characteristic bit of the IP address is 10, the IP address belongs to class B address, and the default subnet mask is 255.255.0.0.

(2) Judge whether two IP addresses are in the same IP network. Only hosts with IP addresses in the same IP network (the same network number) can communicate directly, and hosts with different IP networks must be forwarded through routers to communicate. The router address used to forward data to other IP networks is called the default gateway.

Method: compare the IP address and subnet mask by phase, and the result is the network number. The IP address with the same network number indicates that it is on the same network.

Example 1.4: There are two hosts. Their IP addresses are 201.1.2.3and 201.1.3.3. Use default subnet mask. Can they communicate directly?

$(201)_{10}=(11001001)_2$

Because the characteristic bit of the IP address is 110, therefore,this IP belongs to Class C address,the default subnet mask is 255.255.255.0。

11001001 00000001 00000010 00000011

Sum by phase 11111111 11111111 11111111 00000000

11001001 00000001 00000010 00000000

That is, the network number of the IP address 201.1.2.3 is 201.1.2.0.

11001001 00000001 00000011 00000011

Sum by phase 11111111 11111111 11111111 00000000

11001001 00000001 00000011 00000000

That is, the network number of the IP address 201.1.3.3 is 201.1.3.0.

Because the IP addresses of the two hosts have different network numbers, they cannot communicate directly.

1.2.5.8 Four elements of IP address

When setting the IP address of the host, the subnet mask must be set at the same time to clarify the IP network to which the host belongs. Therefore, the IP address and subnet mask are used in pairs.

After setting the two elements of IP address and subnet mask, as long as the network number of the host IP address is the same, the host can communicate in the LAN.

When the host needs to communicate with other networks (such as the Internet), it must forward the data through the router, which is called the gateway. At this time, the address of the gateway must be added to the IP address setting of the host to realize communication with other networks. It is called the default gateway, which is the third element of the IP address.

Although IP address is used for communication in TCP/IP network (such as Internet), IP address is not convenient for memory, so people have introduced a domain name which is easy to memorize. Domain name is easy to remember, but the computer can not be identified, the use of DNS servers can be analysed for the domain name of the computer can be identified as an IP address, the fourth element of the IP that is DNS servers. The fourth element of IP is the DNS server.

1.3 Project implementation

1.3.1 Installing eNSP

(1) Open the eNSP installation package → Click to install eNSP → Select English → Click Sure, As shown in Figure 1.3.

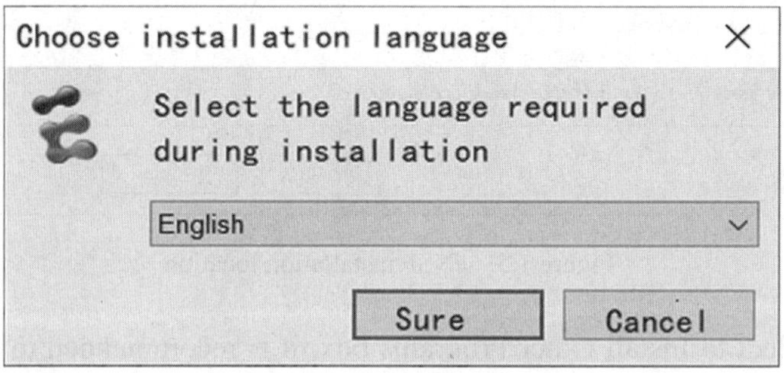

Figure 1.3 Installing eNSP

(2) Click Next to install, as shown in Figure 1.4.

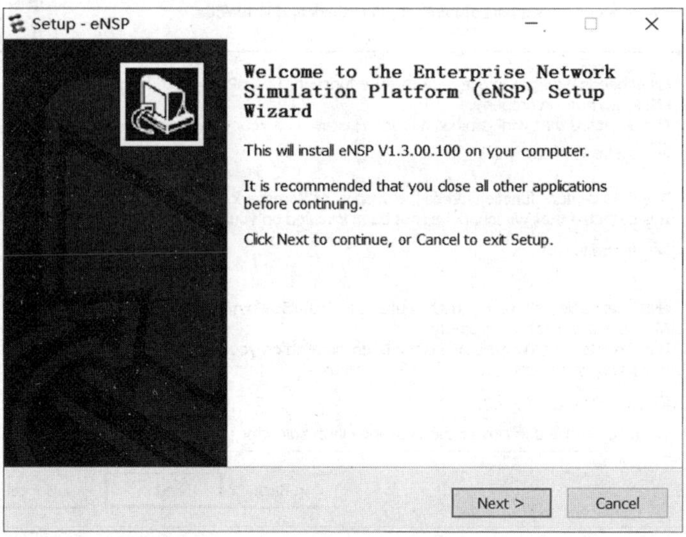

Figure 1.4 eNSP Installation Process

(3) Keep the default selection and click Next, as shown in Figure 1.5.

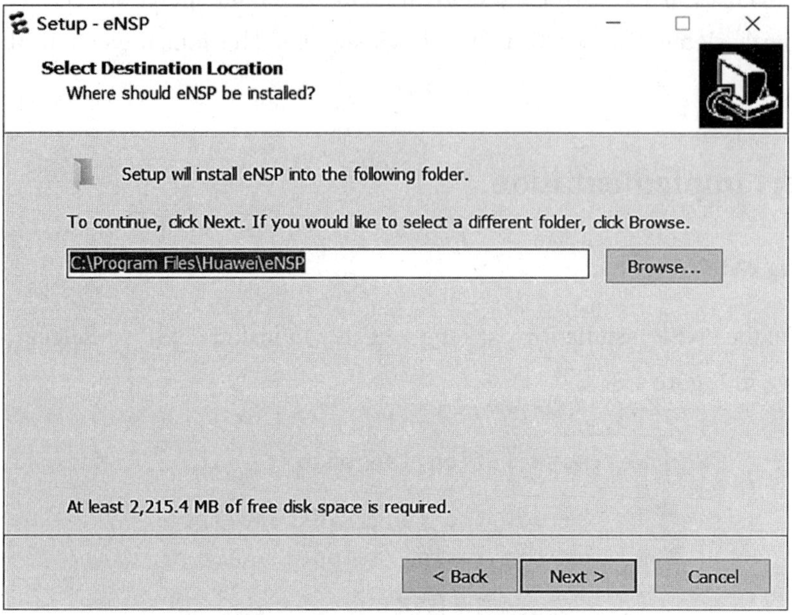

Figure 1.5 eNSP installation location

(4) In the Select to Install Other Programs box, it is recommended to select all the options in order to use eNSP functions better, as shown in Figure 1.6.

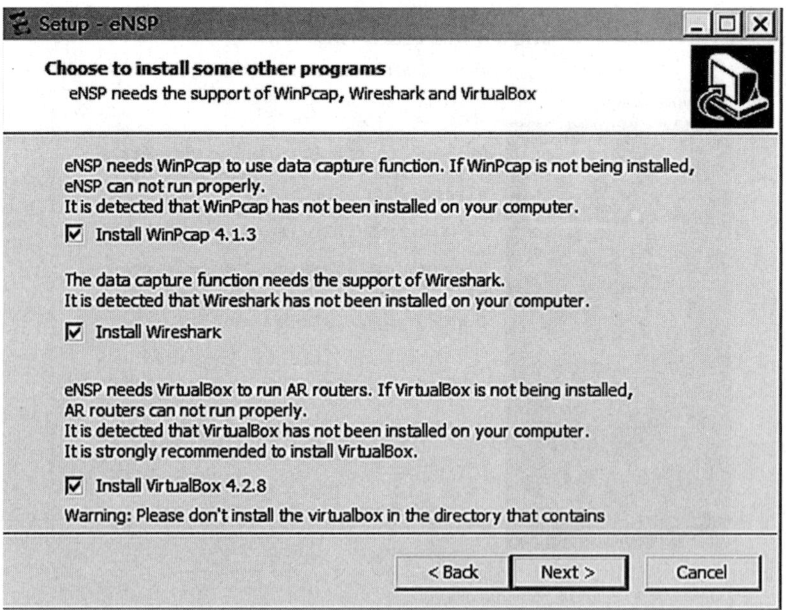

Figure 1.6 eNSP component installation

(5) By default, click Next. Please wait patiently during the installation process, as shown in Figure 1.7.

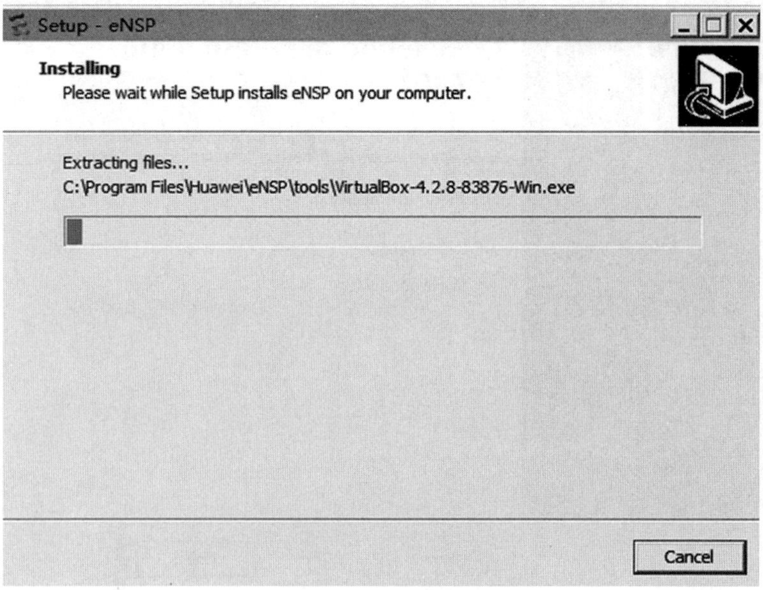

Figure 1.7 eNSP installation process

(6) In the component installation that pops up later, click Next to install in turn, as shown in Figure 1.8.

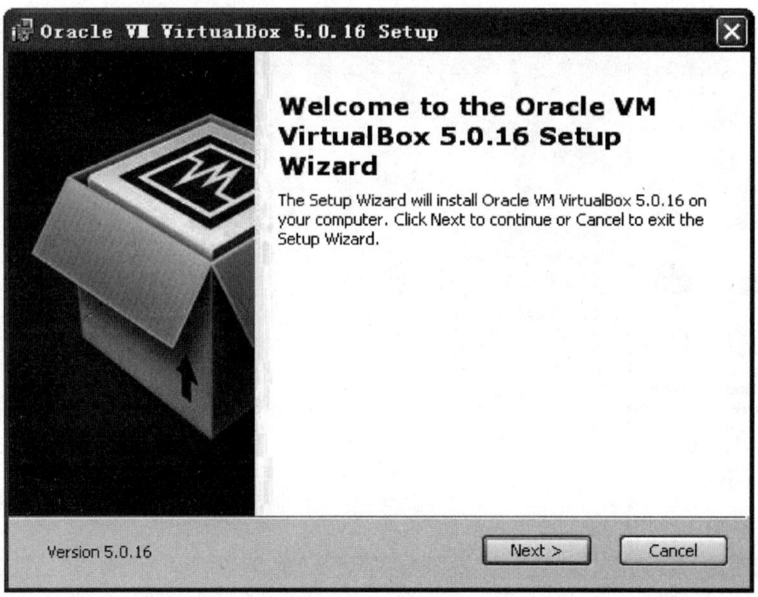

Figure 1.8 Installation of eNSP component VirtualBox

(7) By default, click Next until the installation is completed, as shown in Figure 1.9.

Figure 1.9 eNSP installation complete

(8) The interface after installation is shown in Figure 1.10.

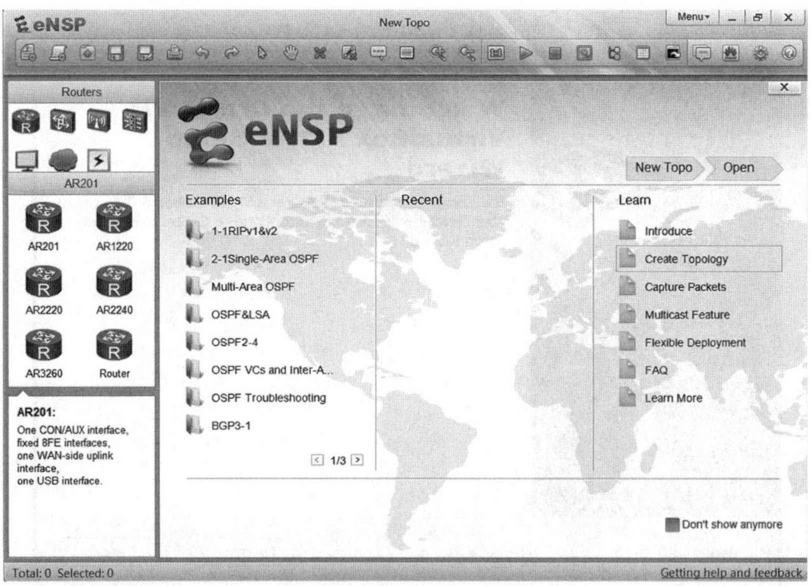

Figure 1.10 eNSP interface

1.3.2　Use eNSP to form a small network project

(1) After selecting the router device, select the appropriate router in the lower device window and drag it to the work area to add a router device, as shown in Figure 1.11.

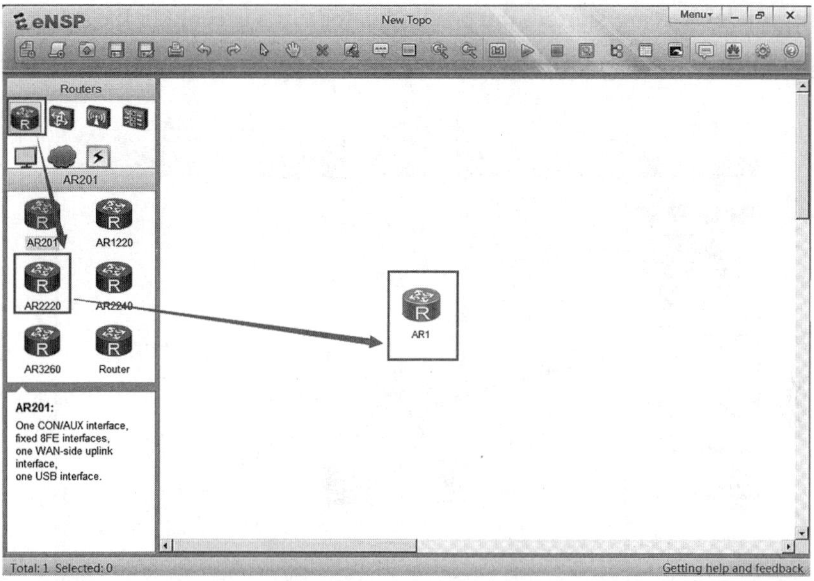

Figure 1.11　Adding routers

(2) In the same way, add switches and PC devices, as shown in Figure 1.12.

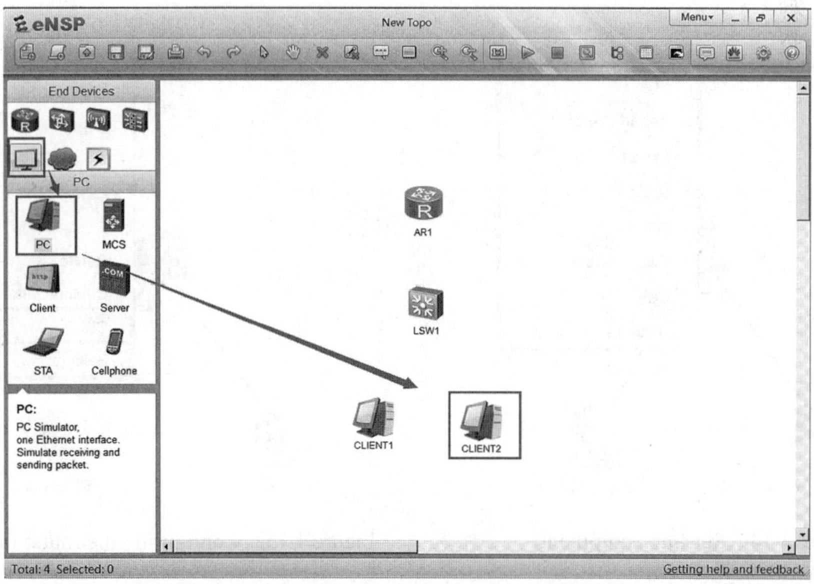

Figure 1.12　Adding switches and PCs

(3) Select the appropriate cable and connect the router, switch and PC that have just been dragged into the work area. Here we choose to select the cable automatically, as shown in Figure 1.13.

(4) Start equipments, as shown in Figure 1.14.

(5) After all devices are started, the port indicator turns green. On the router or switch to be configured, right-click and select CLI to configure the device, as shown in Figure 1.15.

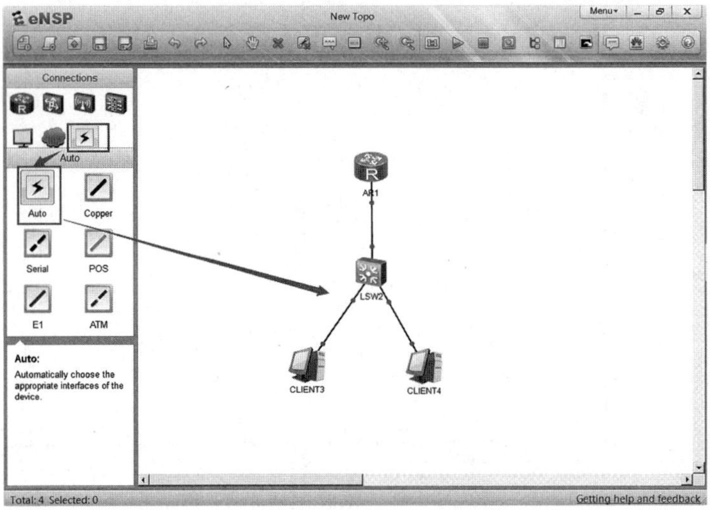

Figure 1.13 Connect equipment with cable

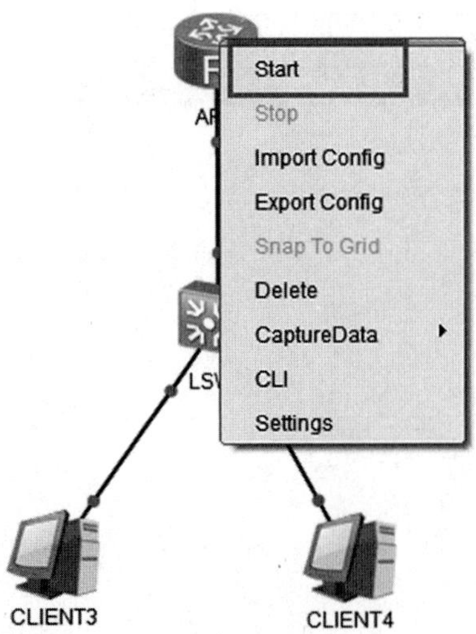

Figure 1.14 Starting equipment

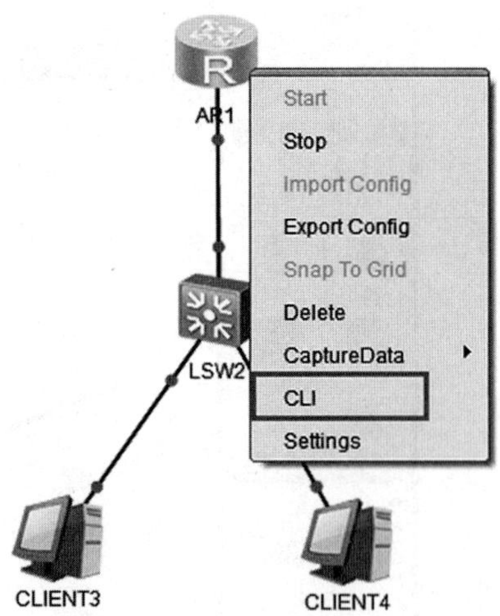

Figure 1.15 Configuring the router through CLI

(6) CLI command configuration interface, as shown in Figure 1.16.

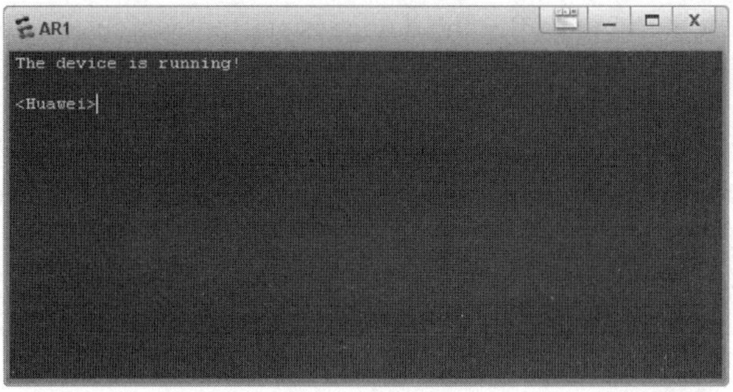

Figure 1.16 Router CLI command configuration interface

(7) Right-click and select Settings. You can set the PC in the pop-up window, as shown in Figure 1.17.

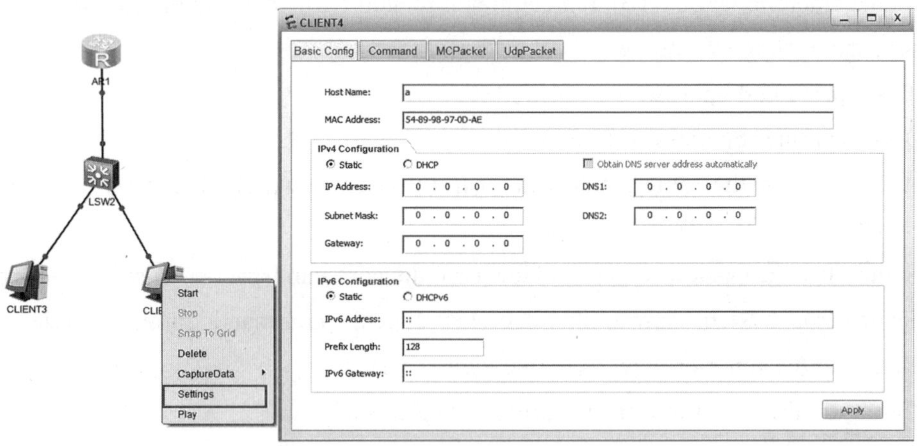

Figure 1.17 PC settings

1.3.3 Familiar with basic VRP operation

1.3.3.1 Switch between different views, as shown in Figure 1.18.

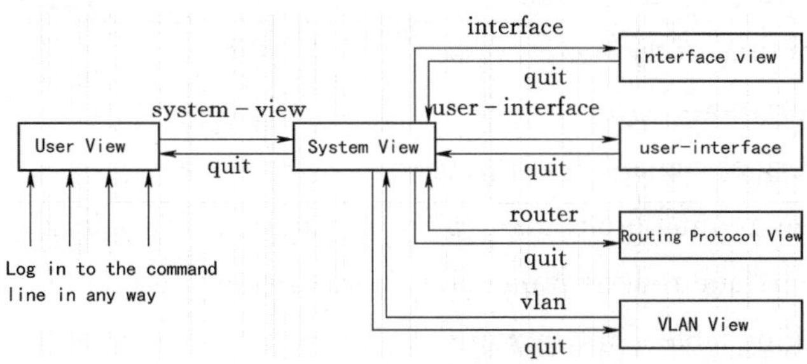

Figure 1.18 Switch between different views

- User view: the default view after the device is started, which can check the basic operation status and statistical information after startup.
- System view: view of configuring global general parameters of the system.
- Routing protocol view: view of configuring routing protocol parameters.
- Interface view: view of configuring interface parameters.
- User interface view: view of configuring various user attributes of the login device.

1.3.3.2 Command level

- Access level (level 0): network diagnostic tool commands, commands to access external devices from this device.
- Monitoring level (level 1): commands for user system maintenance and business fault diagnosis.
- System level (level 2): business configuration command.
- Management level (level 3): the command related to the basic operation of the system and the system support module.

1.3.3.3 Basic command operation

(1) Basic router operations:

- After entering the router, it is located in the user view, which can only complete some basic commands: `<Huawei>`
- Switch from user view to system view,type the command: `system or sys for short`
- Switch from system view to interface view, type the command: `interface s0/0`
- Return to the superior view from one view: `quit`
- Configure the router interface address: `ip address 1.1.1.1 24`
- Check the current router configuration: `dis current`
- Check the saved startup configuration: `dis saved-config`
- Check the version information of the router: `dis version`
- Change the name of the router: `sysname`
- Clear the existing configuration of the router: `reset save`
- Restart the router: `reboot`
- Set the system time of the router: `clock`
- Saving configuration: `save`
- Check the interface status: `dis interface`
- Configure static routing protocols: `ip route-static 192.168.1.0 24 172.16.1.253`
- Configure the default routing protocols: `ip route-static 0.0.0.0 0 172.16.2.253`
- Check the routing table: `dis ip route`
- Check the router load (CPU usage): `dis cpu`

- Cancel the configured command：undo ××××，×××× is a previously configured command

（2）VLAN and related configuration：

- Display all switch ports：dis cur interface
- Entering the switch port：Interface e0/0/3
- Switch port description：description
- Switch port duplex settings：duplex
- Switch port speed setting：speed
- Switch port flow control settings：flow-control
- Switch port aggregation settings（Configure e0/0/1 and e0/0/3 ports as one aggregation group）：
  Link-aggregation e0/0/1 to e0/0/3 both
- Check the port status：dis interface e0/0/5
- Create VLAN：Vlan 3
- Put the port into VLAN：

```
interface e0/0/4
port link-type access
port default  vlan 3
```

- Delete VLAN：undo vlan 3
- Add description for VLAN：description
- Set port 5 as TRUNK port and allow all VLANs to pass：

```
interface e0/0/5
port link-type trunk
port trunk permit vlan all
```

- Setting IP address for VLAN5：

```
interface vlan-interface 5
ip address 5.5.5.5 24
```

- Check VLAN configuration: dis vlan
- Check the VLAN interface status：dis interface vlan-interface
- Nat configuration (on the router or firewall)：

```
nat address-group 1 202.38.160.100 202.38.160.105  # Define public network address pool
acl number 2001   #Define the translated private IP address segment
rule permit source 10.110.10.0 0.0.0.255  #Allow 10.110.10.0/24 address segment translation
quit
interface serial 3/0/0
nat outbound 2001 address-group 1 #NAT at the router's external network port
```

If there is no redundant public address pool, the following options are used on the interface:

```
Int s0/3/0
    nat outbound 2001  # Directly reuse the address of the external network port for NAT
```

- Set the management address for the switch

```
Interface vlan-interface 1
Ip address 172.16.1.2 24        #Configure management IP address
```

- Set the remote management user and password of the switch

```
Local-user huawei               #Create Account
password simple huawei          #Set Password
service-type telnet level 3     #Set permissions
user-interface vty 0 4          #Configure local or remote user name and password authentication
authentication-mode scheme      #Configure local user permissions
```

1.3.4 Small office LAN networking

1.3.4.1 Project requirements

The topology of a small office LAN is shown in Figure 1.19. Please configure the switch and router so that computers in the office LAN can access the Internet smoothly and have simple security isolation functions.

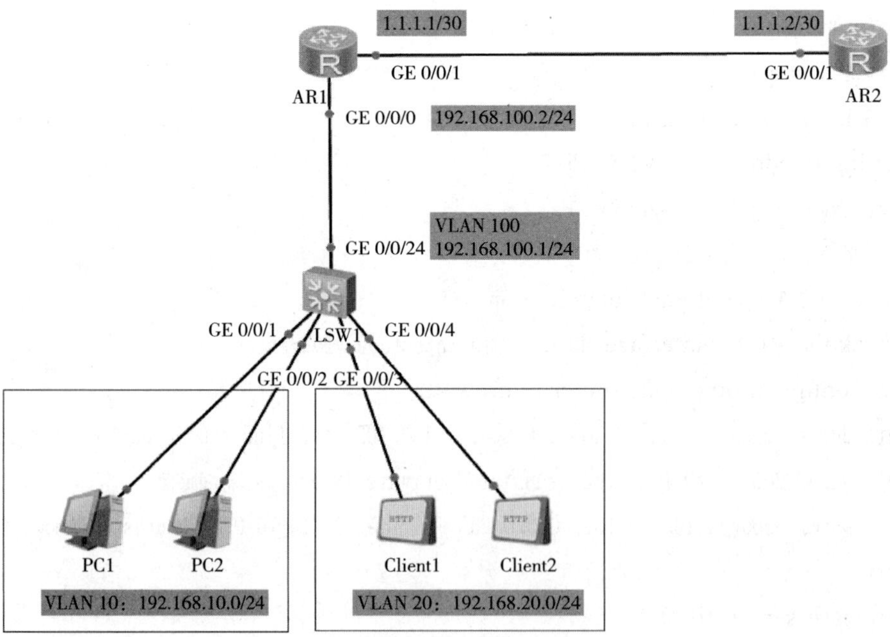

Figure 1.19 Topology of small office LAN

1.3.4.2 Configuration ideas

(1) Switch LSW1 realizes computer access and simple network isolation function, divides VLAN on the switch, and puts corresponding PC into VLAN.

(2) Configure the IP address of each VLAN on the switch LSW1, and configure the default route.

(3) Configure the IP address of each interface of routing AR1.

(4) Configure the default route from router AR1 to AR2, and the static route back to intranet VLAN 10 and VLAN 20.

(5) Router AR1 is configured with PAT to realize network address translation.

1.3.4.3 Configuration Steps

(1) Switch LSW1 VLAN division.

```
<Huawei>system-view                                  #Enter the system view
[Huawei]sysname LSW1                                 #Change the device name to LSW1
[LSW1]undo info-center enable                        #Turn off message prompts
[LSW1]vlan batch 10 20 100                           #Create VLAN 10 20 100
[LSW1]port-group group-member g0/0/1 to g0/0/2       #Add ports G0/0/1 and G0/0/2 to the
                                                      temporary port group
[LSW1-port-group]port link-type access               #Configure the link type of all ports in
                                                      the port group as Access
[LSW1-port-group]port default vlan 10                #Put all ports in the port group into VLAN 10
[LSW1-port-group]quit                                #Exit the current port group mode

[LSW1]port-group group-member g0/0/3 to g0/0/4       #Add ports G0/0/3 and G0/0/4 to the
                                                      temporary port group
[LSW1-port-group]port link-type access               #Configure the link type of all ports in
                                                      the port group as Access
[LSW1-port-group]port default vlan 20                #Put all ports in the port group into VLAN 20
[LSW1-port-group]quit                                #Exit the current port group mode

[LSW1]int g0/0/24                                    #Enter port G0/0/24
[LSW1-GigabitEthernet0/0/24]port link-type access    #The link type of the configured port is Access
[LSW1-GigabitEthernet0/0/24]port default vlan 100    #Put the port into VLAN 100
[LSW1-GigabitEthernet0/0/24]quit                     #Exit the current port group mode
```

(2) Configure each VLAN IP address and default route in the switch LSW1.

```
[LSW1]interface Vlanif 10                            #Create VLANIF10 interface
```

```
[LSW1-Vlanif10]ip address 192.168.10.1 24          #Configure the IP address, which is the gateway address of VLAN10 users
[LSW1-Vlanif10]quit                                #Exit the current mode

[LSW1]interface Vlanif 20                          #Create VLANIF20 interface
[LSW1-Vlanif20]ip address 192.168.20.1 24          #Configure the IP address, which is the gateway address of VLAN10 users
[LSW1-Vlanif20]quit                                # Exit the current mode

[LSW1]interface Vlanif 100                         #Create VLANIF20 interface
[LSW1-Vlanif100]ip address 192.168.100.1 24        #Configure the IP address, which is the gateway address of VLAN10 users
[LSW1-Vlanif100]quit                               # Exit the current mode

[LSW1]ip route-static 0.0.0.0 0 192.168.100.2      #Configure the default route. All the data sent to the unknown network will be sent to the router AR1 for the next hop
```

(3) Configure the IP address of each interface of routing AR1.

```
<Huawei>system-view                                #Enter the system view
[Huawei]sysname AR1                                #Change the device name to AR1

[AR1]interface GigabitEthernet 0/0/1               #Enter port G0/0/1
[AR1-GigabitEthernet0/0/1]ip address 1.1.1.1 30    #Configure IP address as 1.1.1.1, subnet mask as 30 bits
[AR1-GigabitEthernet0/0/1]quit                     #Exit the current interface view

[AR1]interface g0/0/0                              #Enter port G0/0/0
[AR1-GigabitEthernet0/0/0]ip address 192.168.100.2 24  #Configure IP address as 192.168.100.2, subnet mask as 24 bits
[AR1-GigabitEthernet0/0/0]quit                     #Exit the current interface view
```

(4) Configure routing information in router AR1.

```
[AR1]ip route-static 192.168.10.0 24 192.168.100.1 #Configure static routes back to intranet VLAN10
[AR1]ip route-static 192.168.20.0 24 192.168.100.1 #Configure the static route back to
```

```
                                                  the intranet VLAN20

[AR1]ip route-static 0.0.0.0 0 1.1.1.2            #Configure the default route to
                                                  the Internet
```

(5) Router AR1 Network Address Translation Configuration.

```
[AR1]acl number 2000                              #Configure the basic ACL 2000
[AR1-acl-basic-2000]rule  permit source any       #Create a rule, match all source
                                                  addresses
[AR1-acl-basic-2000]quit                          # Exit the current interface view

[AR1]interface g0/0/1                             #Enter Outside Network Port G0/0/1
[AR1-GigabitEthernet0/0/1]nat outbound 2000       #ACL number 2000 matches the ip
                                                  address that can be accessed
                                                  through an external gateway
                                                  G0/0/1Perform address translation
                                                  to access the Internet
[AR1-GigabitEthernet0/0/1]quit                    #Exit the current interface view
```

1.3.4.4 Result Test

Set an address for the PC of the internal network, and then use the PING test to access the Internet, showing that the network connection is normal. As shown in Figure 1.20.

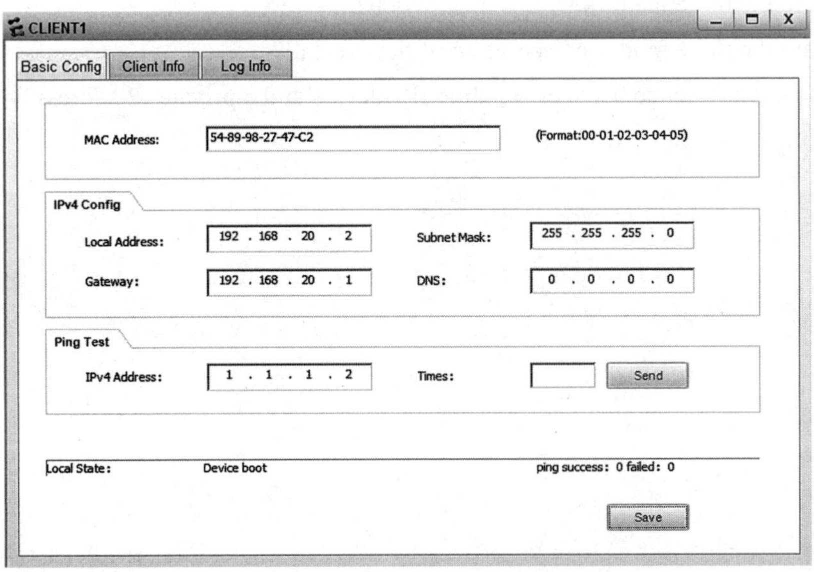

Figure 1.20 Intranet PC Access External Network Test

1.4 Ideological and political integration

In his report to the 20th Party Congress, a macro outlook on the two-step strategic arrangement was given for building a strong socialist modernized country in a comprehensive manner, focusing on the deployment of strategic tasks and major initiatives for the next five years, and presenting a magnificent blueprint that is both desirable and inspiring: looking forward to 2035, China's economic, scientific and technological strengths, and comprehensive national strengths will have jumped dramatically, and its per capita gross domestic product (GDP) will have stepped up to a new level and reached the level of a the level of medium-developed countries; achieve a high level of scientific and technological self-reliance and self-reliance, and enter the forefront of innovative countries. Chinese ICT enterprises innovate independently and overcome difficulties, Chinese scientists and technologists strive diligently and persistently to create one scientific and technological achievement after another. Contemporary college students should shape the soul of hard struggle and hard work, practice the spirit of craftsmanship, and contribute their youthful strength in scientific and technological innovation and network power.

1.5 Thinking and training

1. What layer does the router work on? How does it work?
2. What key information does the routing table include?
3. What are the three main functions of a switch?
4. What are the three modes of operation of the firewall?
5. What is the difference between a public IP address and a private IP address?

项目 2

项目 2　某小型企业局域网网络项目

2.1　项目导入

信息化浪潮风起云涌的今天，企业内部网络的建设已经成为提升企业核心竞争力的关键因素。企业网已经越来越多地被人们提到，利用网络技术，现代企业可以在供应商、合作伙伴、员工之间实现优化的信息沟通。这直接关系到企业能否获得关键竞争优势。近年来越来越多的企业都在加快建设自身信息网络，而其中绝大多数都是中小型企业。

中小型企业局域网通常规模较小，结构相对于简单，对性能的要求则因应用的不同而差别较大。许多中小企业技术人员较少，因而对网络的依赖性很高，要求网络尽可能简单、可靠、易用，可降低网络的使用和维护成本、提高产品的性能价格比就显得尤为重要。

2.2　相关知识

2.2.1　NAT 地址转换

2.2.1.1　NAT 诞生的背景

在网络中 IP 地址有两种：公有地址和私有地址。公有 IP 地址是指在因特网上全球唯一的 IP 地址，所有的公有地址都必须在所属地域 Internet 注册管理机构（RIR）注册，企业从 ISP 租用公有地址。只有公有地址的注册拥有者才能将该地址分配给网络设备。随着接入 Internet 的计算机数量的不断猛增，IP 地址资源也就愈加显得捉襟见肘。

由于公有 IP 地址资源有限，一般在我们企业环境中，使用的都是私有 IP 地址。

RFC 1918 为私有网络预留出了三个 IP 地址块，具体如下：

（1）A 类：10.0.0.0～10.255.255.255。

（2）B 类：172.16.0.0～172.31.255.255。

（3）C 类：192.168.0.0～192.168.255.255。

上述三个范围内的地址不会在因特网上被分配，因此可以不必向 ISP 或注册中心申请而可以在公司或企业内部自由使用。

私有 IP 地址是无法直接访问公网资源的，这个时候就需要用到 NAT 服务。网络地址转换（NAT）将内部私有地址转换为一个或多个公有地址，以便在 Internet 上路由。执行 NAT 功能的网络设备（如路由器、防火墙等）将每个数据包内的私有地址更改为公开注册的公有 IP 地址，然后将数据包发送到 Internet 上。

NAT 服务用于将一个地址段映射到另一个地址段，即可通过 NAT 把内网地址（私有地址）转换成合法的公网 IP 地址，以便内网地址访问公网。

2.2.1.2　NAT 的类型

NAT 的实现方式有静态转换（Static Nat）、动态转换（Dynamic Nat）和端口多路复用（OverLoad）三种。

静态转换是指将内部网络的私有 IP 地址转换为公有 IP 地址，IP 地址对是一对一的，是一成不变的，某个私有 IP 地址只转换为对应的公有 IP 地址。借助于静态转换，可以实现外部网络对内部网络中某些特定设备（如服务器）的访问。

动态转换是指将内部网络的私有 IP 地址转换为公用 IP 地址时，IP 地址是不确定的，是随机的，所有被授权访问上 Internet 的私有 IP 地址可随机转换为任何指定的合法 IP 地址。也就是说，只要指定哪些内部地址可以进行转换，以及用哪些合法地址作为外部地址时，就可以进行动态转换。动态转换可以使用多个合法外部地址集。当 ISP 提供的合法 IP 地址略少于网络内部的计算机数量时，可以采用动态转换的方式。

端口多路复用（Port Address Translation，PAT）是指改变外出数据包的源端口并进行端口转换，即端口地址转换（Port Address Translation，PAT）采用端口多路复用方式。内部网络的所有主机均可共享一个合法外部 IP 地址实现对 Internet 的访问，从而可以最大限度地节约 IP 地址资源。同时，又可隐藏网络内部的所有主机，有效避免来自 Internet 的攻击。因此，目前网络中应用最多的就是端口多路复用方式。

2.2.1.3　NAT 的基本配置

（1）静态 NAT 配置。静态 NAT 为一对一并不能节省 IP 地址，而是为了让映射内网的服务器。现有两个公网地址 202.106.1.1/32 和 202.106.1.2/32；内网 PC 的私有地址为 192.168.10.254/24 和 192.168.20.254/24，则路由器 R1 的静态 NAT 配置为：

```
<R1>system-view
[R1]int g0/0/0
[R1-GigabitEthernet0/0/0]nat static global 202.106.1.1 inside 192.168.10.254  #将这个公网地址映射到内部 IP 为：192.168.10.254 的主机也就是 PC1
[R1-GigabitEthernet0/0/0]nat static global 202.106.1.2 inside 192.168.20.254 #同上映射到 PC2
```

（2）动态 NAT 配置。动态 NAT 是在出口路由器上做了一个地址池，内网 PC 访问外网时会从地址池内获取一个公网 IP。地址池区有多少个公网 IP，同一时刻只能有多少个内网 PC 能上网。

现有以下公网地址：202.106.1.0/24；内网 PC 的私有地址子网为：192.168.10.0/24 192.168.20.0/24，则路由器 R1 的动态 NAT 配置为：

```
[R1]nat address-group 1 202.106.1.1 202.106.1.254   #创建一个 NAT 地址池
[R1]acl 2000  #定义一个访问控制列表
[R1-acl-basic-2000]rule permit source 192.168.10.0 0.0.0.255  #允许 192.168.10.0/24 的子网
[R1-acl-basic-2000]rule permit source 192.168.20.0 0.0.0.255  #允许 192.168.10.0/24 的子网
[R1-acl-basic-2000]quit
[R1]interface g0/0/0
[R1-GigabitEthernet0/0/0]nat outbound 2000 address-group 1 no-pat  #将 ACL 与地址池关联，no-pat 表示不可反复使用
```

（3）PAT 配置配置。PAT 是将一个公网地址反复使用，所有主机都通过它来上网。

现有一个公网地址 202.106.1.1/29；内网 PC 的私有地址子网为：192.168.10.0/24 192.168.20.0/24，则路由器 R1 的 PAT 配置为：

```
[R1]nat address-group 1 202.106.1.1 202.106.1.1  #创建一个地址池
[R1]acl 2000            #定义一个访问控制列表
[R1-acl-basic-2000]rule permit source 192.168.10.0 0.0.0.255   #匹配地址
[R1-acl-basic-2000]rule permit source 192.168.20.0 0.0.0.255   #匹配地址
[R1-GigabitEthernet0/0/0]nat outbound 2000 address-group 1  #将 ACL 与地址池关联
```

现在一个公网没有 IP 只有一个外网口 G0/0/0IP：12.0.0.1；内网 PC 的私有地址子网为：192.168.10.0/24 192.168.20.0/24，则路由器 R1 的 PAT 配置为：

```
[R1]acl 2000            #定义一个访问控制列表
[R1-acl-basic-2000]rule permit source 192.168.10.0 0.0.0.255  #匹配地址
[R1-acl-basic-2000]rule permit source 192.168.20.0 0.0.0.255  #匹配地址
[R1]interface g0/0/0
[R1-GigabitEthernet0/0/0]ip address 12.0.0.1 24
[R1-GigabitEthernet0/0/0]nat outbound 2000  #反复使用当前接口地址
```

（4）静态端口映射配置。在 PAT 的基础上输入如下命令：

```
[R1-GigabitEthernet0/0/0]nat staticprotocol tcp global current-interface telnet inside 192.168.1.10telnet #将当前接口的 23 端口映射到 192.168.1.10 的 23 端口，这里输入端口号或者协议都可以
```

实现效果如图 2.1 所示。

```
[R1-GigabitEthernet0/0/0]natstatic protocol tcp global current-interface1212inside 192.168.1.20 telnet  #将当前接口的 1212 端口映射到 192.168.1.20 的 23 端口
```

实现效果如图 2.2 所示。

```
<R2>telnet 12.0.0.1 23    端口号
  Press CTRL_] to quit telnet mode
  Trying 12.0.0.1 ...
  Connected to 12.0.0.1 ...


Login authentication


Password:
Info: The max number of VTY users is 5, and the number
      of current VTY users on line is 1.
      The current login time is 2015-07-11 15:40:51.
<SW1>
<SW1>
```

图 2.1　映射端口 23 测试

```
<R2>telnet 12.0.0.1 1212
  Press CTRL_] to quit telnet mode
  Trying 12.0.0.1 ...
  Connected to 12.0.0.1 ...


Login authentication


Password:
Info: The max number of VTY users is 5, and the numbe
      of current VTY users on line is 1.
      The current login time is 2015-07-11 15:42:57.
<SW2>
<SW2>
<SW2>
```

图 2.2　映射端口 1212 测试

2.2.2　静态路由

2.2.2.1　静态路由定义

静态路由是指由用户或网络管理员手工配置的路由信息。当网络的拓扑结构或链路的状态发生变化时，网络管理员需要手工去修改路由表中相关的静态路由信息。静态路由信息在缺省情况下是私有的，不会传递给其他的路由器。通过配置静态路由，网络工程师可以人为指定对某一网络访问时所要经过的路径。

在以下场合中，静态路由还是很有用的：

（1）在网络规模小，而且很少变化，或者没有冗余链路。

（2）企业网有很多小的分支机构，并且只有一条路径到达网络的其他部分。

（3）企业想要将数据包发送到互联网主机上，而不是企业网络的主机上。

2.2.2.2　配置代码

命令格式：

```
[R1]ip route-static 10.10.100.0 24192.168.1.1        #这条命令的意思就是到 10.10.100.0/24 这个段的访问下一跳地址是 192.168.1.1
```

在实际中，有可能某条路由失效，但是还有别的路径可以走，可以配置备份静态路由，通过调整优先级参数实现，如：

```
[R1]iproute-static10.10.100.024192.168.2.1preference100        #配置静态路由的优先级，起备份链路的作用
```

2.2.3　默认路由

2.2.3.1　默认路由定义

默认路由是一种特殊的静态路由，指的是当路由表中与包的目的地址之间没有匹配的表项时路由器能够做出的选择。如果没有默认路由，那么目的地址在路由表中没有匹配表项的包将被丢弃，默认路由在某些时候非常有效，当存在末梢网络时，默认路由会大大简化路由器的配置，减轻管理员的工作负担，提高网络性能。

2.2.3.2　配置代码

命令格式：

```
[R1]ip route-static 0.0.0.0 0 10.0.0.1        #配置默认路由
```

在实际中，有可能默认路由失效，但是还有别的路径可以走，可以配置备份默认路由，通过调整优先级参数实现，如：

```
[R1]ip route-static 0.0.0.0 0 10.0.12.2 preference 100        #配置默认备份路由
```

2.2.4　DHCP

2.2.4.1　DHCP 的作用

DHCP（Dynamic Host Configuration Protocol，动态主机配置协议）是一个应用层协议，通常被应用在大型的局域网络环境中，主要作用是集中的管理、分配 IP 地址，使网络环境中的主机动态的获得 IP 地址、Gateway 地址、DNS 服务器地址等信息，并能够提升地址的使用率。

IP 地址是每台计算机必须配置的参数。可以采用静态手工输入或自动获取的方式向 DHCP 服务器获取。手工设置的方式比较容易出错，而且出错时不易找出问题。与此相反，DHCP 服务器能够从预先设置的 IP 地址池里自动给主机分配 IP 地址，它不仅能够保证 IP 地址不会重复分配，也能及时回收 IP 地址以提高 IP 地址的利用效率。

2.2.4.2　DHCP 的工作原理

DHCP 的前身是 BOOTP 协议（Bootstrap Protocol），BOOTP 被创建出来为连接到网络中的设备自动分配地址，后来被 DHCP 取代了，DHCP 比 BOOTP 更加复杂，功能更强大。其工作原理可归结如下。

1. 寻找 DHCP Server

当 DHCP 客户机第一次登录网络的时候（也就是客户机上没有任何 IP 地址数据时），它会通过 UDP67 端口向网络上发出一个 DHCP discover 数据包（包中包含客户机的 MAC 地址和计算机名等信息）。因为客户机还不知道自己属于哪一个网络，所以封包的源地址为 0.0.0.0，目标地址为 255.255.255.255，然后再附上 DHCP discover 的信息，向网络进行广播。

DHCP discover 的等待时间预设为 1s，也就是当客户机将第一个 DHCP discover 封包送出去之后，在 1s 之内没有得到回应的话，就会进行第二次 DHCP discover 广播。若一直没有得到回应，客户机会将这一广播包重新发送四次（以 2s、4s、8s、16s 为间隔，加上 1～1000ms 随机长度的时间）。如果都没有得到 DHCP Server 的回应，客户机会从 169.254.0.0/16 这个自动保留的私有 IP 地址中选用一个 IP 地址。并且每隔 5min 重新广播一次，如果收到某个服务器的响应，则继续 IP 租用过程。

2. 提供 IP 地址租用

当 DHCP Server 监听到客户机发出的 DHCP discover 广播后，它会从那些还没有租出去的地址中，选择最前面的空置 IP，连同其他 TCP/IP 设定，通过 UDP68 端口响应给客户机一个 DHCP OFFER 数据包（包中包含 IP 地址、子网掩码、地址租期等信息）。此时还是使用广播进行通信，源 IP 地址为 DHCP Server 的 IP 地址，目标地址为 255.255.255.255。同时，DHCP Server 为此客户保留它提供的 IP 地址，从而不会为其他 DHCP 客户分配此 IP 地址。

由于客户机在开始的时候还没有 IP 地址，所以在其 DHCP discover 封包内会带有其 MAC 地址信息，并且有一个 XID 编号来辨别该封包，DHCP Server 响应的 DHCP OFFER 封包则会根据这些资料传递给要求租约的客户。

3. 接受 IP 租约

如果客户机收到网络上多台 DHCP 服务器的响应，只会挑选其中一个 DHCP OFFER（一般是最先到达的那个），并且会向网络发送一个 DHCP REQUEST 广播数据包（包含客户端的 MAC 地址、接受的租约中的 IP 地址、提供此租约的 DHCP 服务器地址等），告诉所有 DHCP Server 它将接受哪一台服务器提供的 IP 地址，所有其他的 DHCP 服务器撤销它们的提供以便将 IP 地址提供给下一次 IP 租用请求。此时，由于还没有得到 DHCP Server 的最后确认，客户端仍然使用 0.0.0.0 为源 IP 地址，255.255.255.255 为目标地址进行广播。

事实上，并不是所有 DHCP 客户机都会无条件接受 DHCP Server 的 OFFER，特别是如果这些主机上安装有其他 TCP/IP 相关的客户机软件。客户机也可以用 DHCP REQUEST 向服务器提出 DHCP 选择，这些选择会以不同的号码填写在 DHCP Option Field 里面。客户机可以保留自己的一些 TCP/IP 设定。

4. 租约确认

当 DHCP Server 接收到客户机的 DHCP REQUEST 之后，会广播返回给客户机一个 DHCP ACK 消息包，表明已经接受客户机的选择，并将这一 IP 地址的合法租用以及其他的配置信息都放入该广播包发给客户机。

客户机在接收到 DHCP ACK 广播后，会向网络发送三个针对此 IP 地址的 ARP 解析请求以执行冲突检测，查询网络上有没有其他机器使用该 IP 地址；如果发现该 IP 地址已经被使用，客户机会发出一个 DHCP DECLINE 数据包给 DHCP Server，拒绝此 IP 地址租约，并重新发送 DHCP discover 信息。此时，在 DHCP 服务器管理控制台中，会显示此 IP 地址为 BAD_ADDRESS。

如果网络上没有其他主机使用此 IP 地址，则客户机的 TCP/IP 使用租约中提供的 IP 地址完成初始化，从而可以和其他网络中的主机进行通信。

5. DHCP 客户机租期续约

客户机会在租期过去 50% 的时候，直接向为其提供 IP 地址的 DHCP Server 发送 DHCP REQUEST 消息包。如果客户机接收到该服务器回应的 DHCP ACK 消息包，客户机就根据包中所提供的新的租期以及其他已经更新的 TCP/IP 参数，更新自己的配置，IP 租用更新完成。如果没有收到该服务器的回复，则客户机继续使用现有的 IP 地址，因为当前租期还有 50%。

如果在租期过去 50% 的时候没有更新，则客户机将在租期过去 87.5% 的时候再次向为其提供 IP 地址的 DHCP 联系。如果还不成功，到租约的 100% 时候，客户机必须放弃这个 IP 地址，重新申请。如果此时无 DHCP 可用，客户机会使用 169.254.0.0/16 中随机的一个地址，并且每隔 5min 再进行尝试。

2.2.4.3　基本的 DHCP 服务配置

（1）DHCP 服务的配置命令如下：

```
[Huawei]dhcp enable                                  #开启 DHCP 功能
[Huawei]interface vlanif 1                           #进入 VLANIF 1 接口视图
[Huawei-Vlanif1]ip address 192.168.1.1 24            #配置 VLANIF 接口的 IP 地址
[Huawei-Vlanif1]dhcp select global                   # 能接口的 DHCP 服务功能，指定 DHCP 服务器从全局
                                                       地址池分配地址（可以加在接口或 VLAN 下）
[Huawei-Vlanif1]quit                                 #离开当前视图
[Huawei]ip pool gxsdxy                               #进入全局地址池 gxsdxy
[Huawei-ip-pool-gxsdxy]network 192.168.1.0 mask 255.255.250.0      # 配置地址池下的 IP 地址范围
[Huawei-ip-pool-gxsdxy]gateway-list 192.168.1.1                    #配置 DHCP 的网关
[Huawei-ip-pool-gxsdxy]dns-list 114.114.114.114 8.8.8.8            #配置主 DNS 和备份 DNS
```

（2）可选配置如下：

```
[Huawei-ip-pool-gxsdxy]lease day 1 hour 1 minute 1
# 配置 IP 地址租期，此配置为一天一小时一分钟，默认为一天，unlimited 为无限
[Huawei-ip-pool-gxsdxy]domain-name ×××
# 配置分配给 DHCP 客户端的 DNS 域名后缀（可选中的可选）
[Huawei-ip-pool-gxsdxy]excluded-ip-address 192.168.1.100 192.168.1.254
[Huawei-ip-pool-gxsdxy]excluded-ip-address 192.168.1.1
# 配置地址池中不参与自动分配的 IP 地址，多次执行该命令，可以配置多个不参与自动分配的 IP 地址段
[Huawei-ip-pool-gxsdxy]static-bind ip-address 172.16.35.253 mac-address 28d2-4469-5a55
# 当一个用户需要固定的 IP 地址时，可以将地址池中没有在使用的 IP 地址与用户的 MAC 地址绑定
```

2.2.4.4 DHCP 中继配置

当客户端与 DHCP 服务器不在同一网段时，通过在 DHCP 中继设备转发客户端到 DHCP 服务器的请求。

```
[Huawei]dhcp enable                                   # 开启 DHCP 功能
[Huawei]interface vlanif 1                            # VLANIF 1 接口视图
[Huawei-Vlanif1]ip address 192.168.1.1 mask 24        # VLANIF 接口的 IP 地址
[Huawei-Vlanif1]dhcp select relay
# 启动 VLANIF 接口的 DHCP 中继功能，请在作为 DHCP 中继的设备上进行以下配置
[Huawei-Vlanif1]dhcp server group group-name
# 创建 DHCP 服务器组并进入 DHCP 服务器组视图
[Huawei-Vlanif1]dhcp-server ip-address [ ip-address-index ]
# 向 DHCP 服务器组中添加 DHCP 服务器，每个 DHCP 服务器组下最多可以配置 20 个 DHCP 服务器。不指定索引时，系统将自动分配一个空闲的索引
```

2.2.4.5 校验

```
[Huawei]display dhcp server group [ group-name ]  # 使用命令查看 DHCP 服务器组成员的配置信息
```

2.2.5 VLAN

2.2.5.1 VLAN 概述

虚拟局域网（Virtual Local Area Network，VLAN）是一组逻辑上的设备和用户，这些设备和用户并不受物理位置的限制，可以根据功能、部门及应用等因素将它们组织起来，相互之间的通信就好像它们在同一个网段中一样，由此得名虚拟局域网。VLAN 是一种比较新的技术，工作在 OSI 参考模型的第二层和第三层，在计算机网络中，一个二层网络可以被划分为多个不同的广播域，一个广播域对应了一个特定的用户组，一个 VLAN 就是一个广播域，默认情况下这些不同的广播域是相互隔离的。不同的广播域之间想要通信，需

要通过第三层的路由器来完成。这样的一个广播域就称为 VLAN。与传统的局域网技术相比较，VLAN 技术更加灵活，它具有以下优点：①网络设备的移动、添加和修改的管理开销减少；②可以控制广播活动；③可提高网络的安全性。VLAN 除了能将网络划分为多个广播域，从而有效地控制广播风暴的发生，以及使网络的拓扑结构变得非常灵活外，还可以用于控制网络中不同部门、不同站点之间的互相访问。

传统的局域网（Local Area Network，LAN）是指在某一区域内由多台计算机互联形成的计算机组。一般是方圆几千米以内。局域网可以实现文件管理、应用软件共享、打印机共享、工作组内的日程安排、电子邮件和传真通信服务等功能。局域网是封闭型的，可以由办公室内的两台计算机组成，也可以由一个公司内的上千台计算机组成。

交换技术的发展，加快了新的交换技术（VLAN）的应用速度。通过将企业网络划分为虚拟网络 VLAN 网段，可以强化网络管理和网络安全，控制不必要的数据广播。在共享网络中，一个物理的网段就是一个广播域。而在交换网络中，广播域可以是由一组任意选定的第二层网络地址（MAC 地址）组成的虚拟网段。这样，网络中工作组的划分可以突破共享网络中的地理位置限制，而完全根据管理功能来划分。这种基于工作流的分组模式，大大提高了网络规划和重组的管理功能。在同一个 VLAN 中的工作站，不论它们实际与哪个交换机连接，它们之间的通信就好像在独立的交换机上一样。同一个 VLAN 中的广播只有 VLAN 中的成员才能听到，而不会传输到其他的 VLAN 中去，这样可以很好地控制不必要的广播风暴的产生。同时，若没有路由的话，不同 VLAN 之间不能相互通信，这样增加了企业网络中不同部门之间的安全性。网络管理员可以通过配置 VLAN 之间的路由来全面管理企业内部不同管理单元之间的信息互访。

2.2.5.2　VLAN 的划分方式

（1）基于端口划分的 VLAN：这是最常应用的一种 VLAN 划分方法，根据交换机的物理端口划分到不同的 VLAN 中，目前绝大多数交换机都支持这种 VLAN 划分方法。

优点：定义 VLAN 成员时非常简单，只要将所有的端口都定义为相应的 VLAN 组即可。

缺点：如果某用户离开原来的端口到一个新的交换机的某个端口，必须重新定义。适合于任何大小的网络。

（2）基于 MAC 地址划分 VLAN：这种划分 VLAN 的方法是根据每个用户主机的 MAC 地址来划分。

优点：当用户物理位置从一个交换机换到其他的交换机时，VLAN 不用重新配置。

缺点：初始化时，所有的用户都必须进行配置。适用于小型局域网。

（3）基于网络层协议划分 VLAN：VLAN 按网络层协议来划分，可分为 IP、IPX、DECnet、AppleTalk 等 VLAN 网络。

优点：用户的物理位置改变了，不需要重新配置所属的 VLAN，而且可以根据协议类型来划分 VLAN，并且可以减少网络通信量，可使广播域跨越多个 VLAN 交换机。

缺点：效率低下。适用于需要同时运行多协议的网络。

（4）根据 IP 组播划分 VLAN：IP 组播实际上也是一种 VLAN 的定义，即认为一个 IP 组播组就是一个 VLAN。

优点：更大的灵活性，而且也很容易通过路由器进行扩展。

缺点：适合局域网，主要是效率不高。适合于不在同一地理范围的局域网。

2.2.5.3 VLAN 的配置

（1）常规配置如下：

```
[Huawei]vlan 2                                  #添加一个 vlan2
[Huawei-vlan2]quit                              #退出
[Huawei]int e0/0/2                              #进入 e0/0/2 接口
[Huawei-Ethernet0/0/2]port link-type access     #端口的链路类型为 access（连接计算机的端口），设
                                                置默认接口是 hybrid 的模式
[Huawei-Ethernet0/0/2]port default vlan 2       #默认所有端口都属于 VLAN1，此命令将这个接口加入
                                                到 vlan2 中
[Huawei-Ethernet0/0/2]quit                      #退出
```

（2）把一串连续的端口放到 vlan 中的方法如下：

```
[Huawei]vlan 3                                  #添加一个 vlan3
[Huawei-vlan3]port e0/9 to e0/16                #指定交换机的 e0/9~e0/16 八个端口属于 VLAN3
[Huawei-vlan3]quit                              #退出
```

（3）校验。在完成配置后，执行 display 命令可以显示配置后 VLAN 的运行情况，通过查看显示信息验证配置效果，方法如下：

```
[Huawei]display vlan                            #通过 display vlan 查看当前 vlan 列表
[Huawei]display vlan vlan-id                    #例如 display vlan 100，可查看 vlan100 的状态，可
                                                检查 vlan100 包含哪些端口。
```

2.2.6 以太网端口的链路类型

2.2.6.1 Tag 和 Untag

（1）Untag 就是普通的 Ethernet 报文，普通 PC 机的网卡可以识别这样的报文进行通信。

（2）Tag 报文结构的变化是在源 MAC 地址和目的 MAC 地址之后，加上了 4 bytes 的 VLAN 信息，也就是 VLAN tag 头；一般来说这样的报文普通 PC 机的网卡是不能识别的，带 802.1Q 的帧是在标准以太网帧上插入了 4 个字节的标识。其中包含：

1）2 个字节的协议标识符（TPID)，当前置 0x8100 的固定值，表明该帧带有 802.1Q 的标记信息。

2）2 个字节的标记控制信息（TCI），包含了三个域：① Priority 域，占 3 bit，表示报文的优先级，取值 0～7，7 为最高优先级，0 为最低优先级，该域被 802.1p 采用；②规范格式指示符（CFI）域，占 1 bit，0 表示规范格式，应用于以太网；1 表示非规范格式，应用于 Token Ring；③ VLAN ID 域，占 12bit，用于标示 VLAN 的归属。

2.2.6.2　PVID

PVID 通常代表端口 ID,VID 代表 VLAN ID, 在没有出现 802.1Q 前 , 都是采用基于 MAC 或端口的 VLAN, 推出基于标记的 802.1Q 后才出现 VID, 如果划分基于端口的 VLAN 时 , 只需配置 PVID 就可以了，如果要划分基于 802.1Q 的加标记的 VLAN 时，就需配置 VID 了，交换机默认 PVID 为 1。

2.2.6.3　以太网端口的链路类型 Access

Access 类型的端口只能属于 1 个 VLAN，一般用于连接计算机的端口；Access 端口只属于 1 个 VLAN，所以它的缺省 VLAN 就是它所在的 VLAN；交换机 Access 接口出入数据处理过程如下：

（1）Access 端口收报文：收到一个报文，判断是否有 VLAN 信息：如果没有则打上端口的 PVID，并进行交换转发；如果有则直接丢弃（缺省）。

（2）Access 端口发报文：收到一个报文，判断报文中 VLAN ID 是否与 PVID 相同，如果相同，则去掉报文中的标签，转发；如果不同，则直接丢弃。

（3）配置 Access 端口，方法如下：

```
[Huawei]int e0/0/3
[Huawei-Ethernet0/3]port link-type access  #配置端口 e0/3 access 端口
[Huawei-Ethernet0/3]port default vlan 10   #把端口放入 VLAN 10 中
```

2.2.6.4　以太网端口的链路类型 Trunk

Trunk 类型的端口可以允许多个 VLAN 通过，可以接收和发送多个 VLAN 的报文，一般用于交换机之间连接的端口。

交换机 Trunk 接口出入数据处理过程如下：

（1）Trunk 端口收报文：收到一个报文，判断是否有 VLAN 信息：如果没有则打上端口的 PVID，并进行交换转发；如果有则判断该 Trunk 端口是否允许该 VLAN 的数据进入。如果可以则转发，否则丢弃。

（2）Trunk 端口发报文：比较端口的 PVID 和将要发送报文的 VLAN ID 信息，如果两者相等则剥离 VLAN 信息再发送；如果不相等则判断该 Trunk 端口是否允许该 VLAN 的数据通过，如果可以则转发，否则丢弃。

（3）配置 Trunk 端口，方法如下：

```
[Huawei]int e0/0/3
[Huawei-Ethernet0/3]port link-type trunk  #配置端口 e0/3 为 trunk 端口
```

```
[Huawei-Ethernet0/3]port trunk allow pass vlan all  #允许所有 vlan 通过
```

一般情况下最好指定端口允许通过哪些具体的 VLAN，不要设置允许所有的 VLAN 通过 例如：

```
[Huawei-Ethernet0/3] port trunk allow pass vlan 10  #只允许 vlan 10 通过
```

2.2.6.5 以太网端口的链路类型 Hybrid

Hybrid 类型的端口可以允许多个 VLAN 通过，可以接收和发送多个 VLAN 的报文，可以用于交换机之间连接，也可以用于连接用户的计算机。

Hybrid 端口和 Trunk 端口在接收数据时，处理方法是一样的，唯一不同之处在于发送数据时 Hybrid 端口可以允许多个 VLAN 的报文发送时不打标签，而 Trunk 端口只允许缺省 VLAN 的报文发送时不打标签。

Hybrid 端口和 Trunk 端口属于多个 VLAN，所以需要设置缺省 VLAN ID。缺省情况下，Hybrid 端口和 Trunk 端口的缺省 VLAN 为 VLAN 1。

如果设置了端口的缺省 VLAN ID，当端口接收到不带 VLAN Tag 的报文后，则将报文转发到属于缺省 VLAN 的端口；当端口发送带有 VLAN Tag 的报文时，如果该报文的 VLAN ID 与端口缺省的 VLAN ID 相同，则系统将去掉报文的 VLAN Tag，然后再发送该报文。

对于交换机缺省 VLAN 被称为 “Pvid Vlan”，对于思科交换机缺省 VLAN 被称为 “Native Vlan”。交换机接口出入数据处理过程如下：

（1）hybrid 端口收报文。收到一个报文，判断是否有 VLAN 信息：如果没有则打上端口的 PVID，并进行交换转发；如果有则判断该 hybrid 端口是否允许该 VLAN 的数据进入，如果可以则转发，否则丢弃（此时端口上的 untag 配置是不用考虑的，untag 配置只对发送报文时起作用）。

（2）hybrid 端口发报文：判断该 VLAN 在本端口的属性（通过 display interface 命令，即可看到该端口对哪些 VLAN 是 untag，哪些 VLAN 是 tag）。如果是 untag 则剥离 VLAN 信息，再发送，如果是 tag 则直接发送。

（3）配置命令。配置端口 E0/1 为 Hybrid 端口，能够接收 VLAN 20、30 和 100 发过来的报文，方法如下：

```
[SwitchA]interface Ethernet 0/1
[SwitchA-Ethernet0/1]port link-type hybrid   #配置端口 E0/0/1 口为 Hybird 接口
[SwitchA-Ethernet0/1]port hybrid vlan 20 30 100 untagged  #能够接受来自 vlan 20 30 10 发来
的报文
```

对于 Hybrid 端口来说，可以同时属于多个 VLAN。这些 VLAN 分别是该 Hybrid 端口的 PVID，以及手工配置的 “untagged” 及 “tagged” 方式的 VLAN。一定要注意对应端口的 VLAN 配置，保证报文能够被端口进行正常的收发处理。

2.3　项目实施

2.3.1　项目需求

（1）ISP 运营商出口为静态 IP 形式。

（2）电脑能自动获取 IP 地址。

（3）所有电脑均能访问 Internet。

2.3.2　设备清单

本项目需要用到出口路由器、核心层交换机、接入层交换机，所用设备清单见表 2.1。

表 2.1　　设备清单表

序号	设备型号	数量	功能	备注
1	AR3260	1	路由器	
2	S5700–28C–HI	1	核心层交换机	
3	S3700–26C–HI	4	接入层交换机	

2.3.3　项目拓扑

本项目拓扑图如图 2.3 所示。

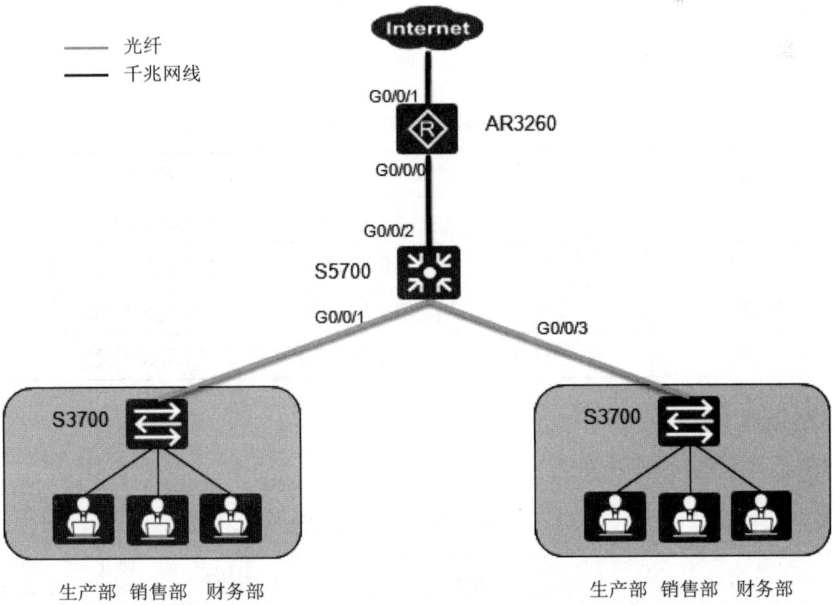

图 2.3　小型企业网络拓扑图

2.3.4　配置思路

项目采用如下的思路配置：

1. 接入层设备

（1）划分 VLAN。

（2）把连接 PC 的端口放到相应 VLAN 中。

（3）把连接核心交换机的端口设置为 Trunk 口，并允许相应的 VLAN 通过。

2. 核心层设备

（1）划分 VLAN。

（2）给 VLAN 配置地址。

（3）把连接接入层交换机的端口设置为 Trunk 口，并允许相应的 VLAN 通过。

（4）把连接出口路由器的端口放到 VLAN 100 中。

（5）配置默认路由指向出口路由器。

（6）配置 DHCP。

3. 出口路由器

（1）给接口配置 IP 地址。

（2）配置局域网用户通过 NAT 转换将私网地址转换为公网地址，进行上网。

（3）配置到 Internet 的默认路由。

2.3.5　相关信息规划

2.3.5.1　核心交换机

核心层交换机的信息规划见表 2.2。

表 2.2　核心层交换机信息规划

序号	使用部门	VLAN 号	IP 地址	掩码	备注
1	生产部	10	10.10.10.1	255.255.255.0	
2	销售部	20	10.10.20.1	255.255.255.0	
3	财务部	30	10.10.30.1	255.255.255.0	
4	互联	100	10.10.100.1	255.255.255.252	与路由器互联

2.3.5.2　出口路由器

出口路由器的信息规划见表 2.3。

表 2.3　出口路由器信息规划

序号	接口	IP 地址	掩码	备注
1	G0/0/0	10.10.100.2	255.255.255.252	与核心交换机互联

2.3.5.3　ISP 出口规划

ISP 出口信息规划见表 2.4。

表 2.4　　ISP 出口信息规划

序号	接口	IP 地址	掩码	网关	备注
1	G0/0/1	1.1.1.1	255.255.255.252	1.1.1.2	静态 IP 地址

2.3.6　操作步骤

2.3.6.1　接入交换机设备

（1）修改设备名，方法如下：

```
<Huawei>system-view
[Huawei]sysname S3700
[S3700]undo info-center enable          #关闭消息提示
```

（2）划分 VLAN，方法如下：

```
[S3700]vlan 10                          #划分 vlan 10
[S3700-vlan10]description shengcanbu    #对 vlan 进行描述
[S3700-vlan10]quit                      #退出
```

同样方法，划分 VLAN 20、30.

```
[S3700]vlan 20
[S3700-vlan20]description xiaoshoubu
[S3700-vlan20]quit

[S3700]vlan 30
[S3700-vlan30]description caiwubu
[S3700-vlan30]quit
```

（3）把连接 PC 的端口放到相应 VLAN 中，方法如下：

```
[S3700]interface Ethernet 0/0/1                   #进入端口 e0/0/1 中
[S3700-Ethernet0/0/1]port link-type access        #把端口设置为 access 模式
[S3700-Ethernet0/0/1]port default vlan 10         #把端口放入 vlan 10 中
[S3700-Ethernet0/0/1]quit                         #退出
```

同样方法，对其他接口进行配置：

```
[S3700]interface Ethernet 0/0/2
```

```
[S3700-Ethernet0/0/2]port link-type access
[S3700-Ethernet0/0/2]port default vlan 20
[S3700-Ethernet0/0/2]quit

[S3700]interface Ethernet 0/0/3
[S3700-Ethernet0/0/3]port link-type access
[S3700-Ethernet0/0/3]port default vlan 30
[S3700-Ethernet0/0/3]quit
```

（4）配置 Trunk，方法如下：

```
[S3700]interface Ethernet 0/0/4
[S3700-Ethernet0/0/4]port link-type trunk          # 在连接核心交换机的上行端口设置为 Trunk
[S3700-Ethernet0/0/4]port trunk allow-pass vlan 10 20 30          # 允许相应的 vlan 通过该接口
[S3700-Ethernet0/0/4]quit
```

2.3.6.2　核心交换机设备

（1）修改设备名，方法如下：

```
<Huawei>system-view
[Huawei]sysname S5700
[S5700]undo info-center enable                     # 关闭消息提示
```

（2）划分 VLAN，方法如下：

```
[S5700]vlan 10                                     # 划分 vlan 10
[S5700-vlan10]description shengcanbu               # 对 vlan 进行描述
[S5700-vlan10]quit                                 # 退出
```

同样方法，划分其他 VLAN：

```
[S5700]vlan 20
[S5700-vlan20]description xiaoshoubu
[S5700-vlan20]quit

[S5700]vlan 30
[S5700-vlan30]description caiwubu
[S5700-vlan30]quit

[S5700]vlan 100
[S5700-vlan100]description hulian
```

```
[S5700-vlan100]quit
```

（3）配置 VLANIF 接口 IP 地址，方法如下：

```
[S5700]interface vlanif 10                    # 进入 vlan 10 接口
[S5700-vlanif10]ip address 10.10.10.1 24      # 配置 ip 地址和掩码信息
[S5700-vlanif10]quit                          # 退出
```

同样方法，对其他 VLAN 进行配置：

```
[S5700]interface vlanif 20
[S5700-vlanif20]ip address 10.10.20.1 24
[S5700-vlanif20]quit

[S5700]interface vlanif 30
[S5700-vlanif30]ip address 10.10.30.1 24
[S5700-vlanif30]quit

[S5700] interface vlanif 100
[S5700-Vlanif100] ip address 10.10.100.1 30
[S5700-Vlanif100] quit
```

（4）配置 Trunk，方法如下：在连接接入层交换机的上行端口设置为 Trunk，并允许相应的 VLAN 通过该接口。

```
[S5700]interface GigabitEthernet 0/0/1                    # 进入端口
[S5700-GigabitEthernet0/0/1]port link-type trunk          # 设置端口为 Trunk
[S5700-GigabitEthernet0/0/1]port trunk allow-pass vlan 10 20 30  # 设置允许通过的 vlan
[S5700-GigabitEthernet0/0/1]quit                          # 退出
```

（5）配置与出口路由器互联接口，方法如下：

```
[S5700]interface GigabitEthernet 0/0/2                    # 进入端口
[S5700-GigabitEthernet0/0/2]port link-type access         # 设置端口为 access 模式
[S5700-GigabitEthernet0/0/2]port default vlan 100         # 把端口划分到 vlan 100 中
[S5700-GigabitEthernet0/0/2]quit                          # 退出
```

（6）配置默认路由，方法如下：

```
[S5700]ip route-static 0.0.0.0 0 10.10.100.2              # 配置默认路由，把未知网络的数据发送给
                                                          路由器
```

（7）配置 DHCP 服务：

1）启用 DHCP 服务，方法如下：

```
[S5700] dhcp enable
```

2）配置接口地址池，方法如下：

```
[S5700] interface vlanif 10                          #进入 vlan10 接口
[S5700-Vlanif10] dhcp select interface               #配置 dhcp 地址获取方式
[S5700-Vlanif10] dhcp server dns-list 202.103.224.68 #配置 dns 地址
[S5700-Vlanif10] quit                                #退出
```

3）同样方法，配置其他接口地址池：

```
[S5700] interface vlanif 20
[S5700-Vlanif20] dhcp select interface
[S5700-Vlanif20] dhcp server dns-list 202.103.224.68
[S5700-Vlanif20] quit

[S5700] interface vlanif 30
[S5700-Vlanif30] dhcp select interface
[S5700-Vlanif30] dhcp server dns-list 202.103.224.68
[S5700-Vlanif30] quit
```

2.3.6.3　出口设备

（1）修改设备名，方法如下：

```
<Huawei>system-view
[Huawei]sysnameAR3260                                #修改设备名
[AR3260]undo info-center enable                      #关闭消息提示
```

（2）给接口配置 IP 地址，方法如下：

```
[AR3260]interface GigabitEthernet 0/0/1              #进入外网口
[AR3260-GigabitEthernet0/0/1]ip address 1.1.1.1 30   #给外网口配置地址
[AR3260-GigabitEthernet0/0/1]quit                    #退出

[AR3260]interface GigabitEthernet 0/0/0              #进入外网口
[AR3260-GigabitEthernet0/0/0]ip address 10.10.100.2 30 #给内网口配置地址
[AR3260-GigabitEthernet0/0/0]quit                    #退出
```

（3）配置局域网用户通过 NAT 转换将私网地址转换为公网地址，进行上网，方法如下：

```
[AR3260]acl number 2000                              #定义一个标准 acl
[AR3260-acl-basic-2000]rule permit source any        #规则是允许源地址为任何人
[AR3260-acl-basic-2000]quit                          #退出
```

```
[AR3260]interface GigabitEthernet 0/0/1                #进入外网口
[AR3260-GigabitEthernet0/0/1]nat outbound 2000         #允许 acl 编号 2000 匹配的计算机能通过 nat 上网
[AR3260-GigabitEthernet0/0/1]quit                      #退出
```

（4）配置到 Internet 的默认路由，方法如下：

```
[AR3260]ip route-static 0.0.0.0 0 1.1.1.2     任何到 Internet 的数据都往 1.1.1.2 地址送
```

（5）配置到回内部网络的静态路由，方法如下：

```
[AR3260]ip route-static 10.10.10.0 24 10.10.100.1     #回 vlan10 的路由网 10.10.100.1 地址送
[AR3260]ip route-static 10.10.20.0 24 10.10.100.1     #回 vlan20 的路由网 10.10.100.1 地址送
[AR3260]ip route-static 10.10.30.0 24 10.10.100.1     #回 vlan30 的路由网 10.10.100.1 地址送
```

2.3.6.4　测试

（1）电脑自动获取 IP 地址，如图 2.4 所示。

图 2.4　电脑设置为通过 DHCP 自动获取地址

（2）查看获取的地址，如图 2.5 所示。

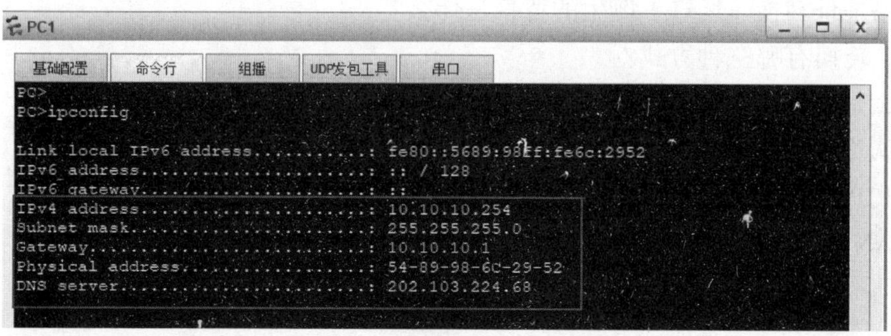

图 2.5　查看电脑获取的 IP 地址

（3）检测访问互联网，如图 2.6 所示。

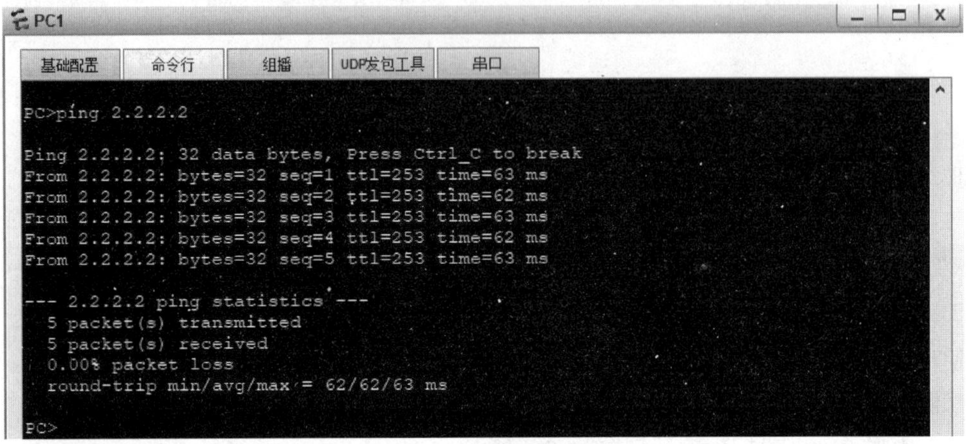

图 2.6　访问互联网检测

2.4　思政融入

互联网 + 时代，国家对新型基础设施建设投入加大，5G 基站、千兆光网、数据中心等基础设施建设需要大量的产业技术工人。课程的学习把不同的网络工程项目融入教材当中，提高实践操作性，企业在构建网络工程项目时，要求设计规范，做到具有实用性、先进性、开放性、可扩展性、安全性和可靠性。在工程项目的学习中，大学生应当树立职业认知，养成良好的职业习惯，秉持严谨的科学态度，对职业敬畏、对工作执着、对项目负责，学思结合、知行合一，增强勇于探索的创新精神、提升解决问题的工程实践能力，树立精益求精的大国工匠精神，通过科技报国实现家国情怀和使命担当。

2.5　思考与训练

1. 网络地址转换（NAT）诞生的背景是什么？
2. NAT 转换有哪三种方式？
3. 什么是静态路由？静态路由适合于什么场合？
4. 默认路由用在什么场合？其命令写法是什么？
5. DHCP 的作用是什么？简述它的工作原理。
6. VLAN 的划分方式有哪几种？请写出划分 VLAN 的配置命令。

Project 2 LAN project of a small enterprise

2.1 Project import

With the information wave surging today, the construction of enterprise internal network has become the key factor to enhance the core competitiveness of enterprises. Enterprise network has been mentioned by more and more people. Using network technology, modern enterprises can achieve optimized information communication among suppliers, partners and employees. This is directly related to whether enterprises can obtain key competitive advantages. In recent years, more and more enterprises are speeding up the construction of their own information networks, most of which are small and medium-sized enterprises.

Small and medium-sized enterprise LANs are usually small in scale, simple in structure, and their requirements for performance vary greatly depending on the application. Many small and medium-sized enterprises have few technicians, so they are highly dependent on the network. They require the network to be as simple, reliable and easy to use as possible, which can reduce the use and maintenance costs of the network and improve the performance price ratio of products.

2.2 Relevant knowledge

2.2.1 NAT address translation

2.2.1.1 Background of NAT birth

There are two kinds of IP addresses in the network: public address and private address. Public IP address refers to the only IP address in the world on the Internet. All public addresses must be registered with the local Internet Registration Authority (RIR). Enterprises rent public addresses from ISPs. Only the registered owner of the public address can assign the address to the network device. With the increasing number of computers connected to the Internet, IP address resources are becoming more and more stretched.

Due to the limited public IP address resources, private IP addresses are generally used in our enterprise environment.

RFC 1918 reserves three IP address blocks for private networks, as follows:

(1) Category A: 10.0.0.0 ~ 10.255.255.255.

(2) Category B: 172.16.0.0 ~ 172.31.255.255.

(3) Category C: 192.168.0.0 ~ 192.168.255.255.

The addresses within the above three ranges will not be assigned on the Internet, so they can be freely used within the company or enterprise without applying to the ISP or registration center.

Private IP addresses cannot directly access public network resources. At this time, we need to use NAT services. Network address translation (NAT) converts internal private addresses to one or more public addresses for routing over the Internet. Network devices (such as routers, firewalls, etc.) that perform NAT function change the private address in each packet to the publicly registered public IP address, and then send the packet to the Internet.

Nat service is used to map one address segment to another, which can convert the intranet address (private address) into a legal public IP address through NAT, so that the intranet address can access the public network.

2.2.1.2 Type of NAT

NAT can be implemented in three ways: static conversion(Static Nat), dynamic conversion(Dynamic Nat) and port multiplexing(OverLoad).

Static conversion refers to converting the private IP address of the internal network into a public IP address. The IP address pair is one-to-one and invariable. A private IP address is only converted into a public IP address. With the help of static transformation, the external network can access some specific devices (such as servers) in the internal network.

Dynamic conversion means that when the private IP address of the internal network is converted to the public IP address, the IP address is uncertain and random. All private IP addresses authorized to access the Internet can be randomly converted to any specified legal IP address. In other words, as long as you specify which internal addresses can be converted and which legal addresses can be used as external addresses, dynamic conversion can be carried out. Dynamic conversion can use multiple legal external address sets. When the legal IP address provided by ISP is slightly less than the number of computers in the network. Dynamic conversion can be adopted.

Port address Translation (PAT) refers to changing the source port of an outgoing packet and performing port translation, i.e., Port Address Translation (PAT). Using port multiplexing. All hosts of the internal network can share a legal external IP address to access the Internet, which can save IP address resources to the maximum extent. At the same time, it can hide all hosts inside the network, effectively avoiding attacks from the Internet. Therefore, port multiplexing is the most widely used method in the current network.

2.2.1.3 Basic configuration of NAT

(1) Static NAT configuration. Static NAT is one-to-one, which can not save IP addresses, but for servers mapped to the intranet. There are two public network addresses 202.106.1.1/32 and 202.106.1.2/32; If the private address of the intranet PC is 192.168.10.254/24 or 192.168.20.254/24, the static NAT configuration of router R1 is:

```
<R1>system-view
[R1]int g0/0/0
[R1-GigabitEthernet0/0/0]nat static global 202.106.1.1 inside 192.168.10.254 #Map this public address to the host with the internal IP address 192.168.10.254, namely PC1
[R1-GigabitEthernet0/0/0]nat static global 202.106.1.2 inside 192.168.20.254 #Same as above mapped to PC2
```

(2) Dynamic NAT configuration. Dynamic NAT makes an address pool on the exit router. When the intranet PC accesses the external network, it will obtain a public IP from the address pool. How many public IPs and how many intranet PCs can access the Internet at the same time.

The following public network addresses are available: 202.106.1.0/24; The private address subnet of the intranet PC is 192.168.10.0/24 192.168.20.0/24, so the dynamic NAT configuration of router R1 is:

```
[R1]nat address-group 1 202.106.1.1 202.106.1.254  #Create a NAT address pool
[R1]acl 2000  #Define an access control list
[R1-acl-basic-2000]rule permit source 192.168.10.0 0.0.0.255  #Allow 192.168.10.0/24 subnets
[R1-acl-basic-2000]rule permit source 192.168.20.0 0.0.0.255  #Allow 192.168.10.0/24 subnets
[R1-acl-basic-2000]quit
[R1]interface g0/0/0
[R1-GigabitEthernet0/0/0]nat outbound 2000 address-group 1 no-pat  #Associate the ACL with the address pool. No-pat indicates that it cannot be used repeatedly
```

(3) PAT configuration. Pat use a public network address repeatedly, and all hosts access the Internet through it.

There is an public address 202.106.1.1/29; The private address subnet of the intranet PC is 192.168.10.0/24 192.168.20.0/24, so the PAT configuration of the router R1 is:

```
[R1]nat address-group 1 202.106.1.1 202.106.1.1  #Create an address pool
[R1]acl 2000          #Define an access control list
[R1-acl-basic-2000]rule permit source 192.168.10.0 0.0.0.255   #Match address
```

[R1-acl-basic-2000]rule permit source 192.168.20.0 0.0.0.255 #Match address

[R1-GigabitEthernet0/0/0]nat outbound 2000 address-group 1 #Associating ACLs with address pools

At present, there is no public IP but one external network interface G0/0/0IP: 12.0.0.1; The private address subnet of the intranet PC is 192.168.10.0/24 192.168.20.0/24, so the PAT configuration of the router R1 is:

[R1]acl 2000 #Define an access control list

[R1-acl-basic-2000]rule permit source 192.168.10.0 0.0.0.255 #Match address

[R1-acl-basic-2000]rule permit source 192.168.20.0 0.0.0.255 #Match address

[R1]interface g0/0/0

[R1-GigabitEthernet0/0/0]ip address 12.0.0.1 24

[R1-GigabitEthernet0/0/0]nat outbound 2000 #Reuse the current interface address

(4) Static port mapping configuration. Enter the following commands based on PAT:

[R1-GigabitEthernet0/0/0]nat static protocol tcp global current-interface telnet inside 192.168.1.10 telnet #Map port 23 of the current interface to port 23 of 192.168.1.10. Enter the port number or protocol here

The implementation effect is shown in Figure 2.1.

```
<R2>telnet 12.0.0.1 23    port number
  Press CTRL_] to quit telnet mode
  Trying 12.0.0.1 ...
  Connected to 12.0.0.1 ...


Login authentication


Password:
Info: The max number of VTY users is 5, and the number
      of current VTY users on line is 1.
      The current login time is 2015-07-11 15:40:51.
<SW1>
<SW1>
```

Figure2.1 Mapping Port 23 Test

[R1-GigabitEthernet0/0/0]nat static protocol tcp global current-interface 1212 inside 192.168.1.20 telnet #Map port 1212 of the current interface to port 23 of 192.168.1.20, and the implementation effect is shown in Figure 2.2.

```
<R2>telnet 12.0.0.1 1212  port number
  Press CTRL_] to quit telnet mode
  Trying 12.0.0.1 ...
  Connected to 12.0.0.1 ...


Login authentication


Password:
Info: The max number of VTY users is 5, and the numbe
      of current VTY users on line is 1.
      The current login time is 2015-07-11 15:42:57.
<SW2>
<SW2>
<SW2>
```

Figure 2.2 Mapping Port 1212 Test

2.2.2 Static routing

2.2.2.1 Static route definition

Static routing refers to the routing information manually configured by users or network administrators. When the network topology or link status changes, the network administrator needs to manually modify the relevant static routing information in the routing table. Static routing information is private by default and will not be passed to other routers. By configuring static routes, network engineers can manually specify the path to be passed when accessing a network.

Static routing is useful in the following situations:

(1) The network scale is small, and there are few changes, or there are no redundant links.

(2) The enterprise network has many small branches, and there is only one path to the rest of the network.

(3) Enterprises want to send data packets to the Internet host, not to the host of the enterprise network.

2.2.2.2 Configuration code

Command format:

[R1]ip route-static 10.10.100.0 24192.168.1.1 #This command means that the next hop address for accessing 10.10.100.0/24 is 192.168.1.1

In practice, it is possible that a route may fail, but there are other paths available. You can configure a backup static route by adjusting the priority parameters, such as:

[R1]ip route-static 10.10.100.0 24 192.168.2.1 preference 100 #Configure the priority of static routes to play the role of backup links

2.2.3 Default route

2.2.3.1 Default route definition

Default route is a special static routing, which refers to the choice that the router can make when there is no matching table entry between the routing table and the destination address of the packet. If there is no default route, the packets whose destination address does not match the table entry in the routing table will be discarded. The default route is very effective in some cases. When there is an end network, the default route will greatly simplify the configuration of the router, reduce the workload of the administrator and improve the network performance.

2.2.3.2 Configuration code

Command format:

```
[R1]ip route-static 0.0.0.0 0 10.0.0.1 #Configure default routes
```

In practice, the default route may be invalid, but there are other paths available. You can configure the backup default route by adjusting the priority parameters, such as:

```
[R1]ip route-static 0.0.0.0 0 10.0.12.2 preference 100 #Configure the default backup route
```

2.2.4 DHCP

2.2.4.1 DHCP role

DHCP (Dynamic Host Configuration Protocol) is an application layer protocol, which is usually applied in large-scale local area network environment. Its main function is to centrally manage and allocate IP address, so that the host in the network environment can dynamically obtain IP address, gateway address, DNS server address and other information, and can improve the utilization rate of address.

IP address is a parameter that must be configured for each computer. It can be obtained from the DHCP server by static manual input or automatic acquisition. Manual setting is easy to make mistakes, and it is not easy to find problems when making mistakes. On the contrary, DHCP server can automatically assign IP address to the host from the preset IP address pool. It can not only ensure that the IP address will not be repeatedly allocated, but also recover the IP address in time to improve the utilization efficiency of IP address.

2.2.4.2 How DHCP works

The predecessor of DHCP was bootstrap protocol. BOOTP was created to automatically assign addresses to devices connected to the network. Later, it was replaced by DHCP. DHCP is more complex and powerful than BOOTP. Its working principle can be summarized as follows:

1. Find DHCP server

When the DHCP client logs in to the network for the first time (that is, there is no IP address data on the client), it will send a DHCP discover packet to the network through UDP 67 port (the packet contains the MAC address, computer name and other information of the client). Because the client does not know which network it belongs to, the source address of the packet is 0.0.0.0 and the destination address is 255.255.255.255, and then attach the information of DHCP discover to broadcast to the network.

The waiting time of DHCP discover is preset to 1 second, that is, when the client sends the first DHCP discover packet and does not get a response within 1 second, it will broadcast the second DHCP discover. If there is no response, the client will resend the broadcast packet four times (at intervals of 2, 4, 8 and 16 seconds, plus a random length of time between 1-1000 milliseconds). If there is no response from the DHCP server, the client will choose an IP address from 169.254.0.0/16, a private IP address automatically reserved. And broadcast again every 5 minutes. If a response is received from one of the servers, the IP lease process continues.

2. Provide IP address rental

When the DHCP server listens to the DHCP discover broadcast sent by the client, it will select the first vacant IP from those addresses that have not been rented out, and respond to a DHCP offer packet (including IP address, subnet mask, address lease term, etc.) to the client through UDP 68 port together with other TCP/IP settings. At this time, broadcast is still used for communication. The source IP address is the IP address of DHCP server and the destination address is 255.255.255.255. At the same time, the DHCP server reserves the IP address it provides for this client, so it will not assign this IP address to other DHCP clients.

Since the client does not have an IP address at the beginning, its MAC address information will be included in its DHCP discover packet, and there will be a XID number to identify the packet. The DHCP offer packet responded by the DHCP server will be passed to the client requesting the lease according to these data.

3. Accept IP lease

If the client receives the response from multiple DHCP servers on the network, it will only select one DHCP offer (generally the one that arrives first), and send a DHCP request broadcast packet to the network (the packet includes the MAC address of the client, the IP address in the accepted lease, the DHCP server address providing this lease, etc.), telling all DHCP servers which server it will accept the IP address provided by. All other DHCP servers revoke their provision in order to provide the IP address to the next IP lease request. At this time, the client still uses 0.0.0.0 as the source IP address and 255.255.255.255 as the target address for broadcasting, because it has not been finally confirmed by the DHCP server.

In fact, not all DHCP clients will unconditionally accept the offer of DHCP server, especially if other TCP/IP related client software is installed on these hosts. The client can also use DHCP request to propose DHCP options to the server. These options will be filled in the DHCP option field with different numbers. Clients can keep some of their own TCP/IP settings.

4. Lease confirmation

After receiving the DHCP request from the client, the DHCP server will broadcast and return a DHCP ACK message packet to the client, indicating that the client' s selection has been accepted, and put the legal lease of this IP, address and other configuration information into the broadcast packet and send it to the client.

After receiving the DHCP ACK broadcast, the client will send three ARP resolution requests for this IP address to the network to perform conflict detection and query whether other machines on the network use this IP address; If the IP address is found to have been used, the client sends a DHCP define packet to the DHCP server, rejects the IP address lease, and resends the DHCP discover information. At this time, in the DHCP server management console, this IP address will be displayed as BAD_ ADDRESS.

If no other host on the network uses this IP address, the TCP/IP of the client completes initialization using the IP address provided in the lease, so that it can communicate with the hosts in other networks.

5. DHCP client lease renewal

When 50% of the lease term has passed, the customer will directly send the DHCP request message packet to the DHCP server that provides the IP address. If the client receives the DHCP ACK message packet responded by the server, the client will update its configuration according to the new lease term and other updated TCP/IP parameters provided in the packet, and the IP lease update is completed. If no reply is received from the server, the client will continue to use the existing IP address because there is still 50% of the current lease term.

If there is no update in the past 50% of the lease term, the client will contact the DHCP that provides it with the IP address again in the past 87.5% of the lease term. If it is not successful, the client must give up the IP address and reapply at 100% of the lease. If DHCP is not available at this time, the client uses a random address in 169.254.0.0/16 and tries again every 5 minutes.

2.2.4.3 Basic DHCP service configuration

(1) The DHCP service configuration command is as follows:

```
[Huawei]dhcp enable                    #Enable DHCP function
[Huawei]interface vlanif 1             #Enter VLANIF 1 interface view
```

[Huawei-Vlanif1]ip address 192.168.1.1 24 #Configure IP address of VLANIF interface

[Huawei-Vlanif1]dhcp select global #The DHCP service function of the interface can specify the DHCP server to allocate addresses from the global address pool (can be added under interface or VLAN)

[Huawei-Vlanif1]quit #Leave the current view

[Huawei]ip pool gxsdxy #Enter the global address pool gxsdxy

[Huawei-ip-pool-gxsdxy]network 192.168.1.0 mask 255.255.250.0 #Configure the IP address range under the address pool

[Huawei-ip-pool-gxsdxy]gateway-list 192.168.1.1 #Configure DHCP gateway

[Huawei-ip-pool-gxsdxy]dns-list 114.114.114.114 8.8.8.8 #Configure Primary DNS and Backup DNS

(2) Optional Configurations

[Huawei-ip-pool-gxsdxy]lease day 1 hour 1 minute 1 #Configure the IP address lease term. This configuration is one hour and one minute a day. The default is one day. Unlimited is unlimited

[Huawei-ip-pool-gxsdxy]domain-name ××× #Configure the DNS domain name suffix assigned to DHCP clients (optional among options)

[Huawei-ip-pool-gxsdxy]excluded-ip-address 192.168.1.100 192.168.1.254

[Huawei-ip-pool-gxsdxy]excluded-ip-address 192.168.1.1

#Configure the IP addresses in the address pool that do not participate in automatic allocation. Execute this command several times to configure multiple IP address segments that do not participate in automatic allocation

[Huawei-ip-pool-gxsdxy]static-bind ip-address 172.16.35.253 mac-address 28d2-4469-5a55

#When a user needs a fixed IP address, the IP address not in use in the address pool can be bound to the user's MAC address

2.2.4.4 DHCP relay configuration

When the client is not in the same network segment as the DHCP server, forward the request from the client to the DHCP server through the DHCP relay device.

[Huawei]dhcp enable #Enable DHCP function

[Huawei]interface vlanif 1 # VLANIF 1 interface view

[Huawei-Vlanif1]ip address 192.168.1.1 mask 24 #IP address of VLANIF interface

[Huawei-Vlanif1]dhcp select relay #Start the DHCP relay function of the VLANIF interface. Please configure the device as a DHCP relay as follows.

[Huawei-Vlanif1]dhcp server group group-name #Create DHCP server group and enter DHCP server group view

[Huawei-Vlanif1]dhcp-server ip-address [ip-address-index] #Add a DHCP server to the DHCP server group. A maximum of 20 DHCP servers can be configured under each DHCP server group. When no index is specified, the system will automatically allocate an idle index.

2.2.4.5 Check

[Huawei]display dhcp server group [group-name] #Use the command to view the configuration information of DHCP server group members

2.2.5 VLAN

2.2.5.1 VLAN overview

Virtual local area network (VLAN) is a group of logical devices and users. These devices and users are not limited by physical location. They can be organized according to functions, departments, applications and other factors. The communication between them is like they are in the same network segment, so it is named virtual LAN. VLAN is a relatively new technology. It works in layers 2 and 3 of OSI reference model. In computer network, a layer 2 network can be divided into multiple different broadcast domains. A broadcast domain corresponds to a specific user group, and a VLAN is a broadcast domain. By default, these different broadcast domains are isolated from each other. Communication between different broadcast domains needs to be completed through layer 3 routers. Such a broadcast domain is called VLAN. Compared with traditional LAN technology, VLAN technology is more flexible. It has the following advantages: the management overhead of moving, adding and modifying network devices is reduced; can control broadcast activities; can improve the security of the network. In addition to the advantages of dividing the network into multiple broadcast domains, so as to effectively control the occurrence of broadcast storms and make the network topology very flexible, VLAN can also be used to control the mutual access between different departments and stations in the network.

Traditional local area network (LAN) refers to a computer group interconnected by multiple computers in a certain area. Usually within a few thousand meters. LAN can realize the functions of file management, application software sharing, printer sharing, schedule within the working group, e-mail and fax communication services. LAN is a closed type, which can be composed of two computers in the office or thousands of computers in a company.

The development of switching technology has also accelerated the application speed of new switching technology (VLAN). By dividing the enterprise network into virtual network VLAN segments, we can strengthen network management and network security and control unnecessary data broadcasting. In a shared network, a physical network segment is a broadcast domain. In the switching network, the broadcast domain can be a virtual network segment composed of a set of arbitrarily selected layer 2 network addresses (MAC addresses). In this way, the division of working groups in the network can break through the geographical location restriction in the shared network and be divided completely according to the management function. This grouping mode based on Workflow greatly improves the management function of network planning and reorganization. Workstations in the same VLAN, no matter which switch they are actually connected to, communicate with each other as if they were on a separate switch. The broadcast in the same VLAN can only be heard by members in the VLAN and will not be transmitted to other VLANs, which can well control the generation of unnecessary broadcast storms. At the same time, if there is no route, different VLANs cannot communicate with each other, which increases the security between different departments in the enterprise network. Network administrators can fully manage the information exchange between different management units within the enterprise by configuring the routing between VLANs.

2.2.5.2 VLAN Division

(1) VLAN based on port division—this is the most commonly used VLAN division method. At present, the vast majority of VLAN protocol switches provide this VLAN configuration method, which is divided into different VLANs according to the physical port of the switch.

Advantages: it is very simple to define VLAN members, as long as all ports are defined as corresponding VLAN groups.

Disadvantages: if a user leaves the original port to a port of a new switch, it must be redefined. Suitable for networks of any size.

(2) VLAN division based on MAC address—this method of VLAN division is based on the MAC address of each user host.

Advantages: when users change their physical location from one switch to another, VLAN does not need to be reconfigured.

Disadvantages: all users must be configured during initialization. Suitable for small LAN.

(3) VLAN division based on network layer protocol—VLAN is divided according to network layer protocol, which can be divided into IP, IPX, DECnet, Apple Talk and other VLAN networks.

Advantages: the physical location of the user has changed, and there is no need to reconfigure the VLAN. In addition, VLANs can be divided according to the protocol type, network traffic can be reduced, and the broadcast domain can span multiple VLAN switches.

Disadvantages: low efficiency. It is suitable for networks that need to run multiple protocols at the same time.

(4) Divide VLANs according to IP Multicast—IP multicast is actually a definition of VLAN, that is, an IP multicast group is a VLAN.

Advantages: greater flexibility and easy expansion through routers.

Disadvantages: suitable for LAN, mainly due to low efficiency. It is suitable for LAN not in the same geographical range.

2.2.5.3 VLAN configuration

(1) The general configuration is as follows:

[Switch]vlan 2 #Add a vlan2

[Switch-vlan2]quit #sign out

[Switch]int e0/0/2 #Enter e0/0/2 interface

[Switch-Ethernet0/0/2]port link-type access #The link type of the port is Access (the port connecting the computer). Huawei's default interface is the hybrid mode

[Switch-Ethernet0/0/2]port default vlan 2 #By default, all ports belong to VLAN1. This command adds this interface to VLAN2

[Switch-Ethernet0/0/2]quit #sign out

(2) The method of putting a series of continuous ports into VLAN is:

[Switch]vlan 3 #Add a vlan2

[Switch-vlan3]port e0/9 to e0/16 #Eight ports from e0/9 to e0/16 of the specified switch belong to VLAN3

[Switch-vlan3]quit #sign out

(3) Check. After the configuration is completed, execute the display command to display the operation of the configured VLAN. Verify the configuration effect by viewing the display information.

[Switch]display vlan #View the current vlan list through display vlan

[Switch]display vlan vlan-id #For example, display vlan 100 to check the status of vlan 100 and to check which ports are included in vlan 100. You can also view the configuration through "display cur", and view the port status "display interface brief"

2.2.6 Link type of Ethernet port

2.2.6.1 Tag and Untag

(1) Untag is an ordinary Ethernet message. The network card of an ordinary PC can recognize such a message for communication.

(2) The change of tag message structure is to add 4 bytes of VLAN information after the source MAC address and destination MAC address, that is, VLAN tag header; Generally speaking, such messages cannot be recognized by the network card of ordinary PC. The frame with 802.1Q inserts a 4-byte identification on the standard Ethernet frame. This includes:

1) The 2-byte protocol identifier (TPID) is currently set to a fixed value of 0x8100,

indicating that the frame has 802.1Q tag information.

2) The 2-byte tag control information (TCI) contains three fields. ① Priority domain, accounting for 3 bits, indicates the priority of the message. The value is 0 to 7. 7 is the highest priority and 0 is the lowest priority. This domain is adopted by 802.1p. ② For Ethernet specification, 1 bit indicates the format (CFI), accounting for 0; 1 refers to non-standard format, which is applied to token ring. ③ VLAN ID domain, accounting for 12 bits, is used to identify the ownership of VLAN.

2.2.6.2 PVID

PVID usually stands for port ID and VID stands for VLAN ID. Before 802.1Q, VLAN based on MAC or port is adopted. VID appears only after tag based 802.1Q is introduced. If port based VLAN is divided, PVID only needs to be configured. If tagged VLAN based on 802.1Q is divided, VID needs to be configured. In summary, a port belongs to one PVID, but can belong to multiple VIDs. The PVID of switch is 1 by default.

2.2.6.3 Access – Link type of Ethernet port access

The access port can only belong to one VLAN, which is generally used to connect the port of the computer. Access port only belongs to one VLAN, so its default VLAN is the VLAN it is in. Access interface access data processing process of switch:

(1) Message received by access port: After receiving a message, judge whether there is VLAN information: if not, mark the PVID of the port and exchange and forward it; If so, discard it directly (default).

(2) A message sent from access port: After receiving a message, judge whether the VLAN ID in the message is the same as the PVID. If the same, remove the tag in the message and forward it; If different, discard it directly.

(3) Configure access port:

```
[Huawei]int e0/0/3
[Huawei-Ethernet0/3]port link-type access  #Configure port e0/3 access port
[Huawei-Ethernet0/3]port default vlan 10   #Put the port into vlan 10
```

2.2.6.4 Trunk–Link type of Ethernet port trunk

Trunk type ports can allow multiple VLANs to pass through, and can receive and send messages from multiple VLANs. They are generally used for the ports connected between switches.

Processing process of incoming and outgoing data of Trunk interface of switch:

(1) Trunk port receiving message: After receiving a message, judge whether there is VLAN information: if not, mark the PVID of the port and exchange and forward it; If yes, judge

whether the Trunk port allows the data of the VLAN to enter. Forward if Possible, otherwise discard.

(2) Send message through Trunk port: Compare the PVID of the port and the VLAN ID information of the message to be sent. If they are equal, peel off the VLAN information and send it again; If they are not equal, judge whether the Trunk port allows the data of the VLAN to pass through. If so, forward it, otherwise discard it.

(3) Configure Trunk port:

```
[Huawei]int e0/0/3
[Huawei-Ethernet0/3]port link-type trunk            #Configure port e0/3 as trunk port
[Huawei-Ethernet0/3]port trunk allow pass vlan all  #Allow all vlans to pass
```

In general, it is better to specify which specific VLANs are allowed to pass through the port. Do not set to allow all VLANs to pass through, for example:

```
[Huawei-Ethernet0/3] port trunk allow pass vlan 10  #Only VLAN 10 is allowed to pass
```

2.2.6.5 Link type of Ethernet port－Hybrid

Hybrid ports can allow multiple VLANs to pass through, receive and send messages from multiple VLANs, and can be used for connection between switches or to connect users' computers.

When receiving data, Hybrid and Trunk ports handle it the same way, except when sending data: Hybrid ports allow multiple VLANs to send messages without labeling, while Trunk ports only allow messages from the default VLANs to be sent without labeling.

The Hybrid port and Trunk port belong to multiple VLANs, so the default VLAN ID needs to be set. By default, the default VLANs for Hybrid and Trunk ports are VLAN 1.

If the default VLAN ID of the port is set, when the port receives a message without VLAN Tag, it forwards the message to the port belonging to the default VLAN; When a message with a VLAN Tag is sent by a port, if the VLAN ID of the message is the same as the default VLAN ID of the port, the VLAN Tag of the message will be removed and sent again.

The default VLAN for switches is called Pvid Vlan, and the default VLAN for Cisco switches is called Native Vlan. Switch interface access data processing:

(1) Message received by hybrid port: Receive a message, determine if there is VLAN information: If not, put on the port's PVID and exchange and forward it; If so, determine if the hybrid port allows data from the VLAN to enter: forward if possible, or discard (the untag configuration on the port is not considered at this time, and the untag configuration only works when sending messages).

(2) Message sent by hybrid port: Determine the properties of the VLAN on the port (you

can see which VLANs are untag and which VLANs are tag for the port by displaying the interface command). If it is untag, the VLAN information is split and then sent. If it is tag, it is sent directly.

(3) Configuration command: Configure port E0/1 to be a Hybrid port capable of receiving messages from VLAN 20, 30, and 100:

```
[SwitchA]interface Ethernet 0/1
[SwitchA-Ethernet0/1]port link-type hybrid  #Configure port E0/0/1 as the Hybrid interface
[SwitchA-Ethernet0/1]port hybrid vlan 20 30 100 untagged  #Can accept the message sent from vlan 20 30 10
```

For Hybrid ports, can belong to multiple VLANs at the same time. These VLANs are the PVIDs of the Hybrid port and the manually configured "untagged" and "tagged" VLANs, respectively. It is important to note the VLAN configuration of the corresponding port to ensure that messages can be sent and received properly by the port.

2.3 Project implementation

2.3.1 Project requirements

(1) ISP operator exports in static IP form.

(2) The PC can automatically obtain the IP address.

(3) All PC computers can access the Internet.

2.3.2 Equipment list

Routers, core layer switches and access layer switches are required for this project. The list of equipment used is shown in Table 2.1.

Table 2.1 Equipment List

Serial No.	Equipment model	Quantity	Function	Remarks
1	AR3260	1	Exit router	
2	S5700–28C–HI	1	Core switch	
3	S3700–26C–HI	4	Access switch	

2.3.3 Project topology

The project topology is shown in Figure 2.3.

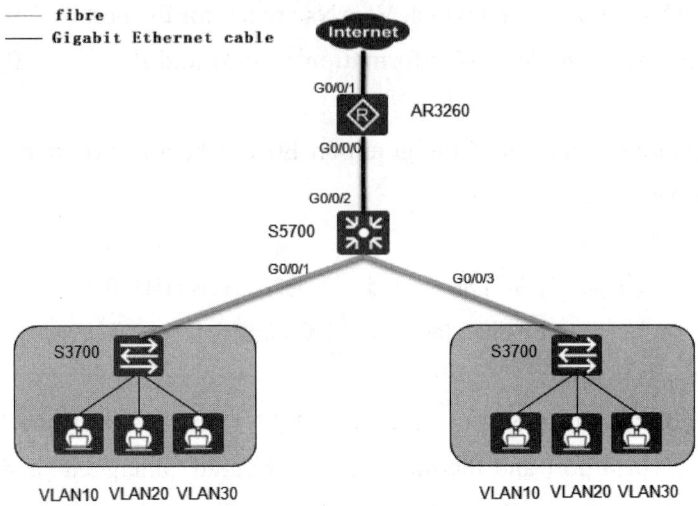

Figure 2.3 Small enterprise network topology

2.3.4 Configuration ideas

The project adopts the following ideas:

1. Access layer equipment

(1) Divide VLANs.

(2) Place the port connected to the PC in the corresponding VLAN.

(3) Set the port connecting the core switch to the Trunk port and allow the corresponding VLAN to pass through.

2. Core layer equipment

(1) Partition vlan.

(2) Configure the address for the vlan.

(3) Set the port connected to the access layer switch as the Trunk port and allow the corresponding vlan to pass through.

(4) Put the port connecting to the exit router in the vlan 100.

(5) Configure the default route to the exit router.

(6) Configure DHCP.

3. Exit router

(1) Configure the IP address for the interface.

(2) Configure LAN users to convert private network addresses to public network addresses through NAT conversion for Internet access.

(3) Configure the default route to the Internet.

2.3.5 Relevant information planning

2.3.5.1 Core switch

The information planning of the core layer switch is shown in Table 2.2.

Table 2.2 Core layer switch information planning

Serial No.	User department	VLAN No.	IP address	Mask	Remarks
1	Production Department	10	10.10.10.1	255.255.255.0	
2	Sales Department	20	10.10.20.1	255.255.255.0	
3	Finance Department	30	10.10.30.1	255.255.255.0	
4	Interconnection	100	10.10.100.1	255.255.255.252	Interconnect with router

2.3.5.2 Exit router

The information planning of the exit router is shown in Table 2.3.

Table 2.3 Exit router information planning

Serial No.	Interface	IP address	Mask	Remarks
1	G0/0/0	10.10.100.2	255.255.255.252	Interconnection with core switch

2.3.5.3 ISP export planning

ISP export information planning is shown in Table 2.4.

Table 2.4 ISP export information planning

Serial No.	Interface	IP address	Mask	Gateway	Remarks
1	G0/0/1	1.1.1.1	255.255.255.252	1.1.1.2	Static IP address

2.3.6 Operating steps

2.3.6.1 Access switch equipment

(1) Modify device name.

```
<Switch>system-view
[Switch]sysname S3700
[S3700]undo info-center enable  #Turn off message prompts
```

(2) Partition VLAN.

```
[S3700]vlan 10                            #Partition vlan10
[S3700-vlan10]description shengchanbu     #Describe vlan
[S3700-vlan10]quit                        #Sign out
```

Divide vlan20, 30 in the same method.

```
[S3700]vlan 20
[S3700-vlan20]description xiaoshoubu
[S3700-vlan20]quit

[S3700]vlan 30
[S3700-vlan30]description caiwubu
[S3700-vlan30]quit
```

(3) Put the port connected to the PC into the corresponding VLAN.

```
[S3700]interface Ethernet 0/0/1                #Enter port e0/0/1
[S3700-Ethernet0/0/1]port link-type access    #Set the port to access mode
[S3700-Ethernet0/0/1]port default vlan 10     #Put the port into vlan 10
[S3700-Ethernet0/0/1]quit                      #Sign out
```

Configure other interfaces in the same way.

```
[S3700]interface Ethernet 0/0/2
[S3700-Ethernet0/0/2]port link-type access
[S3700-Ethernet0/0/2]port default vlan 20
[S3700-Ethernet0/0/2]quit

[S3700]interface Ethernet 0/0/3
[S3700-Ethernet0/0/3]port link-type access
[S3700-Ethernet0/0/3]port default vlan 30
[S3700-Ethernet0/0/3]quit
```

(4) Configure Trunk.

```
[S3700]interface GigabitEthernet 0/0/1
[S3700-GigabitEthernet0/0/1]port link-type trunk  # Set the uplink port of the core switch to Trunk
[S3700-GigabitEthernet0/0/1]port trunk allow-pass vlan 10 20 30  #Allow·the corresponding vlan to pass through the interface
```

```
[S3700-GigabitEthernet0/0/1]quit
```

2.3.6.2 Core switch equipment

(1) Modify device name.

```
<Huawei>system-view
[Huawei]sysname S5700
[S5700]undo info-center enable                 #Turn off message prompts
```

(2) Divide VLAN.

```
[S5700]vlan 10                                 #Partition vlan 10
[S5700-vlan10]description shengchanbu          #Describe vlan
[S5700-vlan10]quit                             #Sign out
```

Divide other vlans in the same way.

```
[S5700]vlan 20
[S5700-vlan20]description xiaoshoubu
[S5700-vlan20]quit

[S5700]vlan 30
[S5700-vlan30]description caiwubu
[S5700-vlan30]quit

[S5700]vlan 100
[S5700-vlan100]description hulian
[S5700-vlan100]quit
```

(3) Configure VLANIF interface IP address.

```
[S5700]interface vlanif 10                     #Enter the vlan 10 interface
[S5700-vlanif10]ip address 10.10.10.1 24       #Configure IP address and mask information
[S5700-vlanif10]quit                           # Sign out
```

Configure other VLANs in the same way.

```
[S5700]interface vlanif 20
[S5700-vlanif20]ip address 10.10.20.1 24
[S5700-vlanif20]quit

[S5700]interface vlanif 30
```

```
[S5700-vlanif30]ip address 10.10.30.1 24
[S5700-vlanif30]quit

[S5700] interface vlanif 100
[S5700-Vlanif100] ip address 10.10.100.1 30
[S5700-Vlanif100] quit
```

(4) Configure Trunk. Set the downlink port of the access layer switch to Trunk, and allow the corresponding VLAN to pass through this interface.

```
[S5700]interface GigabitEthernet 0/0/1                          #Entry port
[S5700-GigabitEthernet0/0/1]port link-type trunk                #Set the port to Trunk
[S5700-GigabitEthernet0/0/1]port trunk allow-pass vlan 10 20 30 #Set the vlans allowed to pass
[S5700-GigabitEthernet0/0/1]quit                                # Sign out
```

(5) Configure the interconnection interface with the exit router.

```
[S5700]interface GigabitEthernet 0/0/2              #Entry port
[S5700-GigabitEthernet0/0/2]port link-type access   #Set the port to access mode
[S5700-GigabitEthernet0/0/2]port default vlan 100   #Set the port to access mode and divide the port into vlan 100
[S5700-GigabitEthernet0/0/2]quit                    # Sign out
```

(6) Configure default routes.

```
[S5700]ip route-static 0.0.0.0 0 10.10.100.2        #Configure the default route and send the data of the unknown network to the router
```

(7) Configure DHCP service.

1) Enable DHCP service.

```
[S5700] dhcp enable
```

2) Configuring the interface address pool.

```
[S5700] interface vlanif 10                          #Enter vlan 10 interface
[S5700-Vlanif10] dhcp select interface               #Configure the dhcp address acquisition method
[S5700-Vlanif10] dhcp server dns-list 202.103.224.68 #Configure dns address
[S5700-Vlanif10] quit                                # Sign out
```

3) Configure other interface address pools in the same way.

```
[S5700] interface vlanif 20
[S5700-Vlanif20] dhcp select interface
[S5700-Vlanif20] dhcp server dns-list 202.103.224.68
[S5700-Vlanif20] quit
```

```
[S5700] interface vlanif 30
[S5700-Vlanif30] dhcp select interface
[S5700-Vlanif30] dhcp server dns-list 202.103.224.68
[S5700-Vlanif30] quit
```

2.3.6.3 Export equipment

(1) Modify device name.

```
<Huawei>system-view
[Huawei]sysname AR3260            #Modify device name
[AR3260]undo info-center enable  #Turn off message prompts
```

(2) Configure the IP address for the interface.

```
[AR3260]interface GigabitEthernet 0/0/1                #Access to external network port
[AR3260-GigabitEthernet0/0/1]ip address 1.1.1.1 30    #Configure the address for the external network port
[AR3260-GigabitEthernet0/0/1]quit                      # Sign out

[AR3260]interface GigabitEthernet 0/0/0                #Access to external network port
[AR3260-GigabitEthernet0/0/0]ip address 10.10.100.2 30  #Configure the address for the intranet port
[AR3260-GigabitEthernet0/0/0]quit                      # Sign out
```

(3) Configure LAN users to convert private network addresses to public network addresses through NAT conversion for Internet access.

```
[AR3260]acl number 2000                                #Define a standard acl
[AR3260-acl-basic-2000]rule permit source any          #The rule is to allow the source address to be anyone
[AR3260-acl-basic-2000]quit                            # Sign out

[AR3260]interface GigabitEthernet 0/0/1                #Access to external network port
[AR3260-GigabitEthernet0/0/1]nat outbound 2000         #Allow computers matching ACL No. 2000 to access the Internet through nat
[AR3260-GigabitEthernet0/0/1]quit                      # Sign out
```

(4) Configure default routes to the Internet.

```
[AR3260]ip route-static 0.0.0.0 0 1.1.1.2            #Any data to the Internet will be sent to the 1.1.1.2 address
```

(5) Configure a static route back to the internal network.

```
[AR3260]ip route-static 10.10.10.0 24 10.10.100.1  #Route back to vlan 10 to 10.10.100.1
[AR3260]ip route-static 10.10.20.0 24 10.10.100.1  #Route back to vlan 20 to 10.10.100.1
[AR3260]ip route-static 10.10.30.0 24 10.10.100.1  #Route back to vlan 30 to 10.10.100.1
```

2.3.6.4 test

(1) The PC automatically obtains the IP address, as shown in Figure 2.4.

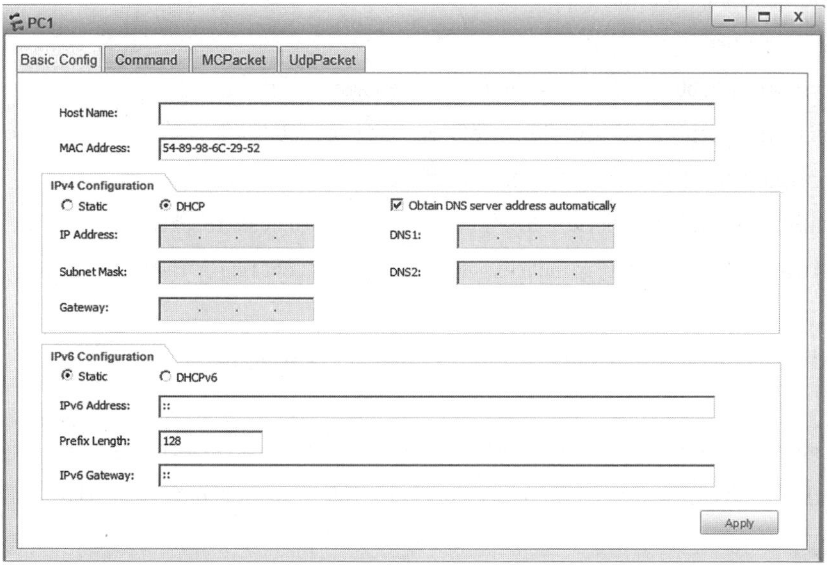

Figure 2.4 The PC is set to automatically obtain the address through DHCP

(2) Check the obtained address, as shown in Figure 2.5.

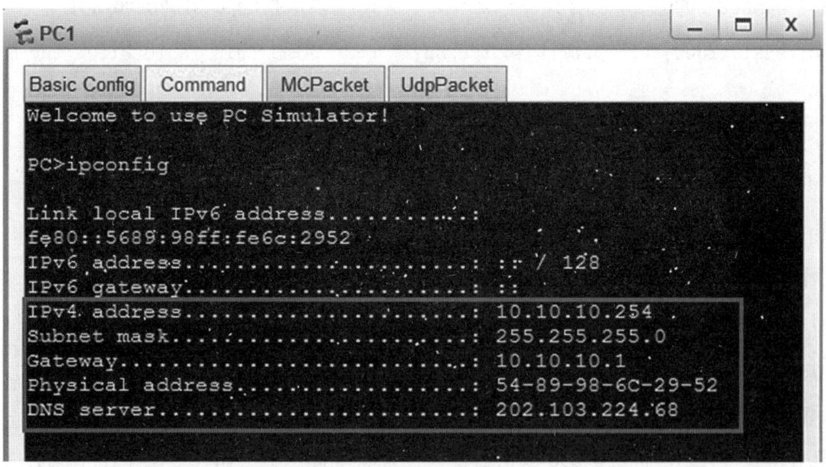

Figure 2.5 Check the IP address obtained by the PC

(3) Detect Internet access, as shown in Figure 2.6.

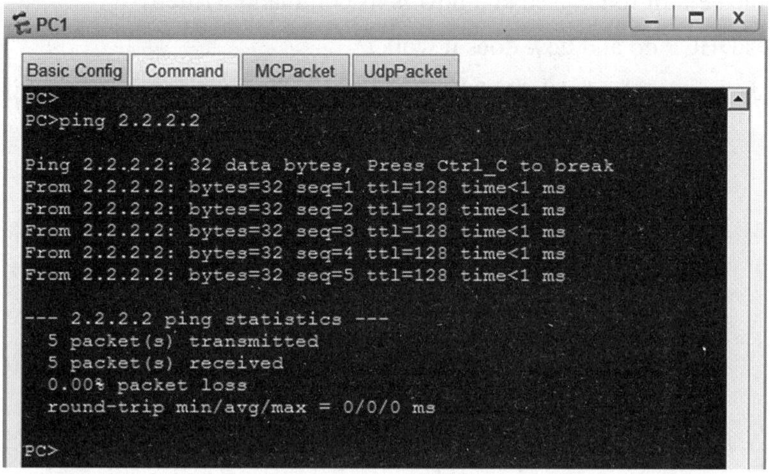

Figure 2.6 Internet access detection

2.4 Ideological and political integration

In the era of Internet plus, the state has increased investment in new infrastructure construction, infrastructure construction such as 5G base stations, gigabit optical networks, and data centers requires a large number of industrial skilled workers.The learning of the course integrates different network engineering projects into the teaching materials to improve the practical operation, and enterprises require design specifications, practicality, advancedness, openness, scalability, security and reliability when building network engineering projects.In the study of engineering projects, college students should establish professional cognition, develop good professional habits, uphold a rigorous scientific attitude, respect for the profession, be persistent to the work, and be responsible for the project. Combine learning and thinking, integrate knowledge and action, enhance the innovative spirit of exploration, improve the engineering practice ability to solve problems, establish the spirit of craftsmen of a great country that strives for excellence, and realize the feelings and mission of the country through science and technology.

2.5 Thinking and training

1. What is the background of the birth of Network Address Translation?
2. What are the three ways of NAT?

3. What is static route? What are the best scenarios for static routes?
4. When is the default route used and how is its command written?
5. What does DHCP do and how does it work?
6. What are the ways to divide VLANs? Write out the configuration command to configure the VLAN.

项目 3

项目 3　某木业集团企业局域网项目

3.1　项目导入

随着近年来企业信息化建设的深入，企业的运作越来越融入计算机网络，企业的沟通、应用、财务、决策、会议等等数据流都在企业网络上传输，构建一个“安全可靠、性能卓越、管理方便”的“高品质”中型企业网络已经成为企业信息化建设成功的关键基石。

某木业集团为了加快信息化建设，新的集团企业网将建设一个以集团办公自动化、电子商务、业务综合管理、多媒体视频会议、远程通信、信息发布及查询为核心，以现代网络技术为依托，技术先进、扩展性强，将集团的各种办公室、多媒体会议室、PC 终端设备、应用系统通过网络连接起来，实现内、外沟通的现代化计算机网络系统。

3.2　相关知识点

3.2.1　单区域 OSPF

3.2.1.1　概述

OSPF（Open Shortest Path First，开放最短链路优先）路由协议是典型的链路状态路由协议。OSPF 中的字母 O 意为 open，也就是开放、公有，任何标准化的设备厂商都能够支持 OSPF。OSPF 由 IETF 在 20 世纪 80 年代末期开发，是 SPF 类路由协议中的开放式版本。最初的 OSPF 规范体现在 RFC1131 中，被称为 OSPF 版本 1，但是版本 1 很快被进行了重大改进的版本所代替，OSPF 版本 2 的最新版体现在 RFC2328 中，OSPFv2 用在 IPv4 网络。而 OSPF 版本 3 是关于 IPv6 的，用在 IPv6 网络。

OSPF 是一个内部网关协议，用于在单一自治系统（Autonomous System,AS）内决策路由。是对链路状态路由协议的一种实现，隶属内部网关协议（IGP），因此运作于自治系统内部。OSPF 是个上层的协议，OSPF 报文封装在 IP 包头中，协议号为 89。

3.2.1.2　OSPF 特征

OSPF 作为一种内部网关协议（Interior Gateway Protocol，IGP），用于在同一个自治系统（AS）中的路由器之间交换路由信息。OSPF 的特性如下 :① 可适应大规模网络；② 收敛速度快；③ 无路由环路；④ 支持 VLSM 和 CIDR；⑤ 支持等价路由；⑥ 支持区域划分，构成结构化的网络；⑦ 提供路由分级管理；⑧ 支持简单口令和 MD5 认证；⑨ 以组播方式传送协议报文；⑩OSPF 路由协议的管理距离是 10；⑪ OSPF 路由协议采用 cost 作为度量标准；⑫OSPF 维护邻居表、拓扑表和路由表。

3.2.1.3 OSPF 的运行过程

（1）每个运行 OSPF 的路由器发送 HELLO 报文到所有启用 OSPF 的接口。如果在共享链路上两个路由器发送的 HELLO 报文内容一致，那么这两个路由器将形成邻居关系。

（2）从这些邻居关系中，部分路由器形成邻接关系。邻接关系的建立由 OSPF 路由器交换 HELLO 报文和网络类型来决定。

（3）形成邻接关系的每个路由器都宣告自己的所有链路状态。

（4）每个路由器都接受邻居发送过来的 LSA，记录在自己的链路数据库中，并将链路数据库的一份复制发送给其他的邻居。

（5）通过在一个区域中泛洪，使得该区域中的所有路由器同步自己的数据库。

（6）当数据库同步之后，OSPF 通过 SPF 算法，计算到目的地的最短路径，并形成一个以自己为根的无自环的最短路径树。

（7）每个路由器根据这个最短路径树建立自己的路由转发表。

3.2.1.4 OSPF 支持的网络类型

OSPF 将网络划分为四种类型：广播多路访问型（BMA）、非广播多路访问型（NBMA）、点到点型（Point-to-Point）、点到多点型（Point-to-MultiPoint）。OSPF 的网络类型决定了邻居邻接关系的形成，以及对 HELLO 报文的处理，使得 OSPF 的适应性和性能得到提高，不同的二层链路的类型需要 OSPF 不同的网络类型来适应。

（1）广播类型（Broadcast）。当链路层协议是 Ethernet、FDDI 时，缺省情况下，OSPF 认为网络类型是 Broadcast。在该类型的网络中通常以组播形式发送 Hello 报文、LSU 报文和 LSAck 报文。其中，224.0.0.5 的组播地址为 OSPF 设备的预留 IP 组播地址；224.0.0.6 的组播地址为 OSPF DR/BDR（Backup Designated Router）的预留 IP 组播地址。

（2）NBMA 类型（Non-Broadcast Multi-Access）。当链路层协议是帧中继、ATM 或者 X.25 时，缺省情况下，OSPF 认为网络类型是 NBMA。在该类型的网络中，以单播形式发送协议报文（Hello 报文、DD 报文、LSR 报文、LSU 报文、LSAck 报文）。

（3）P2MP 类型（Point-to-Multipoint）。没有一种链路层协议会被缺省地认为是 Point-to-Multipoint 类型。点到多点必须是由其他的网络类型强制更改的。常用做法是将非全连通的 NBMA 改为点到多点的网络。在该类型的网络中：① 以组播形式（224.0.0.5）发送 Hello 报文；② 以单播形式发送其他协议报文（DD 报文、LSR 报文、LSU 报文、LSAck 报文）。

（4）P2P 类型（Point-to-Point）。当链路层协议是 PPP、HDLC 和 LAPB 时，缺省情况下，

OSPF 认为网络类型是 P2P。在该类型的网络中，以组播形式（224.0.0.5）发送协议报文（Hello 报文、DD 报文、LSR 报文、LSU 报文、LSAck 报文）。

3.2.1.5　OSPF 选路

最短路径优先（SPF）算法是 OSPF 路由协议的基础。SPF 算法有时也被称为 Dijkstra 算法，这是因为最短路径优先算法（SPF）是 Dijkstra 发明的。OSPF 路由器利用 SPF，独立地计算出到达任意目的地的最佳路由。

著名的迪克斯加算法（Dijkstra）被用来计算最短路径树。与距离矢量路由协议直接交互路由器的路由表不同，OSPF 作为链路状态路由协议，路由器之间交互的是 LSA（链路状态通告），路由器将网络中泛洪的 LSA 搜集到自己的 LSDB（链路状态数据库）中，这有助于 OSPF 理解整张网络拓扑，并在此基础上通过 SPF 最短路径算法计算出以自己为根的、到达网络各个角落的、无环的树，最终，路由器将计算出来的路由装载进路由表中。

OSPF 工作原理：每台路由器通过使用 Hello 报文与它的邻居之间建立邻接关系；每台路由器向每个邻居发送链路状态通告（LSA），有时叫链路状态报文（LSP），每个邻居在收到 LSP 之后要依次向它的邻居转发这些 LSP（泛洪）；每台路由器要在数据库中保存一份它所收到的 LSA 的备份，所有路由器的数据库应该相同；依照拓扑数据库每台路由器使用 Dijkstra 算法（SPF 算法）计算出到每个网络的最短路径，并将结果输出到路由选择表中。

OSPF 的工作原理可简化为：发 Hello 报文→建立邻接关系→形成链路状态数据库→ SPF 算法→形成路由表。

3.2.1.6　OSPF 配置命令

```
[AR1]ospf 1                      #启用 OSPF
[AR1-ospf-1]router-id 1.1.1.1    #手动指定 Router ID
[AR1-ospf-1]area 1               #进区域后再通告接口网段
[AR1-ospf-1-area-0.0.0.1]network 1.1.1.1 0.0.0.0  #宣告接口网段，1.1.1.1 0.0.0.0 是举例的网段
[AR1-ospf-1-area-0.0.0.1]net 10.10.100.0 0.0.0.3  #宣告接口网段，10.10.100.0 0.0.0.3 是举例的网段
[AR1-ospf-1-area-0.0.0.1]quit
[AR1-ospf-1]import-route direct  #将直连路由引入到 OSPF
[AR1-ospf-1]import-route rip 1   #将 RIP 路由引入到 OSPF，默认的类型是 2，也就是 OSPF 域内度量值不累加，可以在引入的时候变成类型 1，这样外部路由在 OSPF 域内度量会累加
```

3.2.2　MSTP

3.2.2.1　概述

MSTP 是多生成树协议，与生成树和快速生成树相比，MSTP 引入了“实例（INSTANCE）”的概念。生成树和快速生成树都是基于交换机端口的技术，在进行生成树计算的时候，所有

VLAN 都共享相同的生成树；而 MSTP 则是基于实例的。所谓“实例”是指多个 VLAN 对应的一个集合，MSTP 把一台设备的一个或多个 VLAN 划分为一个实例，有着相同实例配置的设备就组成了一个 MST 域，运行独立的生成树；这个 MST 就组成了一个大的设备整体，与其他 MST 域在进行生成树算法，得出一个整体的生成树。所有 VLAN 默认映射到实例 0，其他实例则称为多生成树实例。缺省情况下，交换机的工作模式为 MSTP，MSTP 兼容 STP/RSTP。

三种生成树协议的比较见表 3.1。

表 3.1　　三种生成树协议的比较

生成树协议	特　点	应用场景
STP	（1）形成一棵无环路的树，解决广播风暴并实现冗余备份。 （2）收敛速度较慢	无需区分用户或业务流量，所有 VLAN 共享一棵生成树
RSTP	（1）形成一棵无环路的树，解决广播风暴并实现冗余备份。 （2）收敛速度快	
MSTP	（1）形成多棵无环路的树，解决广播风暴并实现冗余备份。 （2）收敛速度快。 （3）多棵生成树在 VLAN 间实现负载均衡，不同 VLAN 的流量按照不同的路径转发	需要区分用户或业务流量，并实现负载分担。不同的 VLAN 通过不同的生成树转发流量，每棵生成树之间相互独立

RSTP 在 STP 基础上进行了改进，实现了网络拓扑快速收敛。但 RSTP 和 STP 还存在同一个缺陷：由于局域网内所有的 VLAN 共享一棵生成树，因此无法在 VLAN 间实现数据流量的负载均衡，链路被阻塞后将不承载任何流量，造成带宽浪费，还有可能造成部分 VLAN 的报文无法转发。

如图 3.1 所示网络中，在局域网内应用 STP 或 RSTP，生成树结构在图中用虚线表示，S6 为根交换设备。S2 和 S5 之间、S1 和 S4 之间的链路被阻塞，除了图中标注了“VLAN 2”或“VLAN 3”的链路允许对应的 VLAN 报文通过外，其他链路均不允许 VLAN 2、VLAN 3 的报文通过。

HostA 和 HostB 同属于 VLAN 2，由于 S2 和 S5 之间的链路被阻塞，S3 和 S6 之间的链路又不允许 VLAN 2 的报文通过，因此 HostA 和 HostB 之间无法互相通信。

为了弥补 STP 和 RSTP 的缺陷，IEEE 于 2002 年发布的 802.1S 标准定义了 MSTP。MSTP 兼容 STP 和 RSTP，既可以快速收敛，又提供了数据转发的多个冗余路径，在数据转发过程中实现 VLAN 数据的负载均衡。

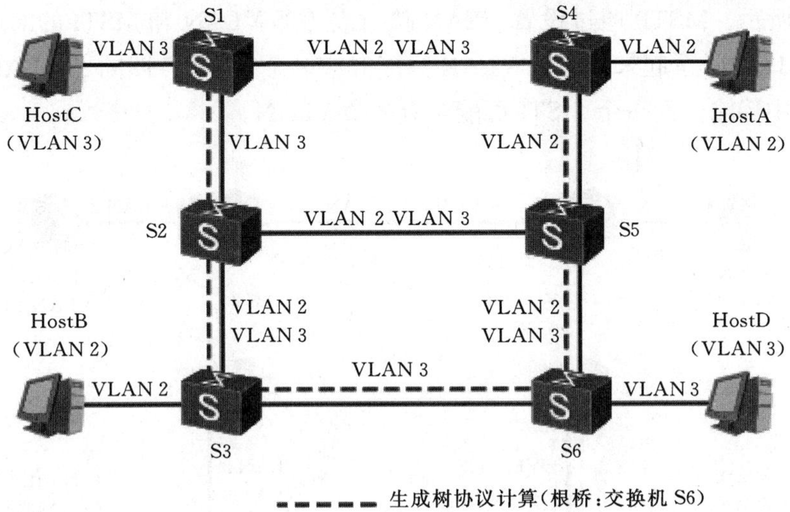

图 3.1　STP/RSTP 的缺陷示意图

如图 3.2 所示，通过 MSTP 把一个交换网络划分成多个域，每个域内形成多棵生成树，生成树之间彼此独立。每棵生成树叫作一个多生成树实例 MSTI（Multiple Spanning Tree Instance），每个域叫作一个 MST 域（MST Region：Multiple Spanning Tree Region）。

所谓实例就是多个 VLAN 的一个集合。映射到一个实例里，这些 VLAN 在端口上的转发状态取决于端口在对应 MSTP 实例的状态。

通过将多个 VLAN 捆绑到一个实例，可以节省通信开销和资源占用率。MSTP 各个实例拓扑的计算相互独立，在这些实例上可以实现负载均衡。

可以通过计算来自动确定 MSTP 的根桥和备份根桥，用户也可以手动配置设备为指定生成树的根桥或备份根桥：设备在各生成树中的角色互相独立，在作为一棵生成树的根桥或备份根桥的同时，也可以作为其他生成树的根桥或备份根桥；但在同一棵生成树中，一台设备不能既作为根桥，又作为备份根桥。

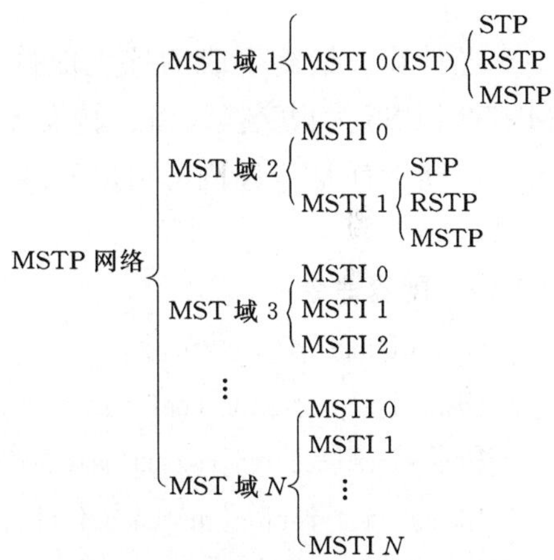

图 3.2　MSTP 网络层次结果关系示例图

在一棵生成树中，生效的根桥只有一个；当两台或两台以上的设备被指定为同一棵生成树的根桥时，系统将选择 MAC 地址最小的设备作为根桥。

可以在每棵生成树中指定多个备份根桥。当根桥出现故障或被关机时，备份根桥可以取代根桥成为指定生成树的根桥；但此时若配置了新的根桥，则备份根桥将不会成为根桥。如果配置了多个备份根桥，则 MAC 地址最小的备份根桥将成为指定生成树的根桥。在配置 MSTP 过程中，建议手动配置根桥和备份根桥。

如图 3.3 所示，MSTP 通过设置 VLAN 映射表（即 VLAN 和 MSTI 的对应关系表），把 VLAN 和 MSTI 联系起来。每个 VLAN 只能对应一个 MSTI，即同一 VLAN 的数据只能在一个 MSTI 中传输，而一个 MSTI 可能对应多个 VLAN。

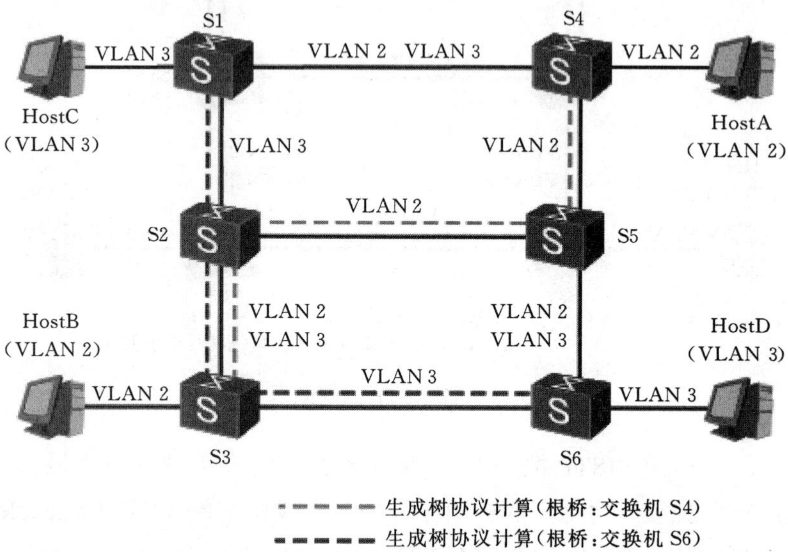

图 3.3　MST 域内的多棵生成树示意图

经计算，最终生成两棵生成树：MSTI1 以 S4 为根交换设备，转发 VLAN 2 的报文；MSTI2 以 S6 为根交换设备，转发 VLAN 3 的报文。

这样所有 VLAN 内部可以互通，同时不同 VLAN 的报文沿不同的路径转发，实现了负载分担。

3.2.2.2　配置命令

[Huawei]stp mode mstp　　#配置 MSTP 工作模式

[Huawei]stp region-configuration　　#进入 MST 域视图

[Huawei-mst-region]region-name gxsdxy　　#配置 MST 域的域名

[Huawei-mst-region]instance 1 vlan 2 to 5　　#配置多生成树实例和 VLAN 的映射关系，此次 VLAN 2 to 5 为举例 VLAN

[Huawei-mst-region]instance 2 vlan 6 to 10　　#配置多生成树实例和 VLAN 的映射关系，此次 VLAN 6 to 10 为举例 VLAN

[Huawei-mst-region]revision-level 0　　#配置 MST 域的 MSTP 修订级别，缺省情况下，MSTP 域的 MSTP 修订级别为 0。

[Huawei-mst-region]active region-configuration　#激活新的 MST 域配置

Info: This operation may take a few seconds. Please wait for a moment...done.

[Huawei-mst-region]quit

```
[Huawei]stp instance 1 root primary          #配置该设备为根桥
[Huawei]stp instance 2 root secondary        #配置该设备为备份根桥
```

3.2.3 VRRP

3.2.3.1 概述 Summary

VRRP（Virtual Router Redundancy Protocol，虚拟路由器冗余协议）将可以承担网关功能的路由器加入到备份组中，形成一台虚拟路由器，由 VRRP 的选举机制决定哪台路由器承担转发任务，局域网内的主机只需将虚拟路由器配置为缺省网关。VRRP 是一种容错协议，在提高可靠性的同时，简化了主机的配置。在具有多播或广播能力的局域网（如以太网）中，借助 VRRP 能在某台设备出现故障时仍然提供高可靠的缺省链路，有效避免单一链路发生故障后网络中断的问题。

在局域网内，主机发往其他网段的报文都由网关进行转发。当网关发生故障时，本网段内所有发往网关的数据将中断。为了避免网络中断，可以在主机上设置多个网关，但是一个主机只允许设置一个默认网关，因此需要管理员手工添加和修改，这样大大增加了网络管理的复杂度。因此通常使用 VRRP 协议（虚拟路由器冗余协议），保证用户快速、不间断、透明地切换到另一个网关。

VRRP 协议首先采用竞选的方法选择主路由器（Master），主路由器负责提供实际的数据转发服务，主路由器选出后，其他设备作为备份路由器（Backup），并通过主路由设备定时发出的 VRRP 报文监测主路由设备的状态。如果组内的备份路由设备在规定的时间内没有收到来自主路由设备的报文，则将自己的状态转为 Master，由于切换非常迅速而且用户终端不需要改变默认网关的 IP 地址和 MAC 地址，所以对用户而言是透明的。

主路由负责转发数据，备份路由不负责转发数据，因此在主路由转发数据的同时，备份路由却一直处于空闲状态，这样势必造成了网络带宽资源的浪费。我们通过在 VRRP 中使用负载均衡技术，创建不同的 VRRP 组，使得路由器在不同的 VRRP 组中担任不同的角色。

VRRP 协议的实现有 VRRPv2 和 VRRPv3 两个版本，VRRPv2 基于 IPv4，VRRPv3 基于 IPv6。

VRRP 路由器在运行过程中有三种状态：

（1）Initialize 状态。系统启动后就进入 Initialize，此状态下路由器不对 VRRP 报文作任何处理，可以理解为初始化。

（2）Master 状态。路由器会发送 VRRP 通告，发送免费 ARP 报文。

（3）Backup 状态。接受 VRRP 通告。

一般主路由器处于 Master 状态，备份路由器处于 Backup 状态。

3.2.3.2 VRRP 选举机制

VRRP 使用选举机制来确定路由器的状态，运行 VRRP 的一组路由器对外构成了一个虚拟路由器，其中一台路由器处于 Master 状态，其他处于 Backup 状态。所以主路由器又叫作 Master 路由器，备份路由器又叫作 Backup 路由器。

（1）路由器启用 VRRP 功能后，会根据优先级确定自己在备份组中的角色。优先级高的路由器成为 Master 路由器，优先级低的成为 Backup 路由器。Master 路由器定期发送 VRRP 通告报文，通知备份组内的其他设备自己工作正常；Backup 路由器则启动定时器等待通告报文的到来。

（2）在抢占方式下，当 Backup 路由器收到 VRRP 通告报文后，会将自己的优先级与通告报文中的优先级进行比较。如果大于通告报文中的优先级，则成为 Master 路由器；否则将保持 Backup 状态。

（3）在非抢占方式下，只要 Master 路由器没有出现故障，备份组中的路由器始终保持 Master 或 Backup 状态，Backup 路由器即使随后被配置了更高的优先级也不会成为 Master 路由器。

（4）如果 Backup 路由器的定时器超时后仍未收到 Master 路由器发送来的 VRRP 通告报文，则认为 Master 路由器已经无法正常工作，此时 Backup 路由器会认为自己是 Master 路由器，并对外发送 VRRP 通告报文。备份组内的路由器根据优先级选举出 Master 路由器，承担报文的转发功能。

3.2.3.3　VRRP 优先级选举的方法

（1）VRRP 组中 IP 拥有者。如果虚拟 IP 地址与 VRRP 组中的某台 VRRP 路由器 IP 地址相同，则此路由器为 IP 地址拥有者，这台路由器将被定位主路由器。

（2）比较优先级。如果没有 IP 地址拥有者，则比较路由器的优先级，优先级的范围是 0～255，大的作为主路由器。

（3）比较 IP 地址。在没有 IP 地址拥有者和优先级相同的情况下，IP 地址大的作为主路由器。

3.2.3.4　VRRP 基本配置

（1）配置 VRRP 组。要启用 VRRP，最基本的配置就是要创建 VRRP 组，并为 VRRP 组配置虚拟 IP 地址，此命令需要在主路由器和备份路由器上配置，方法如下：

[Huawei]interface Ethernet 0/0/1　　　　　　　# 进入端口 e0/0/1 中

[Huawei-Ethernet0/0/1] vrrp ***group-number*** ip ***ip-address***　#***group-number*** 代表 VRRP 组的编号，实际配置中可以调整，取值范围为 1~255。***ip-address*** 代表虚拟组的 IP 地址，例如 192.168.1.1

（2）配置 VRRP 优先级。如果希望指定某台路由器称为主路由器，可以手工调整其优先级，方法如下：

[Huawei]interface Ethernet 0/0/1　　　　　　　# 进入端口 e0/0/1 中

[Huawei-Ethernet0/0/1]vrrp ***group-number*** priority ***number***　# ***group-number*** 代表 VRRP 组的编号，实际配置中可以调整，取值范围为 1~255。***number*** 代表优先级范围，取值 0~255，默认为 100

优先级的配置在没有 IP 地址拥有者的情况下。想让哪台路由器成为主路由器就在哪台路由器上配置。

（3）配置 VRRP 接口跟踪 。VRRP 接口跟踪机制就是检测接口故障的一种机制。配置了接口跟踪机制的路由器，当自己的接口发生故障时会将自己的路由器优先级降低，从而使自己从主路由器变为备份路由器，然后原来的备份路由器此时将成为主路由器。

端口跟踪命令如下：

[Huawei]interface Ethernet 0/0/1 #进入端口 e0/0/1 中

[Huawei-Ethernet0/0/1]vrrp *group-nunmber* track interface *priority-decrement*

group-number 代表 VRRP 组的编号，取值与上面保持一致。*priority-decrement* 表示需要降低的优先级数值

注意：priority–decrement 是降低了多少而不是降低到多少，比如 priority–decrement 为 30，那么此路由器的优先级在原来基础上降低 30。

（4）配置 VRRP 抢占模式 。抢占模式指当原来的路由器从故障中回复并接入到网络层后，配置了 VRRP 抢占模式的路由器将夺回原来属于自己的角色（主路由器），如果没有配置，回复之后将保持备份路由器的状态。

推荐在主路由器中配置该命令。使用启用抢占模式命令如下：

[Huawei]interface Ethernet 0/0/1 #进入端口 e0/0/1 中

[Huawei-Ethernet0/0/1]vrrp group-number preempt {delay [Delay-time] }

Delay #取值范围为 1 ～ 255，如果不配置 delay 时间，那么其默认值为 0s。

delay-time #为延迟抢占的时间即从该路由器发现自己的优先级大于 MASTER 的优先级开始，经过 delay-time 这样长的一段时间之后才允许抢占

（5）配置 VRRP 定时器。VRRP 定时器可以修改通告报文的发送时间，在主路由器上配置，命令如下：

[Huawei]interface Ethernet 0/0/1 #进入端口 e0/0/1 中

[Huawei-Ethernet0/0/1]vrrp group-number timers advertise vrrp-advertise-interval

adver_interval 为设置定时器 adver_timer 的时间间隔。MASTER 每隔这样一个时间间隔，就会发送一个 advertisement 报文以通知组内其他路由器自己工作正常，vrrp–advertise–interval 的取值范围为 0 ～ 254。

配置 VRRP 定时学习功能：配置此命令的路由器会学习发送通告报文时间，进而计算出失效间隔，否则默认 3s，这条命令对于上面的配置 VRRP 定时器，在主路由器中配置了发送时间间隔，那么在备份路由器上就需要配置定时学习功能来计算失效间隔，因为失效间隔是发送时间的 3 倍。

（6）VRRP 负载均衡。在一组 VRRP 组中，主路由器承担数据转发任务的同时，备份路由器的链路将处于空闲状态，这必然造成了带宽资源的浪费。为了避免这种浪费，使用 VRRP 负载均衡。

VRRP 负载均衡是通过实现将路由器加入到多个 VRRP 组实现的，使 VRRP 路由器在不同的组中担任不同的角色。

3.3　项目实施

3.3.1　项目需求

（1）ISP 运营商出口为静态 IP 形式。

（2）所有 PC 电脑均能访问 Internet。

（3）能实现负载均衡，VLAN 10、VLAN 20 的用户数据通过核心交换机 LSW1 访问 Internet，VLAN 30 的用户数据通过核心交换机 LSW2 访问 Internet。

（4）能实现冗余，当任何一个核心设备出故障的时候，网络仍然能够正常运行。

3.3.2　设备清单

本项目设备清单见表 3.2。

表 3.2　设备清单表

序号	设备型号	数　量	功　能	备　注
1	AR3260	1	出口路由器	
2	S5700–28C–HI	2	核心交换机	
3	S3700–26C–HI	2	接入交换机	

3.3.3　项目拓扑

本项目网络拓扑结构如图 3.4 所示。

3.3.4　配置思路

采用如下的思路配置：

3.3.4.1　接入层设备

（1）划分 VLAN。

（2）把连接 PC 的端口放到相应 VLAN 中。

（3）把连接核心交换机的端口设置为 Trunk 口，并允许相应的 VLAN 通过。

3.3.4.2　核心层设备

（1）划分 VLAN。

（2）配置 VLANIF 接口地址。

（3）把连接接入层交换机的端口设置为 Trunk 口，并允许相应的 VLAN 通过。

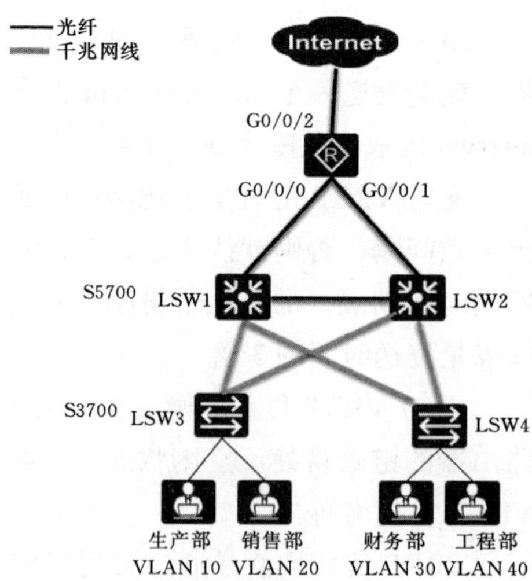

图 3.4　中型企业网络拓扑图

（4）配置 MSTP，把 VLAN 10、VLAN 20 放到实例 1 中，把 VLAN 30、VLAN 40 放到实例 2 中，设置实例 1 的根桥为左边的交换机 LSW1，备份根桥为右边的交换机 LSW2，设置实例 2 的根桥为右边的交换机 LSW2，备份根桥为左边的交换机 LSW1，并对上行端口进行检测。

（5）配置 VRRP。

（6）把连接出口路由器的端口放到 VLAN 100 中，并给 VLAN 100 配置 IP 地址。

（7）配置动态路由 OSPF。

3.3.4.3　出口路由器

（1）给接口配置 IP 地址。

（2）配置到 Internet 的默认路由。

（3）配置动态路由 OSPF，并注入默认路由。

（4）配置局域网用户通过 NAT 转换将私网地址转换为公网地址，进行上网。

3.3.5　相关信息规划

3.3.5.1　核心交换机 LSW1

核心层交换机的信息规划如表 3.3 和表 3.4 所示。

表 3.3　核心层交换机 LSW1 信息规划

序号	使用部门	VLAN 号	IP 地址	掩码	备注
1	生产部	10	10.10.10.1	255.255.255.0	
2	销售部	20	10.10.20.1	255.255.255.0	
3	财务部	30	10.10.30.1	255.255.255.0	
4	工程部	40	10.10.40.1	255.255.255.0	
5	互联	100	10.10.100.1	255.255.255.252	与路由器互联

表 3.4　核心层交换机 LSW2 信息规划

序号	使用部门	VLAN 号	IP 地址	掩码	备注
1	生产部	10	10.10.10.2	255.255.255.0	
2	销售部	20	10.10.20.2	255.255.255.0	
3	财务部	30	10.10.30.2	255.255.255.0	
4	工程部	40	10.10.40.2	255.255.255.0	
5	互联	200	10.10.200.1	255.255.255.252	与路由器互联

3.3.5.2　出口路由器

出口路由器的信息规划见表 3.5。

表 3.5　**出口路由器信息规划**

序号	接口	IP 地址	掩码	备注
1	G0/0/0	10.10.100.2	255.255.255.252	与核心交换机 LSW1 互联
2	G0/0/1	10.10.200.2	255.255.255.252	与核心交换机 LSW2 互联

3.3.5.3　ISP 出口规划

ISP 出口的信息规划见表 3.6。

表 3.6　**ISP 出口信息规划**

序号	接口	IP 地址	掩码	网关	备注
1	G0/0/2	1.1.1.1	255.255.255.252	1.1.1.5	静态 IP 地址

3.3.6　操作步骤

3.3.6.1　接入交换机 LSW3 设备

（1）修改设备名，方法如下：

```
<Huawei>sys
[Huawei]sysname LSW3
[LSW3]undo info-center enable                #关闭消息提示
```

（2）划分 VLAN，方法如下：

```
[LSW3]vlan 10                                #划分 vlan10
[LSW3-vlan10]description shengchanbu         #对 vlan 进行描述
[LSW3-vlan10]quit                            #退出

[LSW3]vlan 20                                #划分 vlan20
[LSW3-vlan20]description xiaoshoubu          #对 vlan 进行描述
[LSW3-vlan20]quit                            #退出

[LSW3]vlan 30                                #划分 vlan30
[LSW3-vlan30]description caiwubu             #对 vlan 进行描述
[LSW3-vlan30]quit                            #退出
```

```
[LSW3]vlan 40                                    # 划分 vlan40
[LSW3-vlan40]description gongchengbu             # 对 vlan 进行描述
[LSW3-vlan40]quit                                # 退出
```

（3）把连接 PC 的端口放到相应 VLAN 中，方法如下：

```
[LSW3]interface Ethernet 0/0/1                   # 进入端口 e0/0/3 中
[LSW3-Ethernet0/0/1]port link-type access        # 把端口设置为 access 模式
[LSW3-Ethernet0/0/1]port default vlan 10         # 把端口放入 vlan 10 中
[LSW3-Ethernet0/0/1]quit                         # 退出
```

同样方法，对其他接口进行配置：

```
[LSW3]interface Ethernet 0/0/2
[LSW3-Ethernet0/0/2]port link-type access
[LSW3-Ethernet0/0/2]port default vlan 20
[LSW3-Ethernet0/0/2]quit
```

（4）配置 Trunk，方法如下：

```
# 在连接核心交换机的上行端口 G0/0/1、G0/0/2 上设置 Trunk，并允许相应的 vlan 通过该接口
[LSW3] interface GigabitEthernet 0/0/1           # 进入端口 G0/0/1 中
[LSW3-GigabitEthernet0/0/1]port link-type trunk  # 把端口设置为 trunk 模式
[LSW3-GigabitEthernet0/0/1]port trunk allow-pass vlan 10 2030 40 # 允许 vlan 10、vlan 20、vlan 30、vlan 40 通过
[LSW3-GigabitEthernet0/0/1]quit                  # 退出

[LSW3]interface GigabitEthernet 0/0/2
[LSW3-GigabitEthernet0/0/2]port link-type trunk
[LSW3-GigabitEthernet0/0/2]port trunk allow-pass vlan 10 20 30 40
[LSW3-GigabitEthernet0/0/2]quit
```

（5）MSTP 配置，方法如下：

```
[LSW3]stp mode mstp        # 配置模式为 MSTP 模式，交换机默认是 MSTP 模式，该步骤可以省略
[LSW3]stp region-configuration                      # 进入 stp 配置
[LSW3-mst-region]region-name sd001                  # 配置域名为 sd001
[LSW3-mst-region]instance 1 vlan 10 20              # 配置 vlan 10、vlan 20 映射为实例 1
[LSW3-mst-region]instance 2 vlan 30 40              # 配置 vlan 30、vlan 40 映射为实例 2
[LSW3-mst-region]active region-configuration        # 激活 mstp 域的配置
Info: This operation may take a few seconds. Please wait for a moment...done.
```

```
[LSW3-mst-region]quit                                  #退出
```

3.3.6.2　接入交换机 LSW4 设备

接入交换机 LSW4 的配置和 LSW3 的配置思路相似，下面给出配置代码。

（1）修改设备名，方法如下：

```
<Huawei>sys
[Huawei]sysname LSW4                                   #修改设备名
[LSW4]undo info-center enable                          #关闭消息提示
```

（2）划分 VLAN，方法如下：

```
[LSW4]vlan 10                                          #划分 vlan10
[LSW4-vlan10]description shengchanbu                   #对 vlan 进行描述
[LSW4-vlan10]quit                                      #退出

[LSW4]vlan 20                                          #划分 vlan20
[LSW4-vlan20]description xiaoshoubu                    #对 vlan 进行描述
[LSW4-vlan20]quit                                      #退出

[LSW4]vlan 30                                          #划分 vlan30
[LSW4-vlan30]description caiwubu                       #对 vlan 进行描述
[LSW4-vlan30]quit                                      #退出

[LSW4]vlan 40                                          #划分 vlan40
[LSW4-vlan40]description gongchengbu                   #对 vlan 进行描述
[LSW4-vlan40]quit                                      #退出
```

（3）把连接 PC 的端口放到相应 VLAN 中，方法如下：

```
[LSW4]interface Ethernet 0/0/1                      #进入端口 e0/0/1 中
[LSW4-Ethernet0/0/1]port link-type access           #把端口设置为 access 模式
[LSW4-Ethernet0/0/1]port default vlan 30            #把端口放入 vlan 30 中
[LSW4-Ethernet0/0/1]quit                            #退出
```

同样方法，对其他接口进行配置：

```
[LSW4]interface Ethernet 0/0/2
[LSW4-Ethernet0/0/2]port link-type access
[LSW4-Ethernet0/0/2]port default vlan 40
[LSW4-Ethernet0/0/2]quit
```

（4）配置 Trunk，方法如下：

```
# 在连接核心交换机的上行端口 G0/0/1、G0/0/2 上设置 trunk，并允许相应的 vlan 通过该接口
[LSW4] interface GigabitEthernet 0/0/1          # 进入端口 G0/0/1 中
[LSW4-GigabitEthernet0/0/1]port link-type trunk  # 把端口设置为 trunk 模式
[LSW4-GigabitEthernet0/0/1]port trunk allow-pass vlan 10 20 30 40   # 允许 vlan 10、vlan 20、vlan 30、vlan 40 通过
[LSW4-GigabitEthernet0/0/1]quit                  # 退出

[LSW4]interface GigabitEthernet 0/0/2
[LSW4-GigabitEthernet0/0/2]port link-type trunk
[LSW4-GigabitEthernet0/0/2]port trunk allow-pass vlan 10 20
[LSW4-GigabitEthernet0/0/2]quit
```

（5）MSTP 配置：

```
[LSW4]stp mode mstp     # 配置模式为 MSTP 模式，交换机默认是 MSTP 模式，该步骤可以省略
[LSW4]stp region-configuration                  # 进入 stp 配置
[LSW4-mst-region]region-name sd001              # 配置域名为 sd001
[LSW4-mst-region]instance 1 vlan 10 20          # 配置 vlan 10、vlan 20 映射为实例 1
[LSW4-mst-region]instance 2 vlan 30 40          # 配置 vlan 30、vlan 40 映射为实例 2
[LSW4-mst-region]active region-configuration    # 激活 mstp 域的配置
Info: This operation may take a few seconds. Please wait for a moment...done.
[LSW4-mst-region]quit                           # 退出
```

3.3.6.3　核心交换机设备 LSW1 的配置

（1）修改设备名，方法如下：

```
<Huawei>system-view
[Huawei]sysname LSW1
[LSW1]undo info-center enable                   # 关闭消息提示
```

（2）划分 VLAN，方法如下：

```
[LSW1]vlan 10                                   # 划分 vlan10
[LSW1-vlan10]description shengchanbu            # 对 vlan 进行描述
[LSW1-vlan10]quit                               # 退出
```

同样方法，划分其他 VLAN：

```
[LSW1]vlan 20
[LSW1-vlan20]description xiaoshoubu
```

```
[LSW1-vlan20]quit

[LSW1]vlan 30
[LSW1-vlan30]description caiwubu
[LSW1-vlan30]quit

[LSW1]vlan 40
[LSW1-vlan40]description gongchengbu
[LSW1-vlan40]quit

[LSW1]vlan 100
[LSW1-vlan100]description hulian
[LSW1-vlan100]quit
```

（3）配置 VLANIF 接口 IP 地址，方法如下：

```
[LSW1]interface vlanif 10                     # 进入 vlan 10 接口
[LSW1-vlanif10]ip address 10.10.10.1 24       # 配置 ip 地址和掩码信息
[LSW1-vlanif10]quit                           # 退出
```

同样方法，对其他 VLAN 进行配置：

```
[LSW1]interface vlanif 20
[LSW1-vlanif20]ip address 10.10.20.1 24
[LSW1-vlanif20]quit

[LSW1]interface vlanif 30
[LSW1-vlanif30]ip address 10.10.30.1 24
[LSW1-vlanif30]quit

[LSW1]interface vlanif 40
[LSW1-vlanif30]ip address 10.10.40.1 24
[LSW1-vlanif30]quit

[LSW1] interface vlanif 100
[LSW1-Vlanif100] ip address 10.10.100.1 30
[LSW1-Vlanif100] quit
```

（4）配置 Trunk，方法如下：

```
# 在连接接入层交换机的下行端口设置为trunk，并允许相应的vlan通过该接口
[LSW1]interface GigabitEthernet 0/0/2                 #进入端口
[LSW1-GigabitEthernet0/0/2]port link-type trunk  #设置端口为trunk
[LSW1-GigabitEthernet0/0/2]port trunk allow-pass vlan 10 20 30 40   #设置允许通过的vlan
[LSW1-GigabitEthernet0/0/2]quit                      #退出

[LSW1]interface GigabitEthernet 0/0/3                 #进入端口
[LSW1-GigabitEthernet0/0/3]port link-type trunk  #设置端口为trunk
[LSW1-GigabitEthernet0/0/3]port trunk allow-pass vlan 10 20 30 40   #设置允许通过的vlan
[LSW1-GigabitEthernet0/0/3]quit                       #退出

[LSW1]interface GigabitEthernet 0/0/4                 #进入端口
[LSW1-GigabitEthernet0/0/3]port link-type trunk  #设置端口为trunk
[LSW1-GigabitEthernet0/0/4]port trunk allow-pass vlan 10 20 30 40  #设置允许通过的vlan
[LSW1-GigabitEthernet0/0/3]quit                       #退出
```

（5）配置 MSTP，方法如下：

```
[LSW1]stp mode mstp       #配置模式为mstp模式，交换机默认是MSTP模式，该步骤可以省略
[LSW1]stp region-configuration                  #进入stp配置
[LSW1-mst-region]region-name sd001              #配置域名为sd001
[LSW1-mst-region]instance 1 vlan 10 20          #配置vlan 10、vlan 20映射为实例1
[LSW1-mst-region]instance 2 vlan 30 40          #配置vlan 30、vlan 40映射为实例2
[LSW1-mst-region]active region-configuration    #激活mstp域的配置
Info: This operation may take a few seconds. Please wait for a moment...done.
[LSW1-mst-region]quit                                #退出

[LSW1]stp instance 1 root primary                    #配置在实例1中交换机LSW1为根桥
[LSW1]stp instance 2 root secondary                  #配置在实例2中交换机LSW1为备份根桥
```

（6）配置 VRRP，方法如下：

```
[LSW1]interface vlan 10                              #进入vlan10接口
[LSW1-Vlanif10]vrrp vrid 1 virtual-ip 10.10.10.254   #创建vrrp组1，虚拟网关为10.10.10.254
[LSW1-Vlanif10]vrrp vrid 1 priority 200              #设置vrrp组1的优先级为200，缺省为
100，用途是和另一个核心交换机LSW2比较，此交换机的vrrp组1优先级较高，工作在Master状态
[LSW1-Vlanif10]vrrp vrid 1 track interface GigabitEthernet 0/0/1 reduced 150
```

```
#监控上行端口 G0/0/1，如果端口 Down 掉则优先级降低 150
[LSW1-Vlanif10]quit                                    #退出
```

同样的方法，创建 VRRP 组 2，3，4：

[LSW1]interface vlan 20

[LSW1-Vlanif20]vrrp vrid 2 virtual-ip 10.10.20.254

[LSW1-Vlanif20]vrrp vrid 2 priority 200 #设置 vrrp 组 2 的优先级为 200，缺省为 100，用途是和另一个核心交换机 LSW2 比较，此交换机的 vrrp 组 2 优先级较高，工作在 Master 状态

[LSW1-Vlanif20]vrrp vrid 2 track interface GigabitEthernet 0/0/1 reduced 150

[LSW1-Vlanif20]quit

[LSW1]interface Vlanif 30

[LSW1-Vlanif30]vrrp vrid 3 virtual-ip 10.10.30.254

[LSW1-Vlanif30]vrrp vrid 3 priority 100 #设置 vrrp 组 3 的优先级为 100，缺省为 100，此命令可以忽略，用途是和另一个核心交换机 LSW2 比较，此交换机的 vrrp 组 3 优先级较低，工作在 Backup 状态

[LSW1-Vlanif30]quit

[LSW1]interface Vlanif 40

[LSW1-Vlanif40]ip address 10.10.40.1 24

[LSW1-Vlanif40]vrrp vrid 4 virtual-ip 10.10.40.254

[LSW1-Vlanif40]vrrp vrid 4 priority 100 #设置 vrrp 组 3 的优先级为 100，缺省为 100，此命令可以忽略，用途是和另一个核心交换机 LSW2 比较，此交换机的 vrrp 组 4 优先级较低，工作在 Backup 状态

[LSW1-Vlanif40]quit

（7）检查 VRRP 配置，方法如下：

```
[LSW1]display vrrp brief
VRID  State      Interface            Type     Virtual IP
-------------------------------------------------------------
1     Master     Vlanif10             Normal   10.10.10.254
# vrrp 组 1 绑定 vlan 10，工作在 Master 状态，虚拟网关是 10.10.10.254
2     Master     Vlanif20             Normal   10.10.20.254
3     Master     Vlanif30             Normal   10.10.30.254
4     Master     Vlanif40             Normal   10.10.40.254
-------------------------------------------------------------
Total:4     Master:4     Backup:0     Non-active:0
```

＊因为 LSW2 还没有进行配置，此时 LSW1 的 4 个 Vlanif 的 vrrp 状态均为 Master

（8）配置与出口路由器互联接口，方法如下：

```
[LSW1]interface GigabitEthernet 0/0/1                  #进入端口
[LSW1-GigabitEthernet0/0/1]port link-type access       #设置端口为 access 模式
[LSW1-GigabitEthernet0/0/1]port default vlan 100       #把端口划分到 vlan100 中
[LSW1-GigabitEthernet0/0/2]quit                        #退出
```

（9）配置动态路由 OSPF，方法如下：

```
[LSW1]ospf 1 router-id 1.1.1.1                         #启用 OSPF 路由，进程号是 1，Router-ID 是 1.1.1.1
[LSW1-ospf-1]import-route direct                       #导入直连路由
[LSW1-ospf-1]area 0                                    #区域 0
[LSW1-ospf-1-area-0.0.0.0]network 10.10.100.1 0.0.0.0  #通告网络
[LSW1-ospf-1-area-0.0.0.0]quit                         #退出
[LSW1-ospf-1]quit                                      #退出
```

3.3.6.4 核心交换机设备 LSW2 的配置

（1）修改设备名，方法如下：

```
<Huawei>system-view
[Huawei]sysname LSW2
[LSW2]undo info-center enable                          #关闭消息提示
```

（2）划分 VLAN，方法如下：

```
[LSW2]vlan 10                                          #划分 vlan10
[LSW2-vlan10]description shengcanbu                    #对 vlan 进行描述
[LSW2-vlan10]quit                                      #退出
```

同样方法，划分其他 VLAN：

```
[LSW2]vlan 20
[LSW2-vlan20]description xiaoshoubu
[LSW2-vlan20]quit

[LSW2]vlan 30
[LSW2-vlan30]description caiwubu
[LSW2-vlan30]quit

[LSW2]vlan 40
[LSW2-vlan40]description gongchengbu
```

```
[LSW2-vlan40]quit

[LSW2]vlan 200
[LSW2-vlan200]description hulian
[LSW2-vlan200]quit
```

（3）配置 VLANIF 接口 IP 地址，方法如下：

```
[LSW2]interface vlanif 10                          # 进入 vlan 10 接口
[LSW2-vlanif10]ip address 10.10.10.2 24            # 配置 ip 地址和掩码信息
[LSW2-vlanif10]quit                                # 退出
```

同样方法，对其他 VLAN 进行配置：

```
[LSW2]interface vlanif 20
[LSW2-vlanif20]ip address 10.10.20.2 24
[LSW2-vlanif20]quit

[LSW2]interface vlanif 30
[LSW2-vlanif30]ip address 10.10.30.2 24
[LSW2-vlanif30]quit

[LSW2]interface vlanif 40
[LSW2-vlanif30]ip address 10.10.40.2 24
[LSW2-vlanif30]quit

[LSW2] interface vlanif 200
[LSW2-Vlanif100] ip address 10.10.200.1 30
[LSW2-Vlanif100] quit
```

（4）配置 Trunk，方法如下：

```
# 在连接接入层交换机的下行端口设置为 trunk，并允许相应的 vlan 通过该接口
[LSW2]interface GigabitEthernet 0/0/2              # 进入端口
[LSW2-GigabitEthernet0/0/2]port link-type trunk    # 设置端口为 trunk
[LSW2-GigabitEthernet0/0/2]port trunk allow-pass vlan 10 20 30 40   # 设置允许通过的 vlan
[LSW2-GigabitEthernet0/0/2]quit                    # 退出

[LSW2]interface GigabitEthernet 0/0/3              # 进入端口
```

```
[LSW2-GigabitEthernet0/0/3]port link-type trunk  #设置端口为trunk
[LSW2-GigabitEthernet0/0/3] port trunk allow-pass vlan 10 20 30 40  #设置允许通过的vlan
[LSW2-GigabitEthernet0/0/3]quit                 #退出

[LSW2]interface GigabitEthernet 0/0/4           #进入端口
[LSW2-GigabitEthernet0/0/3]port link-type trunk  #设置端口为trunk
[LSW2-GigabitEthernet0/0/4]port trunk allow-pass vlan 10 20 30 40  #设置允许通过的vlan
[LSW2-GigabitEthernet0/0/3]quit                 #退出
```

（5）配置 MSTP，方法如下：

```
[LSW2]stp mode mstp  #配置模式为mstp模式，交换机默认是MSTP模式，该步骤可以省略
[LSW2]stp region-configuration                  #进入stp配置
[LSW2-mst-region]region-name sd001              #配置域名为sd001
[LSW2-mst-region]instance 1 vlan 10 20          #配置vlan 10、vlan 20映射为实例1
[LSW2-mst-region]instance 2 vlan 30 40          #配置vlan 30、vlan 40映射为实例2
[LSW2-mst-region]active region-configuration    #激活mstp域的配置
Info: This operation may take a few seconds. Please wait for a moment...done
[LSW2-mst-region]quit                           #退出

[LSW2]stp instance 1 root secondary             #配置在实例1中交换机LSW2为备份根桥
[LSW2]stp instance 2 root primary               #配置在实例2中交换机LSW2为根桥
```

（6）配置 VRRP，方法如下：

```
[LSW2]interface vlan 10                         #进入vlan10接口
[LSW2-Vlanif10]vrrp vrid 1 virtual-ip 10.10.10.254  #创建vrrp组1，虚拟网关为10.10.10.254
[LSW2-Vlanif10]vrrp vrid 1 priority 100  #设置vrrp组1的优先级为100，此命令可以忽略，缺省为100，用途是和另一个核心交换机LSW1比较，此交换机的vrrp组1优先级较低，工作在Backup状态
[LSW2-Vlanif10]quit                             #退出
```

同样的方法，创建 VRRP 组 2、3、4：

```
[LSW2]interface vlan 20
[LSW2-Vlanif20]vrrp vrid 2 virtual-ip 10.10.20.254
[LSW2-Vlanif20]vrrp vrid 2 priority 100  #设置vrrp组2的优先级为100，此命令可以忽略，缺省为100，用途是和另一个核心交换机LSW2比较，此交换机的vrrp组2优先级较低，工作在Backup状态
[LSW2-Vlanif20]quit

[LSW2]interface Vlanif 30
```

[LSW2-Vlanif30]vrrp vrid 3 virtual-ip 10.10.30.254

[LSW2-Vlanif30]vrrp vrid 3 priority 200 # 设置 vrrp 组 3 的优先级为 200，缺省为 100，用途是和另一个核心交换机 LSW1 比较，此交换机的 vrrp 组 3 优先级较高，工作在 Master 状态

[LSW2-Vlanif20]vrrp vrid 2 track interface GigabitEthernet 0/0/1 reduced 150

监控上行端口 G0/0/1，如果端口 Down 掉则优先级降低 150

[LSW2-Vlanif30]quit

[LSW2]interface Vlanif 40

[LSW2-Vlanif40]ip address 10.10.40.1 24

[LSW2-Vlanif40]vrrp vrid 4 virtual-ip 10.10.40.254

[LSW2-Vlanif40]vrrp vrid 4 priority 200 #设置 vrrp 组 3 的优先级为 200，缺省为 100，此命令可以忽略，用途是和另一个核心交换机 LSW1 比较，此交换机的 vrrp 组 4 优先级较高，工作在 Master 状态

[LSW2-Vlanif40]quit

（7）检查 VRRP 配置，方法如下：

[LSW2]display vrrp brief

VRID	State	Interface	Type	Virtual IP
1	Backup	Vlanif10	Normal	10.10.10.254

vrrp 组 1 绑定 vlan 10，工作在 Backup 状态，虚拟网关是 10.10.10.254

2	Backup	Vlanif20	Normal	10.10.20.254
3	Master	Vlanif30	Normal	10.10.30.254

#vrrp 组 3 绑定 vlan 30，工作在 Master 状态，虚拟网关是 10.10.30.254

4	Master	Vlanif40	Normal	10.10.40.254

Total:4 Master:2 Backup:2 Non-active:0

（8）配置与出口路由器互联接口，方法如下：

```
[LSW2]interface GigabitEthernet 0/0/1                 # 进入端口
[LSW2-GigabitEthernet0/0/1]port link-type access      # 设置端口为 access 模式
[LSW2-GigabitEthernet0/0/1]port default vlan 200      # 把端口划分到 vlan200 中
[LSW2-GigabitEthernet0/0/2]quit                       # 退出
```

（9）配置动态路由 OSPF，方法如下：

```
[LSW2]ospf1 router-id 2.2.2.2          #配置OSPF路由，进程号是1，Router-ID是2.2.2.2
[LSW2-ospf-1]import-route direct       # 导入直连路由
[LSW2-ospf-1]area 0                    # 区域 0
```

```
[LSW2-ospf-1-area-0.0.0.0]network 10.10.200.1 0.0.0.0  #通告网络
[LSW2-ospf-1-area-0.0.0.0]quit                         #退出
[LSW2-ospf-1]quit                                      #退出
```

3.3.6.5　出口设备

（1）修改设备名，方法如下：

```
<Huawei>system-view
[Huawei]sysname AR1                                    #修改设备名
[AR1]undo info-center enable                           #关闭消息提示
```

（2）给接口配置 IP 地址，方法如下：

```
[AR1]interface GigabitEthernet 0/0/0                   #进入内网口
[AR1-GigabitEthernet0/0/0]ip address 10.10.100.2 30    #给内网口配置地址
[AR1-GigabitEthernet0/0/0]quit                         #退出

[AR1]interface GigabitEthernet 0/0/1                   #进入内网口
[AR1-GigabitEthernet0/0/1]ip address 10.10.200.2 30    #给内网口配置地址
[AR1-GigabitEthernet0/0/1]quit                         #退出

[AR1]interface GigabitEthernet 0/0/2                   #进入外网口
[AR1-GigabitEthernet0/0/2]ip address 1.1.1.1 29        #给外网口配置地址
[AR1-GigabitEthernet0/0/2]quit                         #退出
```

（3）配置局域网用户通过 NAT 转换将私网地址转换为公网地址，进行上网，方法如下：

```
[AR1]acl number 2000                                   #定义一个标准 acl
[AR1-acl-basic-2000]rule permit source any             #规则是允许源地址为任何人
[AR1-acl-basic-2000]quit                               #退出
[AR1]nat address-group 1 1.1.1.2 1.1.1.5  #创建 nat 转换地址池，编号为 1，范围是 1.1.1.2-1.1.1.5
[AR1]interface GigabitEthernet 0/0/2                   #进入外网口
[AR1-GigabitEthernet0/0/1]nat outbound 2000 address-group 1# 允许 acl 编号 2000 匹配的计算机能通过编号为 1 的地址池上网
[AR1-GigabitEthernet0/0/1]quit                         #退出
```

（4）配置到 Internet 的默认路由，方法如下：

```
[AR1]ip route-static 0.0.0.0 0 1.1.1.6  #任何到 Internet 的数据都往 1.1.1.6 地址送
```

（5）配置动态路由 OSPF，方法如下：

```
[AR1]ospf 1 router-id 3.3.3.3
[AR1-ospf-1]default-route-advertise                    #将缺省路由通告到整个 OSPF 域中
```

```
[AR1-ospf-1]area 0                                    #进入区域0
[AR1-ospf-1-area-0.0.0.0]network 10.10.100.2 0.0.0.0  #宣告网络
[AR1-ospf-1-area-0.0.0.0]network 10.10.200.2 0.0.0.0  #宣告网络
[AR1-ospf-1-area-0.0.0.0]quit                         #退出
[AR1-ospf-1]quit                                      #退出
```

3.3.6.6　测试

（1）给生产部 PC 手工配置 IP 地址（图 3.5）。

图 3.5　手工配置 IP 地址

（2）访问外网（图 3.6）。

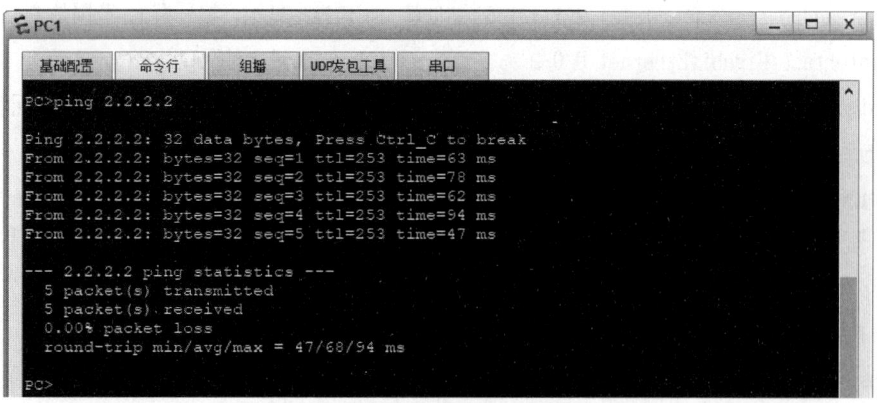

图 3.6　访问外网检测

3.4　思政融入

中国的互联网发展飞速、成果斐然。全球访问量前 20 名的互联网网站，中国占了 7 个；全球市值前 10 位的互联网企业，中国占了 4 位；网民数量规模位居世界第一；光纤宽带用户比例世界第一；4G 商用网络全球覆盖最广；5G 标准必要专利数量全球第一；人工智能、大数据、云计算等新一代信息技术飞速发展。当前的中国已成为名副其实的网络大国，正向着网络强国迈进。实施网络强国战略是立足实践、尊重实践，当代大学生要强化对习近平新时代中国特色社会主义思想的学习，树立爱国情怀，坚定文化自信，践行使命担当，在日常学习和今后的工作中脚踏实地，实事求是，养成良好的职业习惯。

3.5　思考与训练

1. OSPF 路由协议的特征有哪些？
2. 生成树协议 STP、RSTP、MSTP 的特点分别是什么？它们各自的应用场景有哪些？
3. 什么是 VRRP 协议？其工作的过程是什么？
4. VRRP 路由器在运行过程中有哪三种状态？
5. 多实例生成树 MSTP 的配置步骤有哪些？

Project 3

Project 3 LAN project of a wood industry group

3.1 Project import

With the deep–going of enterprise information construction in recent years, the operation of enterprises is increasingly integrated into computer networks. The data streams of enterprise communication, application, finance, decision–making, meetings and so on are transmitted on the enterprise network. Building a "high–quality" medium–sized enterprise network with "safety, reliability, excellent performance and convenient management" has become the key cornerstone of the success of enterprise information construction.

In order to speed up the information construction of a wood industry group, the new group enterprise network will build a group office automation, e–commerce, comprehensive business management, multimedia video conference, remote communication, information release and query as the core, relying on modern network technology, with advanced technology and strong expansibility. It will connect various offices, multimedia conference rooms, PC terminal equipment and application systems of the group through the network to realize internal modern computer network system for external communication.

3.2 Relevant knowledge points

3.2.1 Single area OSPF

3.2.1.1 Summary

OSPF (Open Shortest Path First) routing protocol is a typical link state routing protocol. The letter O in OSPF means open, that is, open, public, and any standardized device manufacturer can support OSPF. OSPF, developed by IETF in the late 1980s, is an open version of SPF-based routing protocols. The original OSPF specification was reflected in RFC113 1, called OSPF version 1, but version 1 was soon replaced by a significantly improved version, the latest version of OSPF version 2 was reflected in RFC 2328, and OSPFv2 was used in IPv4 networks. OSPF version 3 is about IPv6 and is used on IPv6 networks.

OSPF is an internal gateway protocol for routing decisions within a single Autonomous System (AS). It is an implementation of link-state routing protocol, which is subordinate to the Internal Gateway Protocol (IGP), and therefore operates within an autonomous system. OSPF is a top-level protocol. OSPF messages are encapsulated in IP headers with protocol number: 89.

3.2.1.2 OSPF Features

OSPF acts as an Internal Gateway Protocol (IGP) for exchanging routing information between routers in the same autonomous system (AS). The OSPF features are as follows:

(1) Adaptable to large-scale networks.

(2) Fast convergence speed.

(3) No routing loops.

(4) Support VLSM and CIDR.

(5) Support equivalent routing.

(6) Support regional division to form a structured network.

(7) Provide hierarchical routing management.

(8) Support simple passwords and MD5 authentication.

(9) Multicast transmission of protocol messages.

(10) The management distance of the OSPF routing protocol is 10.

(11) OSPF routing protocol uses cost as its measurement standard.

(12) OSPF maintains neighbor, topology, and routing tables.

3.2.1.3 Running process of OSPF

(1) Each router running OSPF sends HELLO messages to all OSPF-enabled interfaces. If the HELLO messages sent by two routers on a shared link are identical in content, the two routers will form a neighborhood relationship.

(2) From these neighborhood relationships, some routers form adjacency relationships. The establishment of adjacency relationship is determined by the exchange of HELLO messages and network type between OSPF routers.

(3) Each router that forms an adjacency relationship announces all its link states.

(4) Each router accepts LSAs sent by its neighbors, records them in its own link database, and sends a copy of the link database to other neighbors.

(5) By flooding an area, all routers in the area are synchronized with their own database.

(6) When the databases are synchronized, OSPF calculates the shortest path to the destination using the SPF algorithm and forms a self-looping shortest path tree with its own root.

(7) Each router establishes its own routing forwarding table based on this shortest path tree.

3.2.1.4 Network types supported by OSPF

OSPF divides networks into four types: broadcast multiple access (BMA), non-broadcast multiple access (NBMA), point-to-point, and point-to-multipoint. The network type of OSPF determines the formation of neighborhood relationship and the processing of HELLO messages, which improves the adaptability and performance of OSPF. Different types of layer-2 links require different network types of OSPF to adapt.

(1) Broadcast Type (Broadcast)

When the Link Layer Protocol is Ethernet or FDDI, by default, OSPF considers the network type to be Broadcast. Hello, LSU, and LSAck messages are usually sent as multicast in this type of network. The multicast address of 224.0.0.5 is the reserved IP multicast address for OSPF devices. The multicast address of 224.0.0.6 is the reserved IP multicast address of OSPF DR/BDR (Backup Designated Router).

(2) NBMA type (Non-Broadcast Multi-Access)

When the Link Layer Protocol is frame relay, ATM, or X.25, OSPF considers the network type to be NBMA by default. In this type of network, protocol messages (Hello, DD, LSR, LSU, LSAck) are sent unicast.

(3) P2MP type (Point-to-Multipoint)

No link-layer protocol is considered by default a Point-to-Multipoint type. Point-to-multipoint must be forced to change by another network type. A common practice is to change an incompletely connected NBMA to a point-to-multipoint network. In this type of network: ① Send Hello messages as multicast (224.0.0.5); ② Send other protocol messages as unicast (DD, LSR, LSU, LSAck).

(4) P2P type (Point-to-Point). When the link-layer protocols are PPP, HDLC, and LAPB, by default, the OSPF default network type is P2P. In this type of network, protocol messages (Hello, DD, LSR, LSU, LSAck) are sent as multicast (224.0.0.5).

3.2.1.5 OSPF routing

Shortest path priority (SPF) algorithm: is the basis of OSPF routing protocol. The SPF algorithm is sometimes called the Dijkstra algorithm because the shortest path priority algorithm (SPF) was invented by Dijkstra. OSPF routers use SPF to independently calculate the best route to any destination.

The well-known Dijkstra algorithm is used to calculate the shortest path tree. Unlike distance vector routing protocol direct interactive router routing tables, OSPF is a link state routing protocol. LSDA (link state notification) interacts with routers. Routers collect LSA flooded in the network into their own LSDB (link state database). This helps OSPF understand the entire network topology, and on this basis, calculates the network rooted by the SPF shortest path algorithm, which reaches all corners of the network. Loopless tree, and eventually, the router loads the calculated route into the routing table.

OSPF works: each router establishes an adjacency relationship with its neighbours by using Hello messages; each router sends link state announcements (LSAs), sometimes called link state packets (LSPs), to each neighbour, and each neighbour forwards these LSPs to its neighbours in turn after receiving the LSPs (flooding); each router keeps a backup copy of the LSAs it receives in a database; all routers should have the same database; according to the topology database each router uses Dijkstra's algorithm (SPF algorithm) to calculate the shortest path to each network and outputs the result to the network database. Each router should keep a backup copy of the LSAs it receives in its database, which should be the same for all routers; in accordance with the topology database each router uses Dijkstra's algorithm (SPF algorithm) to compute the shortest path to each network and outputs the result to the routing table.

OSPF works by simply sending Hello messages→establishing adjacency relationships→forming link state databases→SPF algorithms→forming routing tables.

3.2.1.6 OSPF configuration command

```
[AR1]ospf 1                       #Enable OSPF
[AR1-ospf-1]router-id 1.1.1.1     #Specify Router ID manually
[AR1-ospf-1]area 1  #After entering the area, the interface network segment will be announced
[AR1-ospf-1-area-0.0.0.1]network 1.1.1.1 0.0.0.0 #Declare the interface network segment. 1.1.1.1 0.0.0.0 is an example network segment
[AR1-ospf-1-area-0.0.0.1]net 10.10.100.0 0.0.0.3 #Declare the interface network segment. 10.10.100.0 0.0.0.3 is an example network segment
[AR1-ospf-1-area-0.0.0.1]quit
[AR1-ospf-1]import-route direct #Introduce direct route to OSPF
[AR1-ospf-1]import-route rip 1     #When RIP routes are introduced to OSPF, the default type is 2. That is, the measurement values in the OSPF domain are not cumulative. They can be changed to type 1 when they are introduced. In this way, the measurement values of external routes in the OSPF domain will be cumulative.
```

3.2.2 MSTP

3.2.2.1 Summary

MSTP is a multiple spanning tree protocol. Compared with spanning trees and fast spanning trees, MSTP introduces the concept of Instance. Both spanning tree and fast spanning tree are switch port-based technologies, and all VLANs share the same spanning tree when making spanning tree calculations. MSTP is instance-based. The so-called "instance" refers to a collection of VLANs. MSTP divides one or more VLANs of a device into an instance. Devices with the same instance configuration form an MST domain and run an independent spanning tree. This MST constitutes a large equipment as a whole, and spanning tree algorithm is used with other MST domains to get a whole spanning tree. All VLANs are mapped to instance 0 by default, and other instances are called multiple spanning tree instances. By default, the working mode of the switch is MSTP, which is STP/RSTP compliant.

A comparison of three spanning tree protocols is shown in Table 3.1 below.

Table 3.1 Comparison of three spanning tree protocols

Spanning tree protocol	Characteristic	Application scenario
STP	(1) Form a loop-free tree, resolves broadcast storms and enables redundant backups. (2) Convergence rate is slow	All VLANs share a spanning tree without differentiating between users or business traffics
RSTP	(1) Form a loop-free tree, resolves broadcast storms and enables redundant backups. (2) Fast convergence	
MSTP	(1) Form loop-free trees to resolve broadcast storms and achieve redundant backups. (2) Fast convergence. (3) Multiple spanning trees achieve load balancing between VLANs, and traffic from different VLANs is forwarded along different paths	Users or business traffics need to be differentiated and load-sharing implemented. Different VLANs forward traffic through different spanning trees, each of which is independent of each other

RSTP is improved on the basis of STP to achieve fast convergence of network topology. However, RSTP and STP have the same drawback: because all VLANs in the local area network share a spanning tree, it is not possible to achieve load balancing of data traffic between VLANs. When the link is blocked, it will not carry any traffic, causing a waste of bandwidth, and possibly causing some VLANs to fail to forward their messages.

In the network shown in Figure 3.1, STP or RSTP is applied within the local area network. The spanning tree structure is represented by a dashed line in the graph, and S6 is the root switching device. Links between S2 and S5, S1 and S4 are blocked. No messages from VLAN 2 or VLAN 3 are allowed to pass except for the link labeled VLAN 2 or VLAN 3.

HostA and HostB are both VLAN 2. Because the link between S2 and S5 is blocked, and the link between S3 and S6 does not allow messages from VLAN 2 to pass through, HostA and HostB cannot communicate with each other.

To remedy the shortcomings of STP and RSTP, the 802.1S standard issued by IEEE in 2002 defined MSTP. MSTP is compatible with STP and RSTP, which provides fast convergence and multiple redundant paths for data forwarding, and achieves load balancing of VLAN data during data forwarding.

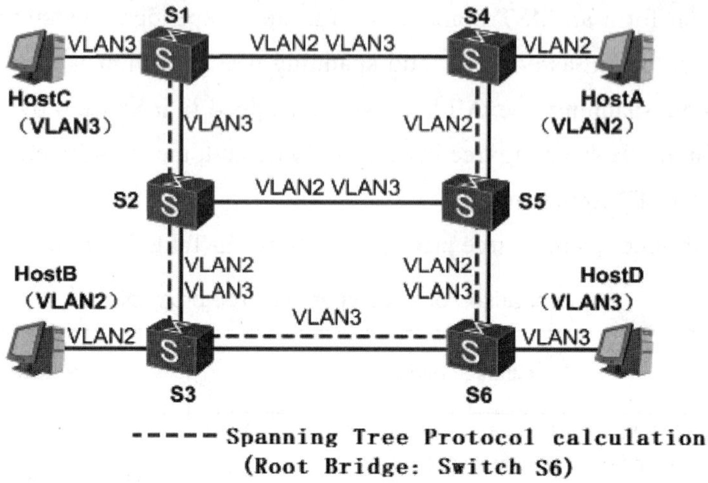

Figure 3.1 Defect Diagram of STP/RSTP

As shown in Figure 3.2, a switching network is divided into domains by MSTP, in which multiple spanning trees are formed, and the spanning trees are independent of each other. Each spanning tree is called a multiple spanning tree instance MSTI, and each domain is called an MST domain (MST Region: Multiple Spanning Tree Region).

An instance is a collection of multiple VLANs. Mapped to an instance, the forwarding status of these VLANs on the port depends on the status of the port on the corresponding MSTP instance.

By bundling multiple VLANs into one instance, you can save communication costs and resource usage. The calculation of MSTP instance topologies is independent, and load balancing can be achieved on these instances.

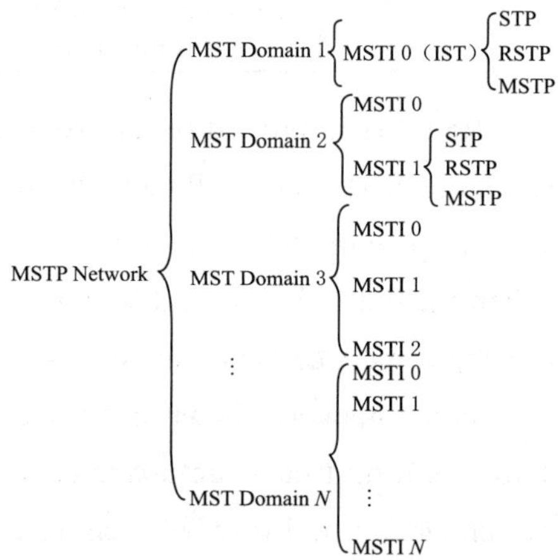

Figure 3.2 MSTP Network Hierarchy Result Diagram Example

By calculating, the root bridge and backup root bridge of MSTP can be determined automatically. Users can also manually configure the device to be the root bridge or backup root bridge of a specified spanning tree: the devices have independent roles in each spanning tree, and can be used as the root bridge or backup root bridge of a spanning tree as well as other spanning trees. But in the same spanning tree, a device cannot be used as either a root bridge or a backup root bridge.

There is only one valid root bridge in a spanning tree. When two or more devices are designated as the root bridge of the same spanning tree, the system will select the device with the smallest MAC address as the root bridge.

Multiple backup root bridges can be specified in each spanning tree. When the root bridge fails or is shut down, the backup root bridge can replace the root bridge as the root bridge for the specified spanning tree; however, if a new root bridge is configured at this time, the backup root bridge will not become the root bridge. If more than one backup root bridge is configured, the backup root bridge with the smallest MAC address will become the root bridge of the specified spanning tree. During the configuration of MSTP, it is recommended to manually configure the root bridge and the backup root bridge.

As shown in Figures 3.3, MSTP associates VLANs with MSTIs by setting up VLAN mapping tables, which are the corresponding relationship tables between VLANs and MSTIs. Each VLAN can only correspond to one MSTI, that is, data from the same VLAN can only be transmitted in one MSTI, and an MSTI may correspond to multiple VLANs.

After calculation, two spanning trees are generated: MSTI1 uses S4 as the root exchange device to forward messages from VLAN 2; MSTI2 forwards messages from VLAN 3 using S6 as the root switching device.

This allows interoperability within all VLANs while forwarding messages from different VLANs along different paths to achieve load sharing.

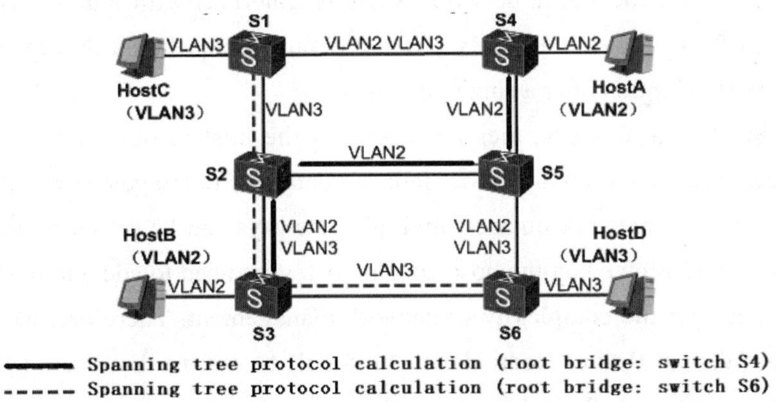

Figure 3.3 Multiple spanning trees in MST domain

3.2.2.2 Configuration command

```
[Huawei]stp mode mstp #Configure MSTP working mode
[Huawei]stp region-configuration #Enter MST domain view
[Huawei-mst-region]region-name gxsdxy #Configure the domain name of the MST domain
[Huawei-mst-region]instance 1 vlan 2 to 5 #Configure the mapping relationship between multiple spanning tree instances and VLANs. This time, VLAN 2 to 5 is an example VLAN
```

```
[Huawei-mst-region]instance 2 vlan 6 to 10  #Configure the mapping relationship between multiple spanning tree instances and VLANs. This time, VLAN 6 to 10 is an example VLAN
[Huawei-mst-region]revision-level0 #Configure the MSTP revision level of the MST domain. By default, the MSTP revision level of the MSTP domain is 0.
[Huawei-mst-region]active region-configuration   #Activate new MST domain configuration
Info: This operation may take a few seconds. Please wait for a moment...done.
[Huawei-mst-region]quit

[Huawei]stp instance 1 root primary   #Configure the device as the root bridge
[Huawei]stp instance 2 root secondary  #Configure the device as a backup root bridge
```

3.2.3 VRRP

3.2.3.1 Summary

VRRP (Virtual Router Redundancy Protocol) adds routers that can take on gateway functions to the backup group to form a virtual router. The voting mechanism of VRRP determines which router should take on forwarding tasks. Hosts in the LAN only need to configure the virtual router as the default gateway. VRRP is a fault-tolerant protocol that improves reliability while simplifying host configuration. In a local area network (such as Ethernet) with multicast or broadcasting capabilities, VRRP can provide a highly reliable default link when a device fails, effectively avoiding network interruption after a single link fails.

Within a local area network, messages sent by the host to other network segments are forwarded by the gateway. When a gateway fails, all data sent to the gateway in this segment will be interrupted. To avoid network outages, multiple gateways can be set up on the host, but only one default gateway is allowed on the host, so administrators need to add and modify it manually, which greatly increases the complexity of network management. Therefore, we usually use the VRRP protocol (Virtual Router Redundancy Protocol) to ensure that users switch to another gateway quickly, uninterruptedly and transparently.

The VRRP protocol first chooses the Master router by election. The Master router is responsible for providing the actual data forwarding service. After the Master router is selected, other devices are used as Backup routers, and the status of the Master router is monitored by the VRRP messages sent periodically by the Master router. If the Backup routing device in the group does not receive a message from the Master routing device within the specified time, it will change its status to Master, which is transparent to the user because the switch is very fast and the user terminal does not need to change the IP address and MAC address of the default gateway.

The Master router is responsible for forwarding data, while the Backup router is not responsible for forwarding data. Therefore, while the Master router forwards data, the Backup router remains idle, which will inevitably result in a waste of network bandwidth resources. We create different VRRP groups by using load balancing technology in VRRP, which enables routers to play different roles in different VRRP groups.

The VRRP protocol is implemented in two versions: VRRPv2 and VRRPv3. VRRPv2 is based on IPv4 and VRRPv3 is based on IPv6.

VRRP routers have three states during operation:

(1) Initialize status: Initialize is entered after the system is started. In this state, the router does not handle VRRP messages anywhere, which can be interpreted as initialization.

(2) Master status: The router sends VRRP notifications and free ARP messages.

(3) Backup status: Accept VRRP notifications.

Normally the Master router is in the Master state and the Backup router is in the Backup state.

3.2.3.2 Election mechanism

VRRP uses an election mechanism to determine the state of the router. A set of routers running VRRP constitutes a virtual router, one of which is in Master state and the other is in Backup state. So the main router is also called the Master router, backing up routers are also called Backup routers.

(1) When a router has VRRP enabled, it determines its role in the backup group based on priority. The router with the highest priority becomes the Master router, and the router with the lower priority becomes the Backup router. Master routers regularly send VRRP notification messages to inform other devices in the backup group that they are working properly; The Backup router starts the timer and waits for the announcement message to arrive.

(2) In preemptive mode, when the Backup router receives a VRRP notification message, it compares its priority with that in the notification message. Become a Master router if it is more preferred than in an announcement message; Otherwise it will remain Backup.

(3) In a non-preemptive manner, as long as the Master router is not failing, the router in the backup group will always remain Master or Backup, and the Backup router will not become a Master router even if it is subsequently configured with a higher priority.

(4) If the timer of the Backup router does not receive the VRRP notification message sent by the Master router after the timeout, the Master router is considered to be no longer functioning properly, and the Backup router considers itself a Master router and sends out the VRRP notification message. Routers in the backup group elect Master routers based on priority to forward messages.

3.2.3.3 Method of VRRP priority election

(1) IP owner in VRRP group. If the virtual IP address is the same as the IP address of a

VRRP router in the VRRP group, the router is the IP address owner and the router will be located on the master router.

(2) Compare priorities. If there is no IP address owner, then compare the priority of the router, the priority range is 0~255, the larger is the master router.

(3) Compare IP addresses. Without the same IP address owner and priority, an IP address with a large size acts as the master router.

3.2.3.4 VRRP basic configuration

(1) Configure VRRP group. To enable VRRP, the basic configuration is to create a VRRP group and configure a virtual IP address for the VRRP group, which needs to be configured on both the master and Backup routers.

[Huawei]interface Ethernet 0/0/1 #Enter port e0/0/1

[Huawei-Ethernet0/0/1] vrrp *group-number* ip *ip-address* # The *group-number* represents the VRRP group number, which can be adjusted in the actual configuration. The value range is 1~255. *ip-address* represents the IP address of the virtual group, such as 192.168.1.1

(2) Configure VRRP priority. If you want to specify a router to be called a Master router, you can manually adjust its priority.

[Huawei]interface Ethernet 0/0/1 # Enter port e0/0/1

[Huawei-Ethernet0/0/1]vrrp *group-number* priority *number* # The *group-number* represents the VRRP group number, which can be adjusted in the actual configuration. The value range is 1~255. *number* represents the priority range, which is between 0~255. The default value is 100.

Priority is configured on which router you want to make the primary router without an IP address owner.

(3) Configure VRRP interface tracing. The VRRP interface tracing mechanism is one of the mechanisms for detecting interface failures. A router configured with an interface tracing mechanism will lower its router priority when its interface fails, changing itself from a master router to a Backup router, and then the original Backup router will become the master router.

Port tracking command:

[Huawei]interface Ethernet 0/0/1 # Enter port e0/0/1

[Huawei-Ethernet0/0/1]vrrp *group-number* track interface *priority-decrement*

The *group-number* represents the VRRP group number, and the value is consistent with the above. *Priority decrement* indicates the priority value to be lowered

Note: priority-decrement is reduced by how much, not how much, such as priority-decrement of 30, then the priority of this router is reduced by 30 from the original.

(4) Configure VRRP preemption mode.Preemptive mode: When the original router is recovered from the failure and connected to the network layer, the router configured with VRRP

preemptive mode will reclaim its original role (master router), if not configured, it will keep the Backup router after the reply.

It is recommended to configure this command in the master router. Use the enable preemption mode command:

```
[Huawei]interface Ethernet 0/0/1                # Enter port e0/0/1
[Huawei-Ethernet0/0/1]vrrp group-number preempt  {delay [Delay-time] }
```

Delay #The value range is 1～255. If the delay time is not configured, the default value is 0 seconds.

delay-time #It is the time to delay preemption, that is, from the time when the router finds that its priority is higher than that of the MASTER, to allow preemption after such a long delay time.

（5）Configure VRRP Timer. VRRP timer can modify the sending time of notification message and configure it on the master router.

command：

```
[Huawei]interface Ethernet 0/0/1                #Enter port e0/0/1
[Huawei-Ethernet0/0/1]vrrp group-number timers advertise vrrp-advertise-interval
```

adver_interval entries for setting the timer interval. MASTER will send an advertisement message every such interval to notify other routers in the group that they are working normally. The value range of vrrp advertisement interval is 0 ～ 254.

Configure VRRP timer learning function：The router that configures this command will learn to send notification message time and calculate the failure interval, otherwise default 3s. This command configures the send interval on the primary router for the VRRP timer configured above, then on the backup router you need to configure the timer learning function to calculate the failure interval because the failure interval is three times the send time.

（6）VRRP load balancing. In a VRRP group, while the master router undertakes the data forwarding task, the link of the Backup router will be idle, which will inevitably lead to a waste of bandwidth resources. To avoid this waste, VRRP load balancing is used.

VRRP load balancing is realized by adding routers to multiple VRRP groups, so that VRRP routers play different roles in different groups.

3.3 Project implementation

3.3.1 Project requirements

（1）The ISP operator's export is in the form of static IP.

（2）All PC computers can access the Internet.

(3) It can realize load balancing. The user data of VLAN 10 and VLAN 20 accesses the Internet through the core switch LSW1, and the user data of VLAN 30 and VLAN 30 accesses the Internet through the core switch LSW2.

(4) Redundancy can be achieved. When any core equipment fails, the network can still operate normally.

3.3.2 Equipment list

The equipment list of this project is shown in Table 3.2.

Table 3.2 Equipment list

Serial No.	Equipment model	Quantity	Function	Remarks
1	AR3260	1	Exit router	
2	S5700–28C–HI	2	Core switch	
3	S3700–26C–HI	2	Access switch	

3.3.3 Access switch project topology

The network topology of the project is shown in Figure 3.4.

3.3.4 Configuration ideas

The following ideas are adopted for configuration:

3.3.4.1 Access layer equipment

(1) Partition VLAN.

(2) Put the port connected to the PC in the corresponding VLAN.

(3) Set the port connecting the core switch to the Trunk port and allow the corresponding VLAN to pass through.

3.3.4.2 Core layer equipment

(1) Partition VLAN.

(2) Configure the VLANIF interface address.

(3) Set the port connected to the

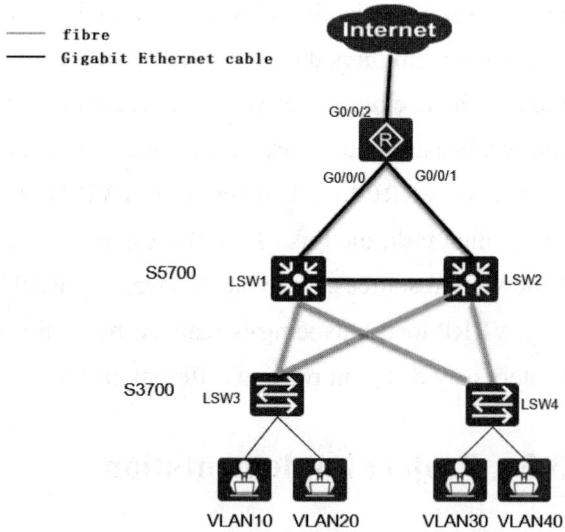

Figure 3.4 Network topology of medium-sized enterprises

access layer switch as the Trunk port and allow the corresponding VLAN to pass through.

(4) Configure MSTP, put VLANs 10 and 20 in instance 1, put VLANs 30 and 40 in instance 2, set the root bridge of instance 1 as the left switch LSW1, the backup root bridge as the right switch LSW2, set the root bridge of instance 2 as the right switch LSW2, and the backup root bridge as the left switch LSW1, and detect the uplink port.

(5) Configure VRRP.

(6) Put the port connecting to the exit router in the VLAN 100, and configure the IP address for the VLAN 100.

(7) Configure dynamic routing OSPF.

3.3.4.3 Exit router

(1) Configure the IP address for the interface.

(2) Configure the default route to the Internet.

(3) Configure dynamic routing OSPF and inject default routing.

(4) Configure LAN users to convert private network addresses to public network addresses through NAT conversion for Internet access.

3.3.5 Relevant information planning

3.3.5.1 Core switch LSW1

The information planning of the core layer switch is shown in Table 3.3 and Table 3.4.

Table 3.3 LSW1 information planning of core layer switch

Serial No.	User department	VLAN No.	IP address	Mask	Remarks
1	Production Department	10	10.10.10.1	255.255.255.0	
2	Sales Department	20	10.10.20.1	255.255.255.0	
3	Finance Department	30	10.10.30.1	255.255.255.0	
4	Engineering Department	40	10.10.40.1	255.255.255.0	
5	Interconnection	100	10.10.100.1	255.255.255.252	Interconnect with router

Table 3.4 LSW2 information planning of core layer switch

Serial No.	User department	VLAN No.	IP address	Mask	Remarks
1	Production Department	10	10.10.10.2	255.255.255.0	
2	Sales Department	20	10.10.20.2	255.255.255.0	

Continue

Serial No.	User department	VLAN No.	IP address	Mask	Remarks
3	Finance Department	30	10.10.30.2	255.255.255.0	
4	Engineering Department	40	10.10.40.2	255.255.255.0	
5	Interconnection	200	10.10.200.1	255.255.255.252	Interconnect with router

3.3.5.2 Exit router

The information planning of the exit router is shown in Table 3.5.

Table 3.5 Information planning of the exit router

Serial No.	Interface	IP address	Mask	Remarks
1	G0/0/0	10.10.100.2	255.255.255.252	Interconnection with core switch LSW1
2	G0/0/1	10.10.200.2	255.255.255.252	Interconnection with core switch LSW2

3.3.5.3 ISP export planning

ISP export information planning is shown in Table 3.6.

Table 3.6 ISP Export Information Planning

Serial No.	Interface	IP address	Mask	Gateway	Remarks
1	G0/0/2	1.1.1.1	255.255.255.252	1.1.1.5	Static IP address

3.3.6 Operating steps

3.3.6.1 LSW3 equipment of access switch

(1) Modify device name:

```
<Huawei>sys
[Huawei]sysname LSW3                    #Modify Device Name
[LSW3]undo info-center enable           #Turn off message prompts
```

(2) Partition VLAN:

```
[LSW3]vlan 10                           #Partition vlan 10
[LSW3-vlan10]description shengchanbu    #Describe vlan
[LSW3-vlan10]quit                       #sign out
```

```
[LSW3]vlan 20                                   #Partition vlan 20
[LSW3-vlan20]description xiaoshoubu             # Describe vlan
[LSW3-vlan20]quit                                # sign out

[LSW3]vlan 30                                   #Partition vlan 30
[LSW3-vlan30]description caiwubu                # Describe vlan
[LSW3-vlan30]quit                               # sign out

[LSW3]vlan 40                                   #Partition vlan 40
[LSW3-vlan40]description gongchengbu            # Describe vlan
[LSW3-vlan40]quit                               # sign out
```

（3）Put the port connected to the PC into the corresponding VLAN:

```
[LSW3]interface Ethernet 0/0/1                  #Enter port e0/0/3
[LSW3-Ethernet0/0/1]port link-type access       #Set the port to access mode
[LSW3-Ethernet0/0/1]port default vlan 10        #Put the port into vlan 10
[LSW3-Ethernet0/0/1]quit                        # sign out
```

Configure other interfaces in the same way:

```
[LSW3]interface Ethernet 0/0/2
[LSW3-Ethernet0/0/2]port link-type access
[LSW3-Ethernet0/0/2]port default vlan 20
[LSW3-Ethernet0/0/2]quit
```

（4）Configure Trunk:

#Set Trunk on the uplink ports G0/0/1 and G0/0/2 connected to the core switch, and allow the corresponding vlan to pass through this interface

```
[LSW3] interface GigabitEthernet 0/0/1                 # Enter port G0/0/1
[LSW3-GigabitEthernet0/0/1]port link-type trunk        #Set the port to Trunk mode
[LSW3-GigabitEthernet0/0/1]port trunk allow-pass vlan 10 2030 40  #Allow vlan 10, 20, 30, 40 to pass
[LSW3-GigabitEthernet0/0/1]quit                        # Sign out

[LSW3]interface GigabitEthernet 0/0/2
[LSW3-GigabitEthernet0/0/2]port link-type trunk
[LSW3-GigabitEthernet0/0/2]port trunk allow-pass vlan 10 20 30 40
```

```
[LSW3-GigabitEthernet0/0/2]quit
```

(5) MSTP Configuration:

```
[LSW3]stp mode mstp#The configuration mode is MSTP. switches are MSTP by default. This step can be omitted
[LSW3]stp region-configuration  #Enter stp configuration
[LSW3-mst-region]region-name sd001  #Configure the domain name as sd001
[LSW3-mst-region]instance 1 vlan 10 20  #Configure vlan 10 and vlan 20 to map to instance 1
[LSW3-mst-region]instance 2 vlan 30 40  #Configure vlan 30 and vlan 40 to map to instance 2
[LSW3-mst-region]active region-configuration  #Activate the configuration of the mstp domain
Info: This operation may take a few seconds. Please wait for a moment...done.
[LSW3-mst-region]quit                          # sign out
```

3.3.6.2 LSW4 equipment of access switch

The configuration of access switch LSW4 is similar to that of LSW3. The configuration code is given below.

(1) Modify device name:

```
<Huawei>sys
[Huawei]sysname LSW4                           #Modify device name
[LSW4]undo info-center enable                  #Turn off message prompts
```

(2) Partition VLAN:

```
[LSW4]vlan 10                                  #Partition vlan 10

[LSW4-vlan10]description shengchanbu           # Describe vlan
[LSW4-vlan10]quit                              # Sign out

[LSW4]vlan 20                                  #Partition vlan 20
[LSW4-vlan20]description xiaoshoubu            # Describe vlan
[LSW4-vlan20]quit                              # Sign out

[LSW4]vlan 30                                  # Partition vlan 30
[LSW4-vlan30]description caiwubu               # Describe vlan
[LSW4-vlan30]quit                              # Sign out

[LSW4]vlan 40                                  # Partition vlan 40
[LSW4-vlan40]description gongchengbu           # Describe vlan
```

```
[LSW4-vlan40]quit                                   # Sign out
```

(3) Put the port connected to the PC into the corresponding VLAN:

```
[LSW4]interface Ethernet 0/0/1                      # Enter port e0/0/1
[LSW4-Ethernet0/0/1]port link-type access           # Set the port to access mode
[LSW4-Ethernet0/0/1]port default vlan 30            # Put the port into vlan 30
[LSW4-Ethernet0/0/1]quit                            # Sign out
```

Configure other interfaces in the same way:

```
[LSW4]interface Ethernet 0/0/2
[LSW4-Ethernet0/0/2]port link-type access
[LSW4-Ethernet0/0/2]port default vlan 40
[LSW4-Ethernet0/0/2]quit
```

(4) Trunk Configuration:

Set Trunk on the uplink ports G0/0/1 and G0/0/2 connected to the core switch, and allow the corresponding vlan to pass through this interface

```
[LSW4] interface GigabitEthernet 0/0/1              #Enter port G0/0/1
[LSW4-GigabitEthernet0/0/1]port link-type trunk  # Set the port to Trunk mode
[LSW4-GigabitEthernet0/0/1]port trunk allow-pass vlan 10 20 30 40  # Allow vlan 10, 20, 30, 40 to pass
[LSW4-GigabitEthernet0/0/1]quit                   # Sign out

[LSW4]interface GigabitEthernet 0/0/2
[LSW4-GigabitEthernet0/0/2]port link-type trunk
[LSW4-GigabitEthernet0/0/2]port trunk allow-pass vlan 10 20
[LSW4-GigabitEthernet0/0/2]quit
```

(5) MSTP Configuration:

```
[LSW4]stp mode mstp  # The configuration mode is MSTP. switches are MSTP by default. This step can be omitted
[LSW4]stp region-configuration                    # Enter stp configuration
[LSW4-mst-region]region-name sd001                # Configure the domain name as sd001
[LSW4-mst-region]instance 1 vlan 10 20            # Configure vlan 10 and vlan 20 to map to instance 1
[LSW4-mst-region]instance 2 vlan 30 40            # Configure vlan 30 and vlan 40 to map to instance 2
[LSW4-mst-region]active region-configuration  # Activate the configuration of the mstp domain
Info: This operation may take a few seconds. Please wait for a moment...done.
```

```
[LSW4-mst-region]quit                    # Sign out
```

3.3.6.3 Configuration of core switch equipment LSW1

(1) Modify device name:

```
<Huawei>system-view
[Huawei]sysname LSW1
[LSW1]undo info-center enable            #Turn off message prompts
```

(2) Partition VLAN:

```
[LSW1]vlan 10                            # Partition vlan 10
[LSW1-vlan10]description shengchanbu     # Describe vlan
[LSW1-vlan10]quit                        # Sign out
```

In the same way, partition other VLANs:

```
[LSW1]vlan 20
[LSW1-vlan20]description xiaoshoubu
[LSW1-vlan20]quit

[LSW1]vlan 30
[LSW1-vlan30]description caiwubu
[LSW1-vlan30]quit

[LSW1]vlan 40
[LSW1-vlan40]description gongchengbu
[LSW1-vlan40]quit

[LSW1]vlan 100
[LSW1-vlan100]description hulian
[LSW1-vlan100]quit
```

(3) Configure VLANIF interface IP address:

```
[LSW1]interface vlanif 10                #Enter the vlan 10 interface
[LSW1-vlanif10]ip address 10.10.10.1 24  #Configure IP address and mask information
[LSW1-vlanif10]quit                      # Sign out
```

Configure other VLANs in the same way:

```
[LSW1]interface vlanif 20
```

```
[LSW1-vlanif20]ip address 10.10.20.1 24
[LSW1-vlanif20]quit

[LSW1]interface vlanif 30
[LSW1-vlanif30]ip address 10.10.30.1 24
[LSW1-vlanif30]quit

[LSW1]interface vlanif 40
[LSW1-vlanif30]ip address 10.10.40.1 24
[LSW1-vlanif30]quit

[LSW1] interface vlanif 100
[LSW1-Vlanif100] ip address 10.10.100.1 30
[LSW1-Vlanif100] quit
```

(4) Configure Trunk:

#Set the downlink port of the access layer switch to Trunk, and allow the corresponding vlan to pass through this interface

```
[LSW1]interface GigabitEthernet 0/0/2            # Enter port G0/0/2
[LSW1-GigabitEthernet0/0/2]port link-type trunk  # set the port to Trunk mode
[LSW1-GigabitEthernet0/0/2]port trunk allow-pass vlan 10 20 30 40  #Allow vlan 10, 20, 30, 40 to pass
[LSW1-GigabitEthernet0/0/2]quit                  # Sign out

[LSW1]interface GigabitEthernet 0/0/3            # Enter port Ge0/0/3
[LSW1-GigabitEthernet0/0/3]port link-type trunk  # Set the port to Trunk mode
[LSW1-GigabitEthernet0/0/3]port trunk allow-pass vlan 10 20 30 40  # Allow vlan 10, 20, 30, 40 to pass
[LSW1-GigabitEthernet0/0/3]quit                  # Sign out

[LSW1]interface GigabitEthernet 0/0/4            # Enter port G0/0/4
[LSW1-GigabitEthernet0/0/3]port link-type trunk  # Set the port to Trunk mode
[LSW1-GigabitEthernet0/0/4]port trunk allow-pass vlan 10 20 30 40  # Allow vlan 10, 20, 30, 40 to pass
[LSW1-GigabitEthernet0/0/3]quit                  # Sign out
```

(5) MSTP Configuration:

[LSW1]stp mode mstp # The configuration mode is MSTP. switches are MSTP by default. This step can be omitted

[LSW1]stp region-configuration # Enter stp configuration

[LSW1-mst-region]region-name sd001 # Configure the domain name as sd001

[LSW1-mst-region]instance 1 vlan 10 20 # Configure VLAN 10 and VLAN 20 to map to instance 1

[LSW1-mst-region]instance 2 vlan 30 40 # Configure VLAN 30 and VLAN 40 to map to instance 2

[LSW1-mst-region]active region-configuration # Activate the configuration of the mstp domain

Info: This operation may take a few seconds. Please wait for a moment...done.

[LSW1-mst-region]quit # Sign out

[LSW1]stp instance 1 root primary #Configure the switch LSW1 as the root bridge in instance 1

[LSW1]stp instance 2 root secondary #Configure the switch LSW1 in instance 2 as the backup root bridge

(6) Configure VRRP:

[LSW1]interface vlan 10 #Enter the vlan 10 interface

[LSW1-Vlanif10]vrrp vrid 1 virtual-ip 10.10.10.254 #Create VRRP group 1, virtual gateway is 10.10.10.254

[LSW1-Vlanif10]vrrp vrid 1 priority 200 #Set the priority of VRRP group 1 to 200, and the default is 100. The purpose is to compare with another core switch LSW2. This switch has a higher priority of VRRP group 1, which works in the master state.

[LSW1-Vlanif10]vrrp vrid 1 track interface GigabitEthernet 0/0/1 reduced 150

#Monitor the uplink port G0/0/1. If the port is down, the priority will be reduced by 150

[LSW1-Vlanif10]quit # Sign out

Create VRRP groups 2, 3, 4 in the same way:

[LSW1]interface vlan 20

[LSW1-Vlanif20]vrrp vrid 2 virtual-ip 10.10.20.254

[LSW1-Vlanif20]vrrp vrid 2 priority 200 #Set the priority of VRRP group 2 to 200, and the default is 100. The purpose is to compare with another core switch LSW2. This switch has a higher priority of VRRP group 2, which works in the Master state.

[LSW1-Vlanif20]vrrp vrid 2 track interface GigabitEthernet 0/0/1 reduced 150

[LSW1-Vlanif20]quit

[LSW1]interface Vlanif 30

[LSW1-Vlanif30]vrrp vrid 3 virtual-ip 10.10.30.254

[LSW1-Vlanif30]vrrp vrid 3 priority 100 #Set the priority of VRRP group 3 to 100. The default is 100. This command can be ignored. It is used to compare with another core switch LSW2. The priority of VRRP group 3 of this switch is lower and it works in the Backup state.

[LSW1-Vlanif30]quit

[LSW1]interface Vlanif 40

[LSW1-Vlanif40]ip address 10.10.40.1 24

[LSW1-Vlanif40]vrrp vrid 4 virtual-ip 10.10.40.254

[LSW1-Vlanif40]vrrp vrid 4 priority 100 #Set the priority of VRRP group 3 to 100, and the default is 100. This command can be ignored. It is used to compare with another core switch LSW2. The priority of VRRP group 4 of this switch is lower, and it works in the Backup state.

[LSW1-Vlanif40]quit

(7) **Check VRRP configuration:**

```
[LSW1]display vrrp brief
VRID  State        Interface              Type      Virtual IP
----------------------------------------------------------------
1     Master       Vlanif10               Normal    10.10.10.254
```

vrrp group 1 is bound to vlan 10 and works in the master state. The virtual gateway is 10.10.10.254

```
2     Master       Vlanif20               Normal    10.10.20.254
3     Master       Vlanif30               Normal    10.10.30.254
4     Master       Vlanif40               Normal    10.10.40.254
----------------------------------------------------------------
Total:4    Master:4    Backup:0    Non-active:0
```

* Since LSW2 has not been configured, the vrrp status of the four vlanifs of LSW1 is Master

(8) **Configure the interconnection interface with the exit router:**

```
[LSW1]interface GigabitEthernet 0/0/1                # Enter port G0/0/1
[LSW1-GigabitEthernet0/0/1]port link-type access     # Set the port to access mode
[LSW1-GigabitEthernet0/0/1]port default vlan 100     #Divide ports into vlan 100
[LSW1-GigabitEthernet0/0/2]quit                       # Sign out
```

(9) **Configure dynamic routing OSPF:**

[LSW1]ospf 1 router-id 1.1.1.1 #Enable OSPF routing. The process number is 1 and the Router ID is 1.1.1.1

```
[LSW1-ospf-1]import-route direct                    #Import direct route
[LSW1-ospf-1]area 0                                 #Region 0
[LSW1-ospf-1-area-0.0.0.0]network 10.10.100.1 0.0.0.0 #Announcement network
[LSW1-ospf-1-area-0.0.0.0]quit                      # Sign out
[LSW1-ospf-1]quit                                   # Sign out
```

3.3.6.4 Configuration of core switch equipment LSW2

(1) Modify device name:

```
<Huawei>system-view
[Huawei]sysname LSW2
[LSW2]undo info-center enable                       #Turn off message prompts
```

(2) Partition VLAN:

```
[LSW2]vlan 10                                       # Partition vlan 10
[LSW2-vlan10]description shengcanbu                 # Describe vlan
[LSW2-vlan10]quit                                   # Sign out
```

In the same way, divide other VLANs:

```
[LSW2]vlan 20
[LSW2-vlan20]description xiaoshoubu
[LSW2-vlan20]quit

[LSW2]vlan 30
[LSW2-vlan30]description caiwubu
[LSW2-vlan30]quit

[LSW2]vlan 40
[LSW2-vlan40]description gongchengbu
[LSW2-vlan40]quit

[LSW2]vlan 200
[LSW2-vlan200]description hulian
[LSW2-vlan200]quit
```

(3) Configure VLANIF interface IP address:

```
[LSW2]interface vlanif 10                      #Enter the vlan 10 interface
[LSW2-vlanif10]ip address 10.10.10.2 24        #Configure ip address and mask information
```

```
[LSW2-vlanif10]quit                              # Sign out
```

In the same way, configure other VLANs:

```
[LSW2]interface vlanif 20
[LSW2-vlanif20]ip address 10.10.20.2 24
[LSW2-vlanif20]quit

[LSW2]interface vlanif 30
[LSW2-vlanif30]ip address 10.10.30.2 24
[LSW2-vlanif30]quit

[LSW2]interface vlanif 40
[LSW2-vlanif30]ip address 10.10.40.2 24
[LSW2-vlanif30]quit

[LSW2] interface vlanif 200
[LSW2-Vlanif100] ip address 10.10.200.1 30
[LSW2-Vlanif100] quit
```

(4) Configure Trunk:

Set the downlink port of the access layer switch to trunk, and allow the corresponding vlan to pass through this interface

```
[LSW2]interface GigabitEthernet 0/0/2                # Enter port G0/0/2
[LSW2-GigabitEthernet0/0/2]port link-type trunk      # Set the port to trunk mode
[LSW2-GigabitEthernet0/0/2]port trunk allow-pass vlan 10 20 30 40  # Allow vlan 10, 20, 30, 40 to pass
[LSW2-GigabitEthernet0/0/2]quit                      # Sign out

[LSW2]interface GigabitEthernet 0/0/3                # Enter port G0/0/3
[LSW2-GigabitEthernet0/0/3]port link-type trunk      # Set the port to trunk mode
[LSW2-GigabitEthernet0/0/3] port trunk allow-pass vlan 10 20 30 40  # Allow vlan 10, 20, 30, 40 to pass
[LSW2-GigabitEthernet0/0/3]quit                      # Sign out

[LSW2]interface GigabitEthernet 0/0/4                # Enter port G0/0/4
[LSW2-GigabitEthernet0/0/3]port link-type trunk      # Set the port to Trunk mode
```

[LSW2-GigabitEthernet0/0/4]port trunk allow-pass vlan 10 20 30 40 # Allow vlan 10, 20, 30, 40 to pass

[LSW2-GigabitEthernet0/0/3]quit # Sign out

（5）**MSTP Configuration**:

[LSW2]stp mode mstp # The configuration mode is MSTP. Switches are MSTP by default. This step can be omitted

[LSW2]stp region-configuration # Enter stp configuration

[LSW2-mst-region]region-name sd001 # Configure the domain name as sd001

[LSW2-mst-region]instance 1 vlan 10 20 # Configure vlan 10 and vlan 20 to map to instance 1

[LSW2-mst-region]instance 2 vlan 30 40 # Configure vlan 30 and vlan 40 to map to instance 2

[LSW2-mst-region]active region-configuration # Activate the configuration of the mstp domain

Info: This operation may take a few seconds. Please wait for a moment...done.

[LSW2-mst-region]quit # Sign out

[LSW2]stp instance 1 root secondary # Configure the switch LSW2 in instance 1 as the backup root bridge

[LSW2]stp instance 2 root primary # Configure the switch LSW2 as the root bridge in instance 2

（6）**Configure VRRP**:

[LSW2]interface vlan 10 #Enter the vlan 10 interface

[LSW2-Vlanif10]vrrp vrid 1 virtual-ip 10.10.10.254 # Create VRRP group 1, virtual gateway is 10.10.10.254

[LSW2-Vlanif10]vrrp vrid 1 priority 100 #Set the priority of VRRP Group 1 to 100. This command can be ignored. The default is 100. It is used to compare with another core switch, LSW1. The VRRP Group 1 of this switch has a lower priority and works in the Backup state.

[LSW2-Vlanif10]quit # Sign out

Create VRRP groups 2, 3, 4 in the same way:

[LSW2]interface vlan 20

[LSW2-Vlanif20]vrrp vrid 2 virtual-ip 10.10.20.254

[LSW2-Vlanif20]vrrp vrid 2 priority 100 #Set the priority of vrrp group 2 to 100. This command can be ignored. The default is 100. It is used to compare with another core switch, LSW2. The vrrp group 2 of this switch has a lower priority and works in the Backup state.

[LSW2-Vlanif20]quit

```
[LSW2]interface Vlanif 30
[LSW2-Vlanif30]vrrp vrid 3 virtual-ip 10.10.30.254
[LSW2-Vlanif30]vrrp vrid 3 priority 200 #Set the priority of vrrp group 3 to 200, and the default is 100. The purpose is to compare with another core switch LSW1. The vrrp group 3 of this switch has a higher priority and works in the master state.
[LSW2-Vlanif20]vrrp vrid 2 track interface GigabitEthernet 0/0/1 reduced 150
#Monitor the uplink port G0/0/1. If the port is down, the priority will be reduced by 150
[LSW2-Vlanif30]quit

[LSW2]interface Vlanif 40
[LSW2-Vlanif40]ip address 10.10.40.1 24
[LSW2-Vlanif40]vrrp vrid 4 virtual-ip 10.10.40.254
[LSW2-Vlanif40]vrrp vrid 4 priority 200 #Set the priority of vrrp group 3 to 200, and the default is 100. This command can be ignored. It is used to compare with another core switch LSW1. vrrp group 4 of this switch has a higher priority and works in the master state.
[LSW2-Vlanif40]quit
```

(7) **Configure VRRP:**

```
[LSW2]display vrrp brief
VRID State       Interface              Type     Virtual IP
----------------------------------------------------------------
1    Backup      Vlanif10               Normal   10.10.10.254
#vrrp group 1 is bound to vlan 10 and works in the Backup state. The virtual gateway is 10.10.10.254
2    Backup      Vlanif20               Normal   10.10.20.254
3    Master      Vlanif30               Normal   10.10.30.254
#vrrp group 3 is bound to vlan 30 and works in the master state. The virtual gateway is 10.10.30.254
4    Master      Vlanif40               Normal   10.10.40.254
----------------------------------------------------------------
Total:4    Master:2    Backup:2    Non-active:0
```

(8) **Configure the interconnection interface with the exit router:**

```
[LSW2]interface GigabitEthernet 0/0/1                 # Enter port G0/0/1
[LSW2-GigabitEthernet0/0/1]port link-type access      # Set the port to access mode
[LSW2-GigabitEthernet0/0/1]port default vlan 200      #Divide ports into vlan 200
```

[LSW2-GigabitEthernet0/0/2]quit # Sign out

(9) Configure dynamic routing OSPF:

[LSW2]ospf1 router-id 2.2.2.2 #Configure OSPF route. The process number is 1 and the Router ID is 2.2.2.2

[LSW2-ospf-1]import-route direct #Import direct route

[LSW2-ospf-1]area 0 #Region 0

[LSW2-ospf-1-area-0.0.0.0]network 10.10.200.1 0.0.0.0 #Announcement network

[LSW2-ospf-1-area-0.0.0.0]quit # Sign out

[LSW2-ospf-1]quit # Sign out

3.3.6.5 Export equipment

(1) Modify device name:

<Huawei>system-view

[Huawei]sysname AR1 #Modify device name

[AR1]undo info-center enable #Turn off message prompts

(2) Configure the IP address for the interface:

[AR1]interface GigabitEthernet 0/0/0 #Enter the intranet port

[AR1-GigabitEthernet0/0/0]ip address 10.10.100.2 30 #Configure the address for the intranet port

[AR1-GigabitEthernet0/0/0]quit # Sign out

[AR1]interface GigabitEthernet 0/0/1 # Enter the intranet port

[AR1-GigabitEthernet0/0/1]ip address 10.10.200.2 30 # Configure the address for the intranet port

[AR1-GigabitEthernet0/0/1]quit # Sign out

[AR1]interface GigabitEthernet 0/0/2 #Access to external network port

[AR1-GigabitEthernet0/0/2]ip address 1.1.1.1 29 #Configure the address for the external network port

[AR1-GigabitEthernet0/0/2]quit # Sign out

(3) Configure LAN users to convert private network addresses to public network addresses through NAT conversion for Internet access:

[AR1]acl number 2000 #Define a standard acl

[AR1-acl-basic-2000]rule permit source any #The rule is to allow the source address to be anyone

```
[AR1-acl-basic-2000]quit                                  # Sign out
[AR1]nat address-group 1 1.1.1.2 1.1.1.5                  #Create a nat translation address pool with the number of 1 and the range of 1.1.1.2-1.1.1.5
[AR1]interface GigabitEthernet 0/0/2                      #Access to external network port
[AR1-GigabitEthernet0/0/1]nat outbound 2000 address-group 1  #Allow computers matching ACL No. 2000 to access the Internet through the address pool numbered 1
[AR1-GigabitEthernet0/0/1]quit                            # Sign out
```

（4）Configure default routes to the Internet:

```
[AR1]ip route-static 0.0.0.0 0 1.1.1.6                    #Any data to the Internet should be sent to the 1.1.1.6 address
```

（5）Configure dynamic routing OSPF:

```
[AR1]ospf 1 router-id 3.3.3.3
[AR1-ospf-1]default-route-advertise                       #Announce the default route to the entire OSPF domain
[AR1-ospf-1]area 0                                        #Enter area 0
[AR1-ospf-1-area-0.0.0.0]network 10.10.100.2 0.0.0.0      #Announcing the Network
[AR1-ospf-1-area-0.0.0.0]network 10.10.200.2 0.0.0.0      #Announcing the Network
[AR1-ospf-1-area-0.0.0.0]quit                             # Sign out
[AR1-ospf-1]quit                                          # Sign out
```

3.3.6.6 Test

（1）Manually configure IP address for the PC of the production department. Manually Configuring IP Addresses is shown in Figure 3.5.

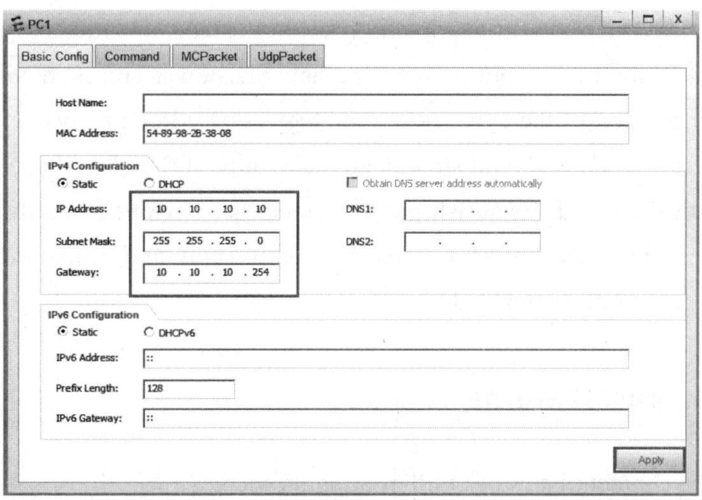

Figure 3.5 Manually Configuring IP Addresses

(2) Access to the Internet. Access to Internet Detection is shown in Figure 3.6.

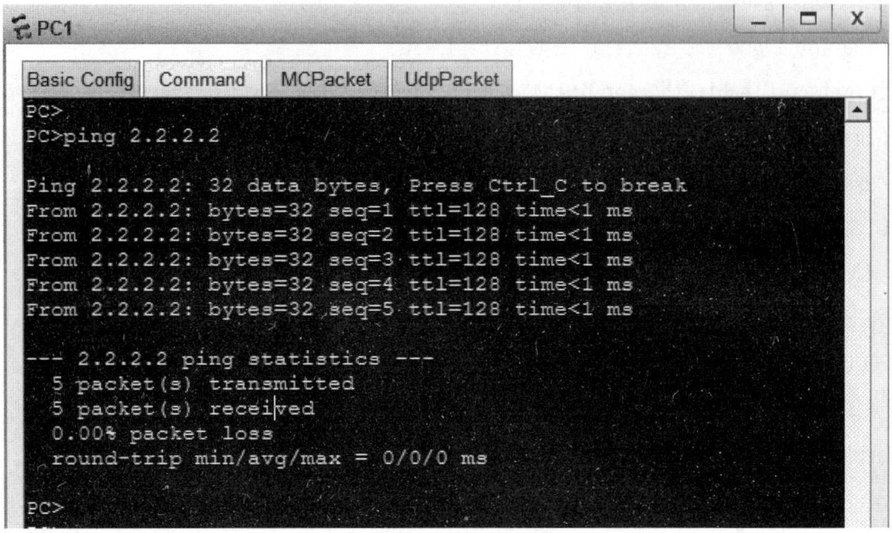

Figure 3.6 Access to Internet Detection

3.4 Ideological and political integration

China's Internet development is rapid and fruitful. China accounts for 7 of the world's top 20 Internet websites by market capitalization, 4 of the world's top 10 Internet companies by market capitalization, the number of netizens ranks first in the world, the proportion of fiber broadband users ranks first in the world, and the 4G commercial network has the widest coverage in the world; The number of 5G standard-essential patents ranks first in the world, and the new generation of information technology such as artificial intelligence, big data, and cloud computing is developing rapidly. At present, China has become a veritable network power and is moving towards a cyber power. The implementation of the cyber power strategy is based on practice and respects practice. Contemporary college students should strengthen their study of the idea of socialism with Chinese characteristics in the new era. Establish patriotic feelings, strengthen cultural self-confidence, practice mission responsibility, be down-to-earth in daily study and future work, seek truth from facts, and develop good professional habits.

3.5 Thinking and training

1. What are the characteristics of OSPF routing protocols?

2. What are the characteristics of spanning-tree protocol STP, RSTP, and MSTP? What are

their respective application scenarios?

3. What is the VRRP protocol? What is the process of its work?

4. What are the three states of VRRP routers during operation?

5. What are the configuration steps for multi-instance spanning-tree MSTP?

项目 4

项目 4　某县公安局网络项目

4.1　项目导入

某县公安局办公大楼对网络系统的要求是极高的，由于该单位职能的特殊性，对网络设计具有一定的特殊要求。除了方便办公外，对网络安全也不同于一般的办公环境。网络的建设应以业务为中心，立足公安系统工作的现实和发展，建成覆盖全县公安系统的信息网络，实现数据的传输交换，建设并不断完善网络系统和运行管理体系，建设网络应用平台、推进业务应用建设，满足公安系统信息化要求及提高工作效率的需要，为整合信息资源奠定基础。

4.2　相关知识点

4.2.1　端口安全

端口安全（Port Security），可根据 MAC 地址来对网络流量的控制和管理，通过 MAC 地址表记录连接到交换机端口的以太网 MAC 地址，并只允许某个 MAC 地址通过本端口通信。其他 MAC 地址发送的数据包通过此端口时，端口安全特性会阻止它。使用端口安全特性可以防止未经允许的设备访问网络，并增强安全性，并可防止 MAC 地址泛洪造成 MAC 地址表填满。

4.2.1.1　MAC 地址表分为三张

（1）静态 MAC 地址表。手工绑定，优先级高于动态 MAC 地址表。

（2）动态 MAC 地址表。交换机收到数据帧后会将源 MAC 学习到 MAC 地址表中。

（3）黑洞 MAC 地址表。手工绑定或自动学习，用于丢弃指定 MAC 地址。

4.2.1.2　MAC 地址表的管理命令

（1）查看 MAC 地址表，方法如下：

```
<Huawei>display mac-address
```

（2）配置静态 MAC 地址表，方法如下：

```
[Huawei] mac-address static 5489-98C0-7E34 GigabitEthernet 0/0/1 vlan 1
```

```
# 将 mac 地址绑定到接口 g0/0/1 在 vlan 1 中有效
```

（3）配置黑洞 MAC 地址表，方法如下：

```
[Huawei] mac-address blackhole 5489-987f-161a vlan 1
# 在 vlan 1 中收到源或目的为此 mac 时丢弃帧
```

（4）禁止端口学习 MAC 地址，可以在端口或者 VLAN 中禁止 MAC 地址学习功能，方法如下：

```
[Huawei-GigabitEthernet0/0/1]mac-address learning disable action discard
```

（5）禁止学习 MAC 地址，并将收到的所有帧丢弃，也可以在 VLAN 中配置，方法如下：

```
[Huawei-GigabitEthernet0/0/1] mac-address learning disable action forward
```

（6）禁止学习 MAC 地址，但是将收到帧以泛洪方式转发（交换机对于未知目的 MAC 地址转发原理），也可以在 VLAN 中配置。

（7）限制 MAC 地址学习数量，可以在端口或者 VLAN 中配置，方法如下：

```
[Huawei-GigabitEthernet0/0/1]mac-limit maximum 9 alarm enable
```

交换机限制 MAC 地址学习数量为 9 个，并在超出数量时发出告警，超过的 MAC 数量将无法被端口学习到，但是可以通过泛洪转发（交换机对于未知目的 MAC 地址转发原理），也可以在 VLAN 中配置。

（8）配置端口安全动态 MAC 地址。此功能是将动态学习到的 MAC 地址设置为安全属性，其他没有被学习到的非安全属性的 MAC 的帧将被端口丢弃，方法如下：

```
[Huawei-GigabitEthernet0/0/3]port-security enable
# 打开端口安全功能
[Huawei-GigabitEthernet0/0/3]port-security max-mac-num 1
# 限制安全 MAC 地址最大数量为 1 个，默认为 1
[Huawei-GigabitEthernet0/0/3]port-security aging-time 300        # 配置安全 MAC 地址的老化时间
300s，默认不老化
```

在端口安全动态 MAC 地址中，配置如上的话，在 g0/0/3 端口学习到的第一个 MAC 地址设置为安全 MAC 地址，此外其他 MAC 地址在接入端口的话都不给予转发，在 300s 后刷新安全 MAC 地址表，并且重新学习安全 MAC 地址，（哪个 MAC 地址）先到就先被学到端口并设置为安全 MAC 地址，但是在交换机重启后安全 MAC 地址会被清空重新学习。

（9）配置端口安全 Sticky 粘贴 MAC 地址。此功能与端口安全动态 MAC 地址一致，唯一不同的是粘贴 MAC 地址不会老化，剪切交换重启后依然存在，动态安全 MAC 地址只能动态学到而安全粘贴 MAC 可以动态学习也可以手工配置，方法如下：

```
[Huawei-GigabitEthernet0/0/3]port-security enable
# 打开端口安全功能
[Huawei-GigabitEthernet0/0/3]port-security mac-address sticky
```

```
#打开安全粘贴mac功能
[Huawei-GigabitEthernet0/0/3]port-security max-mac-num 1
#限制安全mac地址最大数量为1个，默认为1
[Huawei-GigabitEthernet0/0/3]port-security mac-address sticky 5489-98D8-71D5 vlan 1
#手工绑定粘贴MAC地址和所属vlan
[Huawei-GigabitEthernet0/0/3]port-security protect-action  restrict
#配置其他非安全mac地址数据帧的处理动作
```

（10）查看粘贴 MAC 地址状态，方法如下：

```
[Huawei-GigabitEthernet0/0/3] display mac-address
```

4.2.2　MUX VLAN

4.2.2.1　概述

在企业网络中，企业员工和企业客户都可以访问企业的服务器。对于企业来说，希望部分员工之间可以互访，部分员工之间不能够互相交流。为了满足企业的需求，可将企业服务器、可以互访的员工分别划分到不同的 VLAN，不能互相交流的员工分别划分到不同的 VLAN（不能与企业服务器、可互访员工所在的 VLAN 相同）。这不但需要耗费大量的 VLAN ID，还增加了网络管理者的维护量。

为了解决上述问题，可在交换机上部署 MUX VLAN 功能。MUX VLAN 能够提供 VLAN 内的二层流量隔离机制。

MUX VLAN 分为 Principal VLAN （主 VLAN）和 Subordinate VLAN（从 VLAN），从 VLAN 分为 Group VLAN（互通从 VLAN） 和 Separate VLAN（隔离型从 VLAN）。

通信权限如下：

（1）Principal VLAN（主 VLAN）端口可以和所有 VLAN 通信。

（2）Group VLAN（互通型从 VLAN）可以和自己 VLAN 间成员通信，能和 Principal VLAN（主 VLAN）通信。

（3）Separate VLAN（隔离型从 VLAN）只能和 Principal VLAN（主 VLAN）通信，自己 VLAN 的成员也不可通信。

4.2.2.2　配置代码

配置时要注意的是，所有主机必须在同一子网，交换机端口必须为 access 模式加入 VLAN 中，配置 MUX VLAN 不能用于 VLANIF 接口、VLAN Mapping、VLAN Stacking、Super-VLAN、Sub-VLAN 的配置。

配置代码如下：

```
[Huawei]vlan batch 10 20 30
#创建VLAN 10 20 30
```

```
[Huawei]vlan 10
# 进入 VLAN 管理视图
[Huawei-vlan10]mux-vlan
# 配置 VLAN 10 为 Principal VLAN
[Huawei-vlan10]subordinate group 20
# 配置 VLAN 20 为互通型从 VLAN
[Huawei-vlan10]subordinate separate 30
# 配置 VLAN 30 为隔离型从 VLAN
[Huawei-vlan10]quit                # quit
[Huawei]interface g0/0/1           # Enter interface g0/0/1
[Huawei-GigabitEthernet0/0/1]port link-type access
# 配置接口模式为 access
[Huawei-GigabitEthernet0/0/1]port default vlan 10
# 将接口加入 VLAN 10
[Huawei-GigabitEthernet0/0/1]port mux-vlan enable
# 开启接口的 mux-vlan 功能
[Huawei-GigabitEthernet0/0/1]interface g0/0/2
# 进入接口 G0/0/2
[Huawei-GigabitEthernet0/0/2]port link-type access
# 配置接口模式为 access
[Huawei-GigabitEthernet0/0/2]port default vlan 20
# 将接口加入 VLAN 20
[Huawei-GigabitEthernet0/0/2]port mux-vlan enable
# 开启接口的 mux-vlan 功能
[Huawei-GigabitEthernet0/0/2]int g0/0/3
[Huawei-GigabitEthernet0/0/3]port link-type access
[Huawei-GigabitEthernet0/0/3]port default vlan 20
[Huawei-GigabitEthernet0/0/3]port mux-vlan enable
[Huawei-GigabitEthernet0/0/3]int g0/0/4
[Huawei-GigabitEthernet0/0/4]port link-type access
[Huawei-GigabitEthernet0/0/4]port default vlan 30
[Huawei-GigabitEthernet0/0/4]port mux-vlan enable
[Huawei-GigabitEthernet0/0/4]int g0/0/5
[Huawei-GigabitEthernet0/0/5]port link-type access
```

```
[Huawei-GigabitEthernet0/0/5]port default vlan 30
[Huawei-GigabitEthernet0/0/5]port mux-vlan enable
```

4.2.2.3 检验配置

使用 display mux-vlan 查看 mux-vlan 的信息，方法如下：

```
[Huawei]display mux-vlan
Principal Subordinate Type        Interface
--------------------------------------------------------------------
10        -           principal   GigabitEthernet0/0/1
10        30          separate    GigabitEthernet0/0/4 GigabitEthernet0/0/5
10        20          group       GigabitEthernet0/0/2 GigabitEthernet0/0/3
```

从以上信息可以看出，交换机端口 G0/0/1 在主 VLAN 中，端口 G0/0/5 和 G0/0/5 在隔离型从 VLAN 30 中，端口 G0/0/2 和 G0/0/3 在互通型从 VLAN 20 中。

4.3 项目实施

4.3.1 项目需求

（1）全网使用 OSPF 路由协议交互路由信息。

（2）接入交换机上针对于某些 PC 进行 IP+MAC+VLAN 的绑定。

（3）内部网络中，针对于某些地址段限制的上网时间。

（4）同一 VLAN 中大家都可以访问服务器，同一 VLAN 中的部分员工相互隔离，不能相互访问。

4.3.2 设备清单

本项目用的设备主要有三层交换机、二层交换机、出口路由器等，具体型号功能见表 4.1。

表 4.1 网络设备清单

序号	设备型号	数量	功能	备注
1	AR2240	1	出口路由器	
2	S5700	1	核心交换机	
3	S3700-52P-LI-AC	20	接入交换机	

4.3.3 项目拓扑

本项目的网络拓扑如图 4.1 所示。

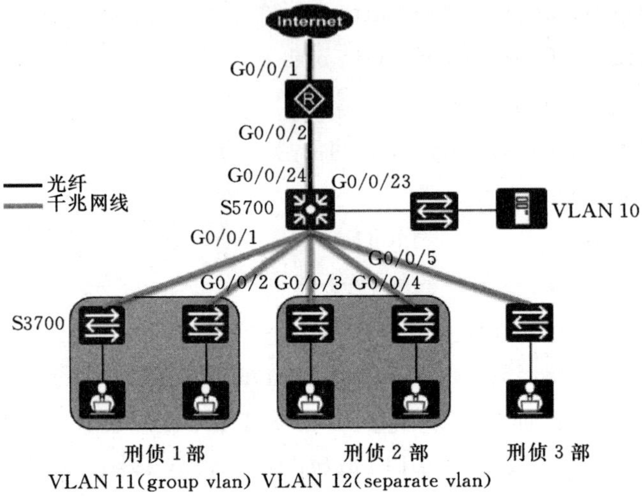

图 4.1　某县公安局网络拓扑图

4.3.4　配置思路

项目采用如下的思路配置：

（1）出口路由器配置 L2TP VPN 服务器。

（2）出口路由器配置基于时间的 ACL 策略限制内部部分 IP 地址的上网时间。

（3）采用 32 位掩码的 lookback 地址作为 OSPF 的 router ID。

（4）接入层交换机配置端口安全机制限制端口终端接入。

（5）在核心交换机采用 MUX VLAN 实现同一 VLAN 中的隔离。

4.3.5　相关信息规划

4.3.5.1　核心交换机

核心交换机信息规划见表 4.2。

表 4.2　核心交换机信息规划表

序号	使用部门	VLAN 号	IP 地址	掩码	备注
1	刑侦服务器	10			192.168.10.250
2	刑侦 1 部	11	192.168.10.1	255.255.255.0	192.168.10.100
3	刑侦 2 部	12			手工指定
4	刑侦 3 部	13	192.168.30.1	255.255.255.0	
5	交换机管理	255	192.168.255.1	255.255.255.0	

续表

序号	使用部门	VLAN 号	IP 地址	掩码	备注
6	互联 1	1000	192.168.250.1	255.255.255.252	与路由器互联
7	Lookback 0		192.168.254.2	255.255.255.255	

其中办公地址段中：192.168.10.250 这个 IP 地址只允许在工作日早上 8 点到 9 点进行因特网访问。

4.3.5.2　出口路由器

出口路由器信息规划见表 4.3。

表 4.3　出口路由器信息规划表

序号	接口	IP 地址	掩　码	备注
1	G0/0/2	192.168.250.2	255.255.255.252	与核心交换机互联
2	Lookback 0	192.168.255.1	255.255.255.255	

4.3.5.3　ISP 出口规划

ISP 出口规划见表 4.4。

表 4.4　ISP 出口信息规划表

序号	接口	IP 地址	掩码	网关	备注
1	G0/0/1	1.1.1.2	255.255.255.192	1.1.1.1	静态 IP 地址

4.3.6　操作步骤

4.3.6.1　接入交换机

（1）基本配置（划分 VLAN、Trunk、绑定地址），方法如下：

```
<Huawei>system-view
[Huawei] YLRXGA _SW_S3700_1
[YLRXGA _SW_S3700_1] vlan batch  10 11 12 13 255 1000
[YLRXGA _SW_S3700_1] interface gigabitethernet 0/0/1
[YLRXGA _SW_S3700_1-GigabitEthernet0/0/1] port link-type trunk
[YLRXGA _SW_S3700_1-GigabitEthernet0/0/1] port trunk allow-pass vlan all
[YLRXGA _SW_S3700_1-GigabitEthernet0/0/1]quit
[YLRXGA _SW_S3700_1] user-bind static ip-address 192.168.10.100 mac-address 5489-
```

```
98D2-0FEB interface Ethernet 0/0/1  #把 IP 地址和 MAC 地址绑定到 E0/0/1 端口中
[YLRXGA _SW_S3700_1] vlan 11
[YLRXGA _SW_S3700_1-vlan11] ip source check user-bind enable  #检查报文功能
```

在网络中经常出现利用仿冒合法用户的源 IP、MAC 等信息的报文访问或攻击网络的情况，导致合法用户无法获得稳定、安全的网络服务。配置 IP 报文检查功能，可以防范上述情况的发生。

当启用了 IP 报文的检查功能后，默认会对 IP 报文中的 IP、MAC、VLAN、接口信息进行绑定表匹配检查。同时也可以执行命令 ip source check user-bind check-item（接口视图）或者 ip source check user-bind check-item（VLAN 视图）灵活配置 IP 报文的检查项。只有与绑定表匹配的 IP 报文才能被转发，否则被丢弃。

其他接入层交换机配置类似，此处省略。

（2）MUX VLAN 配置（只需要在 LSW1-LSW4 中配置），方法如下：

```
[YLRXGA _SW_S3700_1]vlan 10
[YLRXGA _SW_S3700_1-vlan10]mux-vlan    #启用 MUX VLAN，VLAN 10 是主 VLAN
[YLRXGA _SW_S3700_1-vlan10]subordinate group 11   # VLAN 11 是互通型 VLAN
[YLRXGA _SW_S3700_1-vlan10]subordinate separate 12   #VLAN 12 是隔离型 VLAN
[YLRXGA _SW_S3700_1-vlan10]quit
[YLRXGA _SW_S3700_1]interface Ethernet 0/0/1
[YLRXGA _SW_S3700_1-Ethernet0/0/1]port link-type access
[YLRXGA _SW_S3700_1-Ethernet0/0/1]port default vlan 11
[YLRXGA _SW_S3700_1-Ethernet0/0/1]port mux-vlan enable
[YLRXGA _SW_S3700_1-Ethernet0/0/1]quit
```

检查：

```
[YLRXGA _SW_S3700_1]dis mux-vlan
Principal Subordinate Type        Interface
----------------------------------------------------------------------------
10         -          principal
10         12         separate
10         11         group       Ethernet0/0/1
----------------------------------------------------------------------------

[YLRXGA _SW_S3700_1]
```

由检查结果可看出，VLAN 10 属于主 VLAN，VLAN 11 属于互通型 VLAN，VLAN 11 属于隔离型 VLAN。

4.3.6.2　核心交换机

（1）配置设备间的网络互连，方法如下：

```
<Huawei>system-view
[Huawei] sysname YLRXGA _SW
[YLRXGA _SW] vlan batch 1000
[YLRXGA _SW] interface gigabitethernet 0/0/24
[YLRXGA _SW - GigabitEthernet1/0/24] port link-type access
[YLRXGA _SW - GigabitEthernet1/0/24] port default vlan 1000
[YLRXGA _SW] interface vlanif 1000
[YLRXGA _SW - Vlanif1000] ip address 192.168.250.1 30
[YLRXGA _SW -Vlanif1000] quit
```

（2）配置 VLAN 的网关地址，方法如下：

```
[YLRXGA _SW]Vlan batch 10 11 12 13 255 1000
[YLRXGA _SW] interface vlanif 10
[YLRXGA _SW - Vlanif10] ip address 192.168.10.1 24
[YLRXGA _SW - Vlanif10] quit
[YLRXGA _SW] interface vlanif 13
[YLRXGA _SW - Vlanif13] ip address 192.168.30.1 24
[YLRXGA _SW - Vlanif13] quit
[YLRXGA _SW] interface vlanif 255
[YLRXGA _SW - Vlanif255] ip address 192.168.255.1 24
[YLRXGA _SW - Vlanif255] quit
```

*在启用了 MUX VLAN 交换机中启用 VLANIF 的时候，需要 5.130 版本的镜像才能支持，默认如下 eNSP 的版本镜像是 5.110，暂不支持配置了 MUX VLAN 的交换机启用 VLANIF，查看镜像版本的方法如下：

```
<Huawei>dis ver
Huawei Versatile Routing Platform Software
VRP (R) software, Version 5.110 (S5700 V200R00
……
```

*如果采用 eNSP 虚拟仿真软件完成该项目，可使用 CE6800 交换机替代 S5700 交换机

（3）配置 LOOKBACK 地址，方法如下：

```
[YLRXGA _SW] interface loopback 0
[YLRXGA _SW-LoopBack0]ip address  192.168.254.2 32
```

（4）配置 OSPF，方法如下：

```
[YLRXGA _SW] ospf 1 router-id 192.168.254.2
[YLRXGA _SW - ospf-1] import-route direct
```

```
[YLRXGA _SW - ospf-1] area 0
[YLRXGA _SW - ospf-1-area-0.0.0.0] network 192.168.250.1 0.0.0.0
[YLRXGA _SW - ospf-1-area-0.0.0.0] quit
[YLRXGA _SW - ospf-1] quit
```

（5）在各个接口上启用 Trunk 功能，方法如下：

```
[YLRXGA _SW] interface gigabitethernet 0/0/23
[YLRXGA _SW -GigabitEthernet0/0/23] port link-type trunk
[YLRXGA _SW -GigabitEthernet0/0/23] port trunk allow-pass vlan all
[YLRXGA _SW -GigabitEthernet0/0/23] quit
[YLRXGA _SW] interface gigabitethernet 0/0/1
[YLRXGA _SW-GigabitEthernet0/0/1] port link-type trunk
[YLRXGA _SW-GigabitEthernet0/0/1] port trunk allow-pass vlan all
[YLRXGA _SW-GigabitEthernet0/0/1] quit
[YLRXGA _SW] interface gigabitethernet 0/0/2
[YLRXGA_SW-GigabitEthernet0/0/2] port link-type trunk
[YLRXGA_SW-GigabitEthernet0/0/2] port trunk allow-pass vlan all
[YLRXGA_SW-GigabitEthernet0/0/2] quit
[YLRXGA _SW] interface gigabitethernet 0/0/3
[YLRXGA_SW-GigabitEthernet0/0/3] port link-type trunk
[YLRXGA_SW-GigabitEthernet0/0/3] port trunk allow-pass vlan all
[YLRXGA_SW-GigabitEthernet0/0/3] quit
[YLRXGA _SW] interface gigabitethernet 0/0/4
[YLRXGA_SW-GigabitEthernet0/0/4] port link-type trunk
[YLRXGA_SW-GigabitEthernet0/0/4] port trunk allow-pass vlan all
[YLRXGA_SW-GigabitEthernet0/0/4] quit
[YLRXGA _SW] interface gigabitethernet 0/0/5
[YLRXGA_SW-GigabitEthernet0/0/5] port link-type trunk
[YLRXGA_SW-GigabitEthernet0/0/5] port trunk allow-pass vlan all
[YLRXGA_SW-GigabitEthernet0/0/5] quit
```

或者

```
[YLRXGA _SW]port-group group-member g0/0/1 to g0/0/5
[YLRXGA _SW-port-group]port link-type trunk
[YLRXGA _SW-GigabitEthernet0/0/1]port link-type trunk
[YLRXGA _SW-GigabitEthernet0/0/2]port link-type trunk
```

```
[YLRXGA _SW-GigabitEthernet0/0/3]port link-type trunk
[YLRXGA _SW-GigabitEthernet0/0/4]port link-type trunk
[YLRXGA _SW-GigabitEthernet0/0/5]port link-type trunk
[YLRXGA _SW-port-group]port trunk allow-pass vlan all
[YLRXGA _SW-GigabitEthernet0/0/1]port trunk allow-pass vlan all
[YLRXGA _SW-GigabitEthernet0/0/2]port trunk allow-pass vlan all
[YLRXGA _SW-GigabitEthernet0/0/3]port trunk allow-pass vlan all
[YLRXGA _SW-GigabitEthernet0/0/4]port trunk allow-pass vlan all
[YLRXGA _SW-GigabitEthernet0/0/5]port trunk allow-pass vlan all
[YLRXGA _SW-port-group]quit
```

（6）配置 MUX VLAN 及各个 VLAN 间的关系，方法如下：

```
[YLRXGA _SW]vlan 10
[YLRXGA _SW - vlan10] mux-vlan
[YLRXGA _SW - vlan10] subordinate group 11
[YLRXGA _SW - vlan01] subordinate separate 12
[YLRXGA _SW - vlan01] quit
```

4.3.6.3 出口设备相关配置

（1）外网出口配置 IP 地址，方法如下：

```
<Huawei> system-view
[Huawei] sysname LWCD_AR2240
[LWCD_AR2240] interface gigabitethernet 0/0/1
[LWCD_AR2240-GigabitEthernet0/0/1] ip address 1.1.1.2 255.255.255.192
```

（2）基于时间段的 ACL。

1）配置时间段，方法如下：

```
[YLRXGA__AR2240] time-range satime 8:00 to 9:00 working-day
```

2）配置 ACL，方法如下：

```
[YLRXGA__AR2240] acl 3001
[YLRXGA_AR2240-acl-adv-3001] rule 5 permit ip source 192.168.30.250 0.0.0.0time-range satime
[YLRXGA _AR2240-acl-adv-3001]rule 7 deny ip source 192.168.30.250 0.0.0.0
[YLRXGA _AR2240-acl-adv-3001] rule 10 permit ip source 192.168.30.250 0.0.0.255
[YLRXGA _AR2240-acl-adv-3001] rule 25 permit ip source 192.168.10.250 0.0.0.255
```

（3）NAT 地址转换，方法如下：

```
[YLRXGA _AR2240] interface gigabitethernet 0/0/1
[YLRXGA _AR2240-GigabitEthernet0/0/1]nat outbound 3001
```

```
[YLRXGA _AR2240-GigabitEthernet0/0/1] quite
```

（4）默认路由，方法如下：

```
[YLRXGA _AR2240] ip route-static 0.0.0.0 0 1.1.1.1
```

（5）OSPF。

1）配置 LOOKBACK 地址，方法如下：

```
[YLRXGA _AR2240] interface loopback 0
[YLRXGA _AR2240-LoopBack0]ip address  192.168.254.1 255.255.255.255
```

2）配置 ROUTER ID ，方法如下：

```
[YLRXGA_AR2240] router id 192.168.254.1
```

3）配置 OSPF 的基础功能，方法如下：

```
[YLRXGA _AR2240] ospf
[YLRXGA _AR2240-ospf-1] area 0
[YLRXGA _AR2240-ospf-1-area-0.0.0.0] network 192.168.250.2 0.0.0.0
[YLRXGA _AR2240-ospf-1-area-0.0.0.0] quit
```

4）在 OSPF 中引入默认路由，方法如下：

```
[YLRXGA _AR2240] import-route static
```

（6）配置接口互联地址，方法如下：

```
[YLRXGA _AR2240] interface gigabitethernet 0/0/2
[YLRXGA _AR2240-GigabitEthernet0/0/2] ip address 192.168.250.2 30
[YLRXGA _AR2240-GigabitEthernet0/0/2]quit
[YLRXGA _AR2240] interface gigabitethernet 0/0/1
[YLRXGA _AR2240-GigabitEthernet0/0/1] ip address 1.1.1.2 26
[YLRXGA _AR2240-GigabitEthernet0/0/1]quit
```

4.4　思政融入

随着信息技术的快速发展，特别是云计算、万物互联、大数据、智慧城市等热点领域的研究与发展，信息已成为非常重要的战略资源，网络中各类信息数据的安全更是尤为重要。习近平总书记提出“没有网络安全，就没有国家安全”。网络安全事关国家安全与社会稳定，因此必须采取有效措施，保障网络安全。网络安全为人民，网络安全靠人民。大学生是国家发展中的一支重要的人力资源，担负着国家、民族兴旺发达的重任，信息技术专业大学生要做好日常网络安全防护，强化网络安全意识和本领，树立良好专业素养和优秀的品格，遵纪守法，担负起发展国家网络安全的重任。

4.5　思考与训练

1. MUX VLAN 有哪些优点？
2. 什么是主 VLAN、互通型从 VLAN、隔离型从 VLAN？
3. 端口安全的优点有哪些？如何进行配置？
4. 动态端口安全与安全粘贴 MAC 有什么区别？
5. 请写出 MUX VLAN 的配置步骤。

Project 4

Project 4 Network project of a county public security bureau

4.1 Project import

The office building of a County Public Security Bureau has very high requirements for the network system. Because of the special functions of the unit, it has certain special requirements for the network design. In addition to office convenience, network security is different from the general office environment. Network construction should be business-centric, based on the reality and development of the work of the public security system, build an information network covering the county's public security system, realize the transmission and exchange of mould data, build and continuously improve the network system and operation management system, build a network application platform, promote the construction of business applications, meet the requirements of informationization of the public security system and improve work efficiency, and lay a foundation for integrating information resources.

4.2 Relevant knowledge

4.2.1 Port Security

Port Security, which controls and manages network traffic based on the MAC address, records the Ethernet MAC address connected to the switch port through the MAC address table, and allows only one MAC address to communicate through the port. Port security prevents packets sent by other MAC addresses from passing through this port. Using port security features prevents unauthorized devices from accessing the network, enhances security, and prevents MAC address tables from filling up due to flooding of MAC addresses.

4.2.1.1 The MAC address table is divided into three table.

(1) Static MAC address table, manually bound, with higher priority than dynamic MAC address table.

(2) Dynamic MAC address table, the switch will learn the source MAC into the MAC address table when it receives a data frame.

(3) Black hole MAC address table, manual binding or automatic learning, used to discard specified MAC addresses.

4.2.1.2 Management commands of MAC address tables

(1) View the MAC address table.

```
<Huawei>display mac-address
```

(2) Configure static MAC address table.

```
[Huawei] mac-address static 5489-98C0-7E34 GigabitEthernet 0/0/1 vlan 1
#Binding MAC address to interface g0/0/1 is valid in VLAN 1
```

(3) Configure black hole Mac address table.

```
[Huawei] mac-address blackhole 5489-987f-161a vlan 1
#Drop frames when source or purpose is received in vlan 1 for this mac
```

(4) Disable port learning MAC address, you can disable MAC address learning function in port or VLAN.

```
[Huawei-GigabitEthernet0/0/1]mac-address learning disable action discard
```

(5) Forbidden to learn MAC address and discard all frames received, or configure in VLAN.

```
[Huawei-GigabitEthernet0/0/1] mac-address learning disable action forward
```

(6) It is forbidden to learn mac addresses, but the received frames will be forwarded in flood mode (the switch's forwarding principle for unknown destination mac addresses), which can also be configured in VLAN.

(7) Limit the number of MAC addresses to learn, can be configured in a port or VLAN.

```
[Huawei-GigabitEthernet0/0/1]mac-limit maximum 9 alarm enable
```

The switch limits the number of MAC addresses to 9, and sends an alarm when the number exceeds. The number of MAC addresses beyond the limit will not be learned by the port. However, it can be forwarded by flooding (the principle of MAC address forwarding for unknown destination by the switch), or it can be configured in the VLAN.

(8) Configure port security dynamic MAC address.

This function is to set the MAC address dynamically learned as the security attribute, and the frames of other MAC without learned non security attributes will be discarded by the port.

```
[Huawei-GigabitEthernet0/0/3]port-security enable
#Turn on Port Security
[Huawei-GigabitEthernet0/0/3]port-security max-mac-num 1
```

```
#Limit the maximum number of secure MAC addresses to 1. The default is 1.
[Huawei-GigabitEthernet0/0/3]port-security aging-time 300
#Configure the aging time of the security MAC address for 300s, and it is not aging by default
```

In the port security dynamic MAC address, if configured as above, the first MAC address learned at G0/0/3 port is set as the security MAC address. In addition, other MAC addresses are not forwarded at the access port. Refresh the security MAC address table after 300s, and relearn the security MAC address. The switch learns the port first and sets it as the secure MAC address (which MAC address arrives first). However, after the switch is restarted, the security MAC address will be cleared and relearned.

(9) Configure port security sticky stick MAC address.

This function is consistent with the port security dynamic mac address, the only difference is that the pasted MAC address will not age, and still exists after the cut switch is restarted. The dynamic security mac address can only be learned dynamically, while the secure pasted MAC can be dynamically learned or manually configured.

```
[Huawei-GigabitEthernet0/0/3]port-security enable
#Turn on port security
[Huawei-GigabitEthernet0/0/3]port-security mac-address sticky
#Turn on the security paste MAC function
[Huawei-GigabitEthernet0/0/3]port-security max-mac-num 1
#Limit the maximum number of secure MAC addresses to 1. The default is 1.
[Huawei-GigabitEthernet0/0/3]port-security mac-address sticky 5489-98D8-71D5 vlan 1
#Manually bind and paste MAC address and VLAN
[Huawei-GigabitEthernet0/0/3]port-security protect-action restrict
#Configure the processing actions of other non secure MAC address data frames
```

(10) View and paste MAC address status.

```
[Huawei-GigabitEthernet0/0/3] display mac-address
```

4.2.2 MUX VLAN

4.2.2.1 Summary

In the enterprise network, both employees and customers can access the server of the enterprise. For enterprises, they hope that some employees can visit each other, but some employees cannot communicate with each other. To meet the needs of enterprises, enterprise servers and employees who can access each other can be divided into different VLANs, and employees who cannot communicate with each other can be divided into different VLANs (not the same VLAN as enterprise servers and employees who can access each other). This not only requires a lot of VLAN IDs, but also increases the maintenance of network managers.

To solve the above problems, MUX VLAN function can be deployed on the switch. MUX VLAN can provide a two-layer traffic isolation mechanism in VLAN.

MUX VLAN is divided into principal VLAN and subordinate VLAN. Slave VLAN is divided into group VLAN and separate VLAN.

Communication authority:

(1) The principal VLAN port can communicate with all VLANs.

(2) Group VLAN (interworking slave VLAN) can communicate with its own VLAN members and with principal VLAN (master VLAN).

(3) Separate VLAN can only communicate with principal VLAN, and members of their own VLAN can not communicate.

4.2.2.2 Configuration code

During configuration, it should be noted that all hosts must be in the same subnet, and the switch port must be added to the VLAN in access mode. The configuration of MUX VLAN cannot be used for the configuration of VLANIF interface, VLAN Mapping, VLAN Stacking, Super–VLAN and Sub–VLAN.

The configuration code is as follows:

```
[Huawei]vlan batch 10 20 30
# Create VLAN 10 20 30
[Huawei]vlan 10
#Enter VLAN management view
[Huawei-vlan 10]mux-vlan
#Configure vlan 10 as principal VLAN
[Huawei-vlan 10]subordinate group 20
#Configure vlan 20 as an interworking slave VLAN
[Huawei-vlan10]subordinate separate 30
#Configure vlan 30 as isolated slave VLAN
[Huawei-vlan10]quit                #quit
[Huawei]interface g0/0/1           #Enter interface g0/0/1
[Huawei-GigabitEthernet0/0/1]port link-type access
#Configure the interface mode to access
[Huawei-GigabitEthernet0/0/1]port default vlan 10
#Add interface to vlan 10
```

```
[Huawei-GigabitEthernet0/0/1]port mux-vlan enable
#Turn on the MUX VLAN function of the interface
[Huawei-GigabitEthernet0/0/1]interface g0/0/2
#Enter interface g0/0/2
[Huawei-GigabitEthernet0/0/2]port link-type access
#Configure the interface mode to access
[Huawei-GigabitEthernet0/0/2]port default vlan 20
#Add interface to vlan 20
[Huawei-GigabitEthernet0/0/2]port mux-vlan enable
#Turn on the MUX VLAN function of the interface
[Huawei-GigabitEthernet0/0/2]int g0/0/3
[Huawei-GigabitEthernet0/0/3]port link-type access
[Huawei-GigabitEthernet0/0/3]port default vlan 20
[Huawei-GigabitEthernet0/0/3]port mux-vlan enable
[Huawei-GigabitEthernet0/0/3]int g0/0/4
[Huawei-GigabitEthernet0/0/4]port link-type access
[Huawei-GigabitEthernet0/0/4]port default vlan 30
[Huawei-GigabitEthernet0/0/4]port mux-vlan enable
[Huawei-GigabitEthernet0/0/4]int g0/0/5
[Huawei-GigabitEthernet0/0/5]port link-type access
[Huawei-GigabitEthernet0/0/5]port default vlan 30
[Huawei-GigabitEthernet0/0/5]port mux-vlan enable
```

4.2.2.3 Inspection configuration

Use display MUX VLAN to view MUX VLAN information.

```
[Huawei]display mux-vlan
Principal Subordinate Type          Interface
-------------------------------------------------------------------------
10          -         principal     GigabitEthernet0/0/1
10          30        separate      GigabitEthernet0/0/4 GigabitEthernet0/0/5
10          20        group         GigabitEthernet0/0/2 GigabitEthernet0/0/3
```

It can be seen from the above information that the switch port G0/0/1 is in the master VLAN, the ports G0/0/5 and G0/0/5 are in the isolated slave VLAN 30, and the ports G0/0/2 and G0/0/3 are in the interworking slave VLAN 20.

4.3 Project implementation

4.3.1 Project requirements

(1) The whole network uses OSPF routing protocol to exchange routing information.

(2) IP+MAC+VLAN binding for some PCs on the access switch.

(3) In the internal network, the online time limited by certain address segments.

(4) Everyone in the same VLAN can access the server. Some employees in the same VLAN are isolated from each other and cannot access each other.

4.3.2 Equipment list

The equipment used in this project mainly includes layer 3 switch, layer 2 switch, exit router, etc. The specific model and functions are shown in Table 4.1.

Table 4.1 List of Network Equipment

Serial No.	Equipment model	Quantity	Function	Remarks
1	AR2240	1	Exit router	
2	S5700	1	Core switch	
3	S3700–52P–LI–AC	20	access switch	

4.3.3 Access switch project topology

The network topology of the project is shown in Figure 4.1

4.3.4 Configuration ideas

The project adopts the following ideas:

(1) The exit router is configured with L2TP VPN server.

(2) The exit router is configured with a time–based ACL policy to limit the online time of some internal IP addresses.

(3) The lookback address with a 32–bit mask is used as the router ID of OSPF.

(4) The access layer switch is configured with port security mechanism to restrict port terminal access.

(5) MUX VLAN is used in the core switch to achieve isolation in the same VLAN.

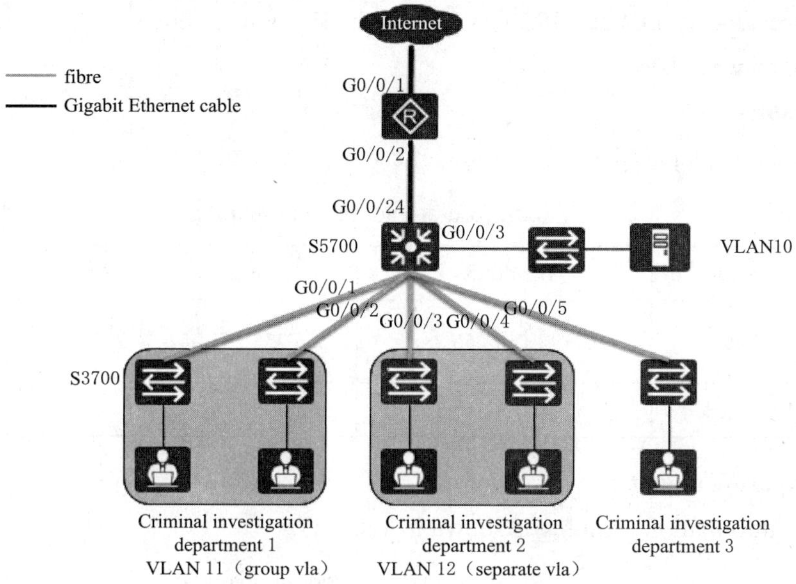

Figure 4.1 Network topology of a county public security bureau

4.3.5 Relevant information planning

4.3.5.1 Core switch

The information planning of the core layer switch is shown in Table 4.2.

Table 4.2 Core Switch Information Planning Table

Serial No.	User department	VLAN No.	IP address	Mask	Remarks
1	Criminal investigation server	10	192.168.10.1	255.255.255.0	192.168.10.250
2	Criminal investigation department 1	11			192.168.10.100
3	Criminal investigation department 2	12			Manually specified
4	Criminal investigation department 3	13	192.168.30.1	255.255.255.0	
5	Switch management	255	192.168.255.1	255.255.255.0	
6	interconnection1	1000	192.168.250.1	255.255.255.252	Interconnect with router
7	Lookback 0		192.168.254.2	255.255.255.255	

In the office address segment: 192.168.10.250, the IP address allows Internet access only from 8 a.m. to 9 a.m. on weekdays.

4.3.5.2 Exit router

The information planning of the exit router is shown in Table 4.3.

Table 4.3 Export Router Information Planning

Serial No.	Interface	IP address	Mask	Remarks
1	G0/0/2	192.168.250.2	255.255.255.252	Interconnection with core switch
2	Lookback 0	192.168.255.1	255.255.255.255	

4.3.5.3 ISP export planning

ISP export information planning is shown in Table 4.4.

Table 4.4 ISP Export Information Planning

Serial No.	Interface	IP address	Mask	Gateway	Remarks
1	G0/0/1	1.1.1.2	255.255.255.192	1.1.1.1	Static IP address

4.3.6 Operating steps

4.3.6.1 Access switch equipment

(1) Basic configuration (VLAN, Trunk, binding address).

```
<Huawei>system-view
[Huawei] YLRXGA _SW_S3700_1
[YLRXGA _SW_S3700_1] vlan batch  10 11 12 13 255 1000
[YLRXGA _SW_S3700_1] interface gigabitethernet 0/0/1
[YLRXGA _SW_S3700_1-GigabitEthernet0/0/1] port link-type trunk
[YLRXGA _SW_S3700_1-GigabitEthernet0/0/1] port trunk allow-pass vlan all
[YLRXGA _SW_S3700_1-GigabitEthernet0/0/1]quit
[YLRXGA _SW_S3700_1] user-bind static ip-address 192.168.10.100 mac-address 5489-98D2-0FEB
interface Ethernet 0/0/1  #Bind the IP address and MAC address to the E0/0/1 port
[YLRXGA _SW_S3700_1] vlan 11
[YLRXGA _SW_S3700_1-vlan11] ip source check user-bind enable  #Check message function
```

In the network, it often occurs to access or attack the network by using the message that counterfeits the source IP, MAC and other information of legitimate users, which leads to the inability of legitimate users to obtain stable and secure network services. Configure the IP message checking function to prevent the above situations.

When the check function of IP message is enabled, the IP, MAC, VLAN and interface information in the IP message will be checked for binding table matching by default. At the same time, you can execute the command ip source check user-bind check-item (interface view) or ip source check user-bind check-item (VLAN view) to flexibly configure the check items of IP messages. Only IP packets matching the binding table can be forwarded, otherwise they will be discarded.

The configuration of other access layer switches is similar, which is omitted here.

(2) MUX VLAN configuration (only need to be configured in LSW1-LSW4).

```
[YLRXGA _SW_S3700_1]vlan 10
[YLRXGA _SW_S3700_1-vlan10]mux-vlan                    #Enable MUX VLAN, VLAN 10 is the primary VLAN
[YLRXGA _SW_S3700_1-vlan10]subordinate group 11        #VLAN 11 is an interworking VLAN
[YLRXGA _SW_S3700_1-vlan10]subordinate separate 12     #VLAN 12 is an isolated VLAN
[YLRXGA _SW_S3700_1-vlan10]quit
[YLRXGA _SW_S3700_1]interface Ethernet0/0/1
[YLRXGA _SW_S3700_1-Ethernet0/0/1]port link-type access
[YLRXGA _SW_S3700_1-Ethernet0/0/1]port default vlan 11
[YLRXGA _SW_S3700_1-Ethernet0/0/1]port mux-vlan enable
[YLRXGA _SW_S3700_1-Ethernet0/0/1]quit
```

inspect:

```
[YLRXGA _SW_S3700_1]dis mux-vlan
Principal Subordinate Type          Interface
-----------------------------------------------------------------
10          -          principal
10          12         separate
10          11         group          Ethernet0/0/1
-----------------------------------------------------------------
[YLRXGA _SW_S3700_1]
```

It can be seen from the inspection results that VLAN 10 belongs to the main VLAN, VLAN 11 belongs to the group VLAN, and VLAN 11 belongs to the separate VLAN.

4.3.6.2 Core switch

(1) Configure network interconnection between devices.

```
<Huawei>system-view
[Huawei] sysname YLRXGA _SW
```

```
[YLRXGA _SW] vlan batch 1000
[YLRXGA _SW] interface gigabitethernet 0/0/24
[YLRXGA _SW - GigabitEthernet1/0/24] port link-type access
[YLRXGA _SW - GigabitEthernet1/0/24] port default vlan 1000
[YLRXGA _SW] interface vlanif 1000
[YLRXGA _SW - Vlanif1000] ip address 192.168.250.1 30
[YLRXGA _SW -Vlanif1000] quit
```

(2) Configure the gateway address of the VLAN.

```
[YLRXGA _SW]Vlan batch 10 11 12 13 255 1000
[YLRXGA _SW] interface vlanif 10
[YLRXGA _SW - Vlanif10] ip address 192.168.10.1 24
[YLRXGA _SW - Vlanif10] quit
[YLRXGA _SW] interface vlanif 13
[YLRXGA _SW - Vlanif13] ip address 192.168.30.1 24
[YLRXGA _SW - Vlanif13] quit
[YLRXGA _SW] interface vlanif 255
[YLRXGA _SW - Vlanif255] ip address 192.168.255.1 24
[YLRXGA _SW - Vlanif255] quit
```

*When VLANIF is enabled in a MUX VLAN switch, the image version 5.130 is required to support it. By default, the version image of eNSP is 5.110. The switch configured with MUX VLAN is temporarily not supported to enable VLANIF. The method to view the image version is as follows:

```
<Huawei>dis ver
Versatile Routing Platform Software
VRP (R) software, Version 5.110 (S5700 V200R00
……
```

*If eNSP virtual simulation software is used to complete the project, CE6800 switch can be used to replace S5700 switch

(3) Configure LOOKBACK address.

```
[YLRXGA _SW] interface loopback 0
[YLRXGA _SW-LoopBack0]ip address  192.168.254.2 32
```

(4) Configure OSPF.

```
[YLRXGA _SW] ospf 1 router-id 192.168.254.2
[YLRXGA _SW - ospf-1] import-route direct
```

```
[YLRXGA _SW - ospf-1] area 0
[YLRXGA _SW - ospf-1-area-0.0.0.0] network 192.168.250.1 0.0.0.0
[YLRXGA _SW - ospf-1-area-0.0.0.0] quit
[YLRXGA _SW - ospf-1] quit
```

(5) Enable Trunk functionality on each interface.

```
[YLRXGA _SW] interface gigabitethernet 0/0/23
[YLRXGA _SW -GigabitEthernet0/0/23] port link-type trunk
[YLRXGA _SW -GigabitEthernet0/0/23] port trunk allow-pass vlan all
[YLRXGA _SW -GigabitEthernet0/0/23] quit
[YLRXGA _SW] interface gigabitethernet 0/0/1
[YLRXGA _SW-GigabitEthernet0/0/1] port link-type trunk
[YLRXGA _SW-GigabitEthernet0/0/1] port trunk allow-pass vlan all
[YLRXGA _SW-GigabitEthernet0/0/1] quit
[YLRXGA _SW] interface gigabitethernet 0/0/2
[YLRXGA_SW-GigabitEthernet0/0/2] port link-type trunk
[YLRXGA_SW-GigabitEthernet0/0/2] port trunk allow-pass vlan all
[YLRXGA_SW-GigabitEthernet0/0/2] quit
[YLRXGA _SW] interface gigabitethernet 0/0/3
[YLRXGA_SW-GigabitEthernet0/0/3] port link-type trunk
[YLRXGA_SW-GigabitEthernet0/0/3] port trunk allow-pass vlan all
[YLRXGA_SW-GigabitEthernet0/0/3] quit
[YLRXGA _SW] interface gigabitethernet 0/0/4
[YLRXGA_SW-GigabitEthernet0/0/4] port link-type trunk
[YLRXGA_SW-GigabitEthernet0/0/4] port trunk allow-pass vlan all
[YLRXGA_SW-GigabitEthernet0/0/4] quit
[YLRXGA _SW] interface gigabitethernet 0/0/5
[YLRXGA_SW-GigabitEthernet0/0/5] port link-type trunk
[YLRXGA_SW-GigabitEthernet0/0/5] port trunk allow-pass vlan all
[YLRXGA_SW-GigabitEthernet0/0/5] quit
```

perhaps

```
[YLRXGA _SW]port-group group-member g0/0/1 to g0/0/5
[YLRXGA _SW-port-group]port link-type trunk
[YLRXGA _SW-GigabitEthernet0/0/1]port link-type trunk
[YLRXGA _SW-GigabitEthernet0/0/2]port link-type trunk
```

```
[YLRXGA _SW-GigabitEthernet0/0/3]port link-type trunk
[YLRXGA _SW-GigabitEthernet0/0/4]port link-type trunk
[YLRXGA _SW-GigabitEthernet0/0/5]port link-type trunk
[YLRXGA _SW-port-group]port trunk allow-pass vlan all
[YLRXGA _SW-GigabitEthernet0/0/1]port trunk allow-pass vlan all
[YLRXGA _SW-GigabitEthernet0/0/2]port trunk allow-pass vlan all
[YLRXGA _SW-GigabitEthernet0/0/3]port trunk allow-pass vlan all
[YLRXGA _SW-GigabitEthernet0/0/4]port trunk allow-pass vlan all
[YLRXGA _SW-GigabitEthernet0/0/5]port trunk allow-pass vlan all
[YLRXGA _SW-port-group]quit
```

(6) Configure MUX VLANs and the relationship between VLANs.

```
[YLRXGA _SW]vlan 10
[YLRXGA _SW - vlan10] mux-vlan
[YLRXGA _SW - vlan10] subordinate group 11
[YLRXGA _SW - vlan01] subordinate separate 12
[YLRXGA _SW - vlan01] quit
```

4.3.6.3 Relevant configuration of export equipment

(1) IP address configured at the internet outlet.

```
<Huawei> system-view
[Huawei] sysname LWCD_AR2240
[LWCD_AR2240] interface gigabitethernet 0/0/1
[LWCD_AR2240-GigabitEthernet0/0/1] ip address 1.1.1.2 255.255.255.192
```

(2) Time period based ACL.

1) Configure time period:

```
[YLRXGA__AR2240] time-range satime 8:00 to 9:00 working-day
```

2) Configure ACL:

```
[YLRXGA__AR2240] acl 3001
[YLRXGA_AR2240-acl-adv-3001] rule 5 permit ip source 192.168.30.250 0.0.0.0time-range satime
[YLRXGA _AR2240-acl-adv-3001]rule 7 deny ip source 192.168.30.250 0.0.0.0
[YLRXGA _AR2240-acl-adv-3001] rule 10 permit ip source 192.168.30.250 0.0.0.255
[YLRXGA _AR2240-acl-adv-3001] rule 25 permit ip source 192.168.10.250 0.0.0.255
```

(3) NAT address translation.

```
[YLRXGA _AR2240] interface gigabitethernet 0/0/1
[YLRXGA _AR2240-GigabitEthernet0/0/1]nat outbound 3001
```

```
[YLRXGA _AR2240-GigabitEthernet0/0/1] quite
```

(4) Default route.

```
[YLRXGA _AR2240] ip route-static 0.0.0.0 0 1.1.1.1
```

(5) OSPF.

1) Configure LOOKBACK address:

```
[YLRXGA _AR2240] interface loopback 0
[YLRXGA _AR2240-LoopBack0]ip address  192.168.254.1 255.255.255.255
```

2) Configure ROUTER ID:

```
[YLRXGA_AR2240] router id 192.168.254.1
```

3) Configure the basic functions of OSPF:

```
[YLRXGA _AR2240] ospf
[YLRXGA _AR2240-ospf-1] area 0
[YLRXGA _AR2240-ospf-1-area-0.0.0.0] network 192.168.250.2 0.0.0.0
[YLRXGA _AR2240-ospf-1-area-0.0.0.0] quit
```

4) Introducing default routes in OSPF:

```
[YLRXGA _AR2240] import-route static
```

(6) Configure interface interconnection address.

```
[YLRXGA _AR2240] interface gigabitethernet 0/0/2
[YLRXGA _AR2240-GigabitEthernet0/0/2] ip address 192.168.250.2 30
[YLRXGA _AR2240-GigabitEthernet0/0/2]quit
[YLRXGA _AR2240] interface gigabitethernet 0/0/1
[YLRXGA _AR2240-GigabitEthernet0/0/1] ip address 1.1.1.2 26
[YLRXGA _AR2240-GigabitEthernet0/0/1]quit
```

4.4 Ideological and political integration

With the rapid development of information technology, especially the research and development of hot areas such as cloud computing, Internet of everything, big data, and smart cities, information has become a very important strategic resource, and the security of all kinds of information and data in the network is particularly important."Without cyber security, there can be no national security."Cyber security is related to national security and social stability, so effective measures must be taken to ensure cyber security.Cyber security is for the people, and cyber security depends on the people.College students are important human resources in the development of the country and take on the heavy responsibilities of the prosperity of the country and the nation.College students majoring in information technology should do a good job in daily network security protection, strengthen network security awareness and skills, establish good professional qualities and excellent character, abide by discipline and law, and undertake the important task of developing national network security.

4.5 Thinking and training

1. What are the advantages of MUX VLANs?
2. What are primary VLANs, interconnected slave VLANs, and isolated slave VLANs?
3. What are the advantages of port security? How to configure it?
4. What is the difference between Dynamic Port Security and Secure Paste MAC?
5. Please write out the configuration steps for the MUX VLAN?

项目 5

项目 5　某集团公司网络项目

5.1　项目导入

随着通信技术、计算机技术、网络技术的应用普及和加深，许多员工的办公不再仅仅局限于同一物理位置上的办公，即使在办事处、分支机构、出差在外、在家中，均可像在公司总部办公一样协同工作。尤其是出现自然因素（如台风等）而导致员工需要远程协同办公，这就需要建立一个安全、快捷、经济、方便的信息交互平台，来传输远程办公员工与公司之间的信息交流。

某集团公司作为国内知名企业，信息的敏感性决定了它们历来都是各种居心叵测者的重要关注对象，甚至也是内部员工十分感兴趣的内容，这提醒我们应该更加注重网络安全的建设。

5.2　相关知识点

5.2.1　IPSec VPN

5.2.1.1　概述

由于 VPN（虚拟专网）比租用专线更加便宜、安全、灵活，所以有越来越多的公司采用 VPN 连接在家工作和出差在外的员工，以及替代连接分公司和合作伙伴的标准广域网。VPN 建在互联网的公共网络架构上，一般通过加密协议，在发送端加密数据、在接收端解密数据，以保证数据的私密性。

IPSec 全称为 Internet Protocol Security，指采用 IPSec 协议来实现远程接入的一种 VPN 技术，是 IETF 制定的三层隧道加密协议，它为 Internet 上传输的数据提供了高质量的、可互操作的、基于密码学的安全保证，是一种传统的实现三层 VPN（Virtual Private Network，虚拟专用网）的安全技术，在特定的通信方之间，通过 IP 层加密与数据源验证来保证数据包在 Internet 上传输时的私有性、完整性和真实性。IPSec VPN 是目前 VPN 技术中使用率非常高的一种技术，同时提供 VPN 和信息加密两项技术。

IPSec 通过两个安全协议来实现对 IP 数据包或上层协议的保护。而且此实现不会对用户主机或其他 Internet 组件造成影响。用户还可以实现不同的加密算法而不会影响其他部分的实现。

5.2.1.2　IPSec 协议类型

IPSec 是一个框架性架构，具体由两类协议组成：

第一类是 AH 协议（Authentication Header）：使用较少，可以同时提供数据完整性确认、数据来源确认、防重放等安全特性；AH 常用摘要算法（单向 Hash 函数）MD5 和 SHA1 实现该特性。由于 AH 无法提供数据加密，所有数据在传输时以明文传输，而 ESP 提供数据加密；而且 AH 因为提供数据来源确认（源 IP 地址一旦改变，AH 校验失败），所以无法穿越 NAT，在实际中较少使用 AH 协议。

第二类是 ESP 协议（Encapsulated Security Payload）：使用较广，可以同时提供数据完整性确认、数据加密、防重放等安全特性；ESP 通常使用 DES、3DES、AES 等加密算法实现数据加密，使用 MD5 或 SHA1 来实现数据完整性。

5.2.1.3　IPSec 工作模式

IPSec 的工作模式主要有两种：

（1）隧道（Tunnel）模式：用户的整个 IP 数据包被用来计算 AH 或 ESP 头，AH 或 ESP 头以及 ESP 加密的用户数据被封装在一个新的 IP 数据包中。通常隧道模式应用在两个安全网关之间的通信。

（2）传输（transport）模式：只有传输层数据被用来计算 AH 或 ESP 头，AH 或 ESP 头以及 ESP 加密的用户数据被放置在原 IP 包头后面。通常，传输模式应用在两台主机之间的通信，或一台主机和一个安全网关之间的通信。

5.2.1.4　IPSec VPN 工作原理

IPSec 提供三种不同的形式来保护通过公有或私有 IP 网络来传送的私有数据。

（1）认证：可以确定所接受的数据与所发送的数据是一致的，同时可以确定申请发送者在实际上是真实发送者，而不是伪装的。

（2）数据完整性：保证数据从原发地到目的地的传送过程中没有任何不可检测的数据丢失与改变。

（3）机密性：使相应的接收者能获取发送的真正内容，而无意获取数据的接收者无法获知数据的真正内容。在 IPSec 由三个基本要素来提供以上三种保护形式：认证协议头（AH）、安全加载封装（ESP）和互联网密钥管理协议（IKMP）。认证协议头和安全加载封装可以通过分开或组合使用来达到所希望的保护等级。

5.2.1.5　IPSec VPN 的应用场景

IPSec 作为一种 VPN 技术，其最显著的特点就是可以为载荷数据提供加密以及数据源验证服务，保护数据的机密性和完整性。同时通过 IKE 可以对密钥进行定时更新维护，增强系统的安全性。鉴于这些特点，IPSec 被广泛应用于传输敏感数据的 VPN 网络中。

Site-to-Site（站点到站点或者网关到网关）：如企业的 3 个机构分布在互联网的 3 个不同的地方，各使用一个网关建立 VPN 隧道，企业内网（若干 PC 机）之间的数据通过这些网关建立的 IPSec 隧道实现安全互联。

End-to-End（端到端或者 PC 机到 PC 机）：两个 PC 机之间的通信由两个 PC 机之间的 IPSec 会话保护，而不是网关。

End-to-Site（端到站点或者 PC 机到网关）：两个 PC 机之间的通信由网关和异地 PC 机之间的 IPSec 进行保护。

5.2.2　DHCP Snooping

5.2.2.1　DHCP Snooping 作用

架设 DHCP 服务器可以为客户端自动分配 IP 地址、掩码、默认网关、DNS 服务器等网络参数，简化了网络配置，提高了管理效率。但在 DHCP 服务的管理上存在一些问题，例如，DHCP Server 的冒充，DHCP Server 的耗竭攻击，某些用户随便指定 IP 地址，造成 IP 地址冲突等，DHCP 监听（DHCP Snooping）是一种 DHCP 安全特性。华为交换机支持在每个 VLAN 基础上启用 DHCP 监听特性。通过这种特性，交换机能够拦截第二层 VLAN 域内的所有 DHCP 报文。

DHCP 监听还有一个非常重要的作用就是建立一张 DHCP 监听绑定表（DHCP Snooping Binding）。一旦一个连接在非信任端口的客户端获得一个合法的 DHCP Offer，交换机就会自动在 DHCP 监听绑定表里添加一个绑定条目，内容包括了该非信任端口的客户端 IP 地址、MAC 地址、端口号、VLAN 编号、租期等信息。

5.2.2.2　DHCP 监听端口

启用了 DHCP 监听的交换机，将交换机端口划分为两类：

（1）非信任端口：通常为连接终端设备的端口，如 PC、网络打印机等。

（2）信任端口：连接合法 DHCP 服务器的端口或者连接汇聚交换机的上行端口。

通过开启 DHCP 监听特性，交换机限制用户端口（非信任端口）只能够发送 DHCP 请求，丢弃来自用户端口的所有其他 DHCP 报文，例如 DHCP Offer 报文等。而且，并非所有来自用户端口的 DHCP 请求都被允许通过，交换机还会比较 DHCP 请求报文的（报文头里的）源 MAC 地址和（报文内容里的）DHCP 客户机的硬件地址（即 CHADDR 字段），只有这两者相同的请求报文才会被转发，否则将被丢弃。这样就防止了 DHCP 耗竭攻击。

信任端口可以接收所有的 DHCP 报文。通过只将交换机连接到合法 DHCP 服务器的端口设置为信任端口，其他端口设置为非信任端口，就可以防止用户伪造 DHCP 服务器来攻击网络。DHCP 监听特性还可以对端口的 DHCP 报文进行限速。通过在每个非信任端口下进行限速，将可以阻止合法 DHCP 请求报文的广播攻击。

5.2.2.3　基本配置代码

```
[Huawei]dhcp enable  #启用 DHCP 功能
[Huawei]dhcp snooping enable  #启用 DHCP Snooping 功能
其他非信任端口：
[Huawei-GigabitEthernet0/0/3]dhcp snooping enable
信任端口：
[Huawei]interface GigabitEthernet 0/0/1
[Huawei-GigabitEthernet0/0/3]dhcp snooping trusted #设置端口 G0/0/1 为信任端口
[Huawei-GigabitEthernet0/0/3]arp dhcp-snooping-detect enable
#配置 ARP 与 DHCP Snooping 的联动功能
[Huawei]dhcp snooping user-offline remove mac-address
#配置用户下线后及时清除对应 MAC 表项功能
[Huawei]dhcp snooping check dhcp-giaddr enable vlan 2
#配置丢弃 GIADDR 字段非零的 DHCP 报文
[Huawei]dhcp server detect #启用 DHCP Server 探测功能
[Huawei]dhcp snooping  check dhcp-rate enable
#防止 DHCP 报文泛洪攻击
```

5.2.3　端口镜像

端口镜像功能通过在交换机或路由器上，将镜像端口的流量复制一份发送到观察端口供观察端口下连的流量分析设备（软件）对复制来的镜像端口的流量进行分析。在企业中用镜像功能，可以很好地对企业内部的网络数据进行监控管理，在网络出故障的时候，可以快速地定位故障。

5.2.3.1　镜像的分类

（1）端口镜像。端口镜像是基于端口的镜像，分为本地端口镜像、二层远程端口镜像、三层远程端口镜像，镜像的流量可以是入向或者出向。

（2）流镜像。流镜像是基于流的镜像，是根据用户配置的流策略匹配的流量进行镜像，只支持（镜像端口的）入方向，不支持出方向。流镜像分为本地流镜像、二层远程流镜像、三层远程流镜像。

（3）VLAN 镜像。VLAN 镜像是基于 VLAN 的镜像，是将制定的 VLAN 内的所有活动接口的入方向的流量复制到观察端口，不支持出方向。VLAN 镜像分为本地 VLAN 镜像、二层远程 VLAN 镜像。

（4）MAC 地址镜像。基于 MAC 地址的镜像，将匹配源或目的的 MAC 地址的入方向的流量复制到观察关口，不支持出方向。MAC 地址镜像支持本地 MAC 地址镜像、二层远程 MAC 地址镜像。

5.2.3.2　端口镜像的配置

（1）本地端口镜像配置，如图 5.1 所示。

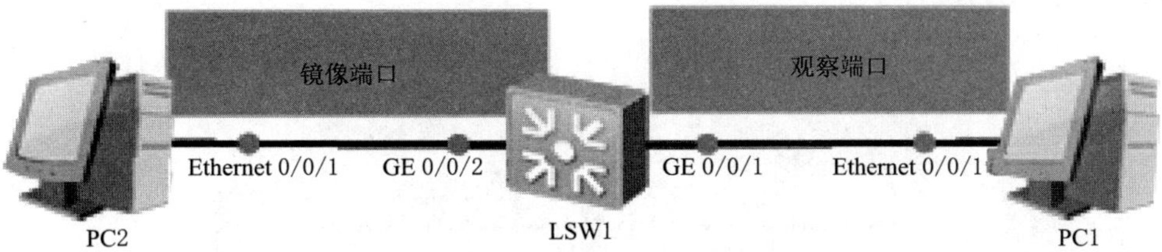

图 5.1　本地端口镜像配置

```
[Huawei]observe-port 1 interface g0/0/1
#配置一个序列号为1的观察端口 g0/0/1
[Huawei]interface g0/0/2
#配置镜像端口
[Huawei-GigabitEthernet0/0/2]port-mirroring to observe-port 1 both
#配置一个镜像端口，将双向流量复制到序列号为1的观察端口
 [Huawei]display observe-port  #查看端口配置状态
----------------------------------------------------------------------------------
Index     : 1
Interface: GigabitEthernet0/0/1
Used      : 2
----------------------------------------------------------------------------------

[Huawei]display port-mirroring #查看端口镜像情况
Port-mirror:
  --------------------------------------------------------------------------------
Mirror-port           Direction        Observe-port
----------------------------------------------------------------------------------
GigabitEthernet0/0/2      Both         GigabitEthernet0/0/1
----------------------------------------------------------------------------------
```

（2）二层远程端口镜像，如图 5.2 所示。端口的二层远程镜像的原理是通过创建一个 VLAN，将镜像端口的流量复制到观察端口中，观察端口在该 VLAN 中进行广播，通过 VLAN 的广播将复制的流量发送到监控设备连接的端口上进行监控。值得注意的是镜像关口和观察端口必须在一台设备上，镜像端口不能在该 VLAN 内（实际中发现镜像端口也可

以在该 VLAN 中），而观察端口因为要收集镜像端口的流量，所以必须在这个 VLAN 内。在图 5.2 拓扑中 LSW1 上的镜像端口将流量复制到观察端口中，由观察端口在 VLAN2 中通过广播将从镜像端口复制的流量发送至监控服务器 Server1 所连接的 GE 0/0/2 接口，供 Server1 进行流量的分析。

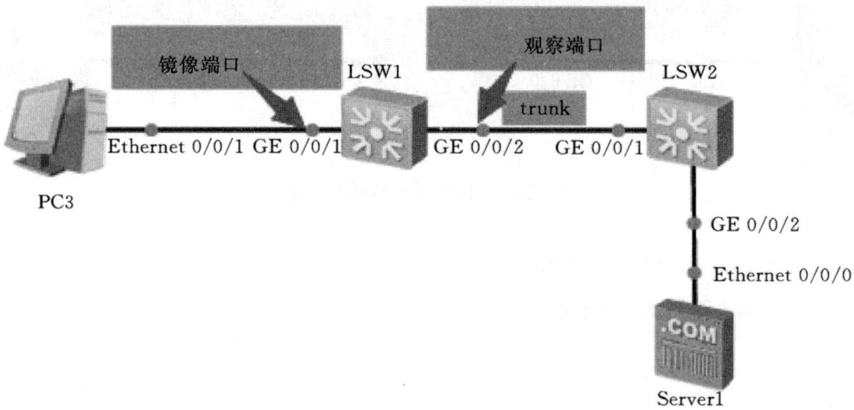

图 5.2　二层远程端口镜像

配置方法：首先创建一个用于广播镜像流量的 VLAN，分别在 LSW 1 与 LSW 2 上创建 VLAN 2，并将 Server 1 连接的接口 GE0/0/2 划分到 VLAN 2 中用于接收复制的镜像流量，而 PC3 默认在 VLAN 1 中不属于 VLAN 2 不用做其他的配置，在 LSW 之间允许 VLAN 2 的流量通过，注意观察 VLAN 中必须关闭 MAC 地址学习功能，因为该功能与镜像功能相冲突。

```
[Huawei]vlan 2
[Huawei-vlan2]mac-address learning disable
#必须在观察 vlan 中关闭 MAC 地址学习功能
[Huawei-vlan2]quit
[Huawei]interface g0/0/2
[Huawei-GigabitEthernet0/0/2]port link-type trunk
[Huawei-GigabitEthernet0/0/2]port trunk allow-pass vlan all
[Huawei-GigabitEthernet0/0/2]q
[Huawei]observe-port 1 interface g0/0/2 vlan 2
 #指定观察端口，并在 vlan2 中广播复制的流量
[Huawei]interface g0/0/1
[Huawei-GigabitEthernet0/0/1]port-mirroring to observe-port 1 both
#指定镜像端口，将流量复制到序列号为 1 的观察端口
```

```
[Huawei-GigabitEthernet0/0/1]q
[Huawei]vlan 2
[Huawei-vlan2]quit
[Huawei]interface g0/0/1
[Huawei-GigabitEthernet0/0/1]port link-type trunk
[Huawei-GigabitEthernet0/0/1]port trunk allow-pass vlan all
[Huawei-GigabitEthernet0/0/1]quit
[Huawei]interface g0/0/2      #将监控设备连接的端口加入 vlan 2 收取从镜像端口复制来的流量
[Huawei-GigabitEthernet0/0/2]port link-type access
[Huawei-GigabitEthernet0/0/2]port default vlan 2
[Huawei-GigabitEthernet0/0/2]quit
```

（3）三层远程镜像端口配置，如图 5.3 所示。端口的三层远程镜像的原理是，通过在 IP 层建立一条 GRE 的隧道将镜像端口的流量复制到观察端口，观察端口在通过 GRE 隧道将流量发送到监控设备所在的端口上，进行流量的分析。在图 5.3 拓扑中交换机的镜像端口将流量复制到观察端口，由观察端口通过 GRE-tunnel 发送至 Server2 监控服务器连接的交换机的 Ethernet 0/0/2 接口，供 Server2 对镜像端口的流量进行分析。

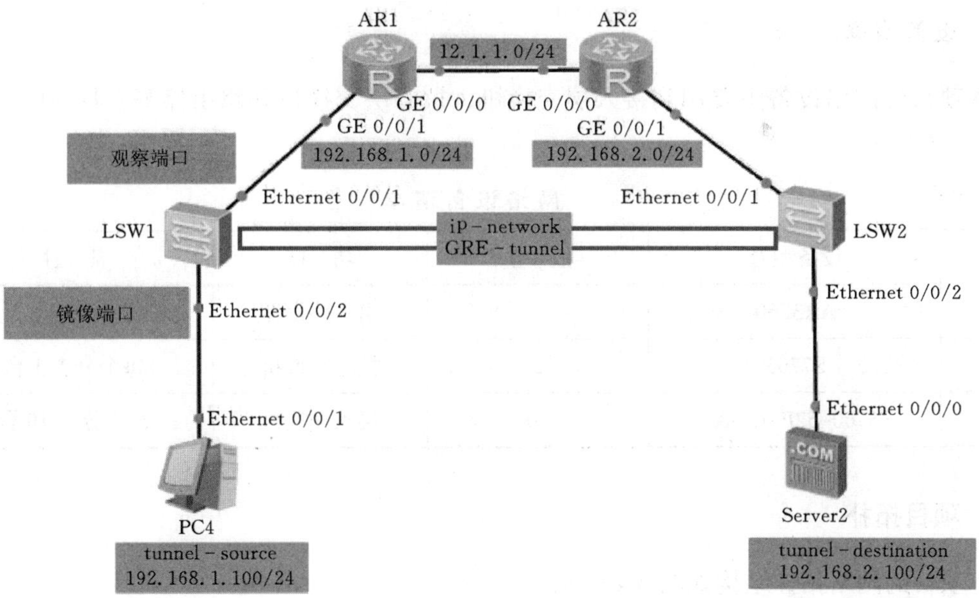

图 5.3　三层远程端口镜像配置

配置方法：首先在 AR1 和 AR2 上分别配置路由网段及静态路由，保证三层互通，然后在交换机 LSW1 上配置镜像端口及观察端口的 gre 隧道。

路由部分省略，在交换机 LSW1 上配置如下：

```
[Huawei]observe-port 1 interface e0/0/1 destination-ip 192.168.2.100 source-ip 192.168.1.100    #创建观察端口及隧道
[Huawei]interface e0/0/2
[Huawei-Ethernet0/0/1]port-mirroring to observe-port 1 both
#指定镜像端口和序列号 1
[Huawei-Ethernet0/0/1]quit
```

5.3　项目实施

5.3.1　项目需求

（1）全网使用静态路由交互信息。

（2）两个分支之间通过 IPSec VPN 构建同一个内网。

（3）需要把服务器的流量镜像到审计设备上。

（4）内部中有人使用小路由器，需要杜绝小路由器的接入，以免影响正常的 DHCP 功能。

5.3.2　设备清单

本项目的网络设备主要用到接入层交换机、核心层交换机和路由器等，具体的型号数量见表 5.1。

表 5.1　　网络设备清单

序号	设备型号	数　量	功　能	备　注
1	AR3260	2	出口路由器	每个分支 1 台
2	S7703	2	核心交换机	每个分支 1 台
3	S5700-52P-LI-AC	20	接入交换机	每个分支 10 台

5.3.3　项目拓扑

本项目的网络拓扑结构如图 5.4 所示。

5.3.4　配置思路

采用如下的思路配置：

（1）出口路由器配置 IP Sec VPN 建立 VPN 隧道。

（2）两个分支内部地址不能相互冲突。

（3）在接入交换机上开启 Snooping 功能，防止私接路由器发送 DHCP 报文。

（4）核心交换机上开启端口镜像功能。

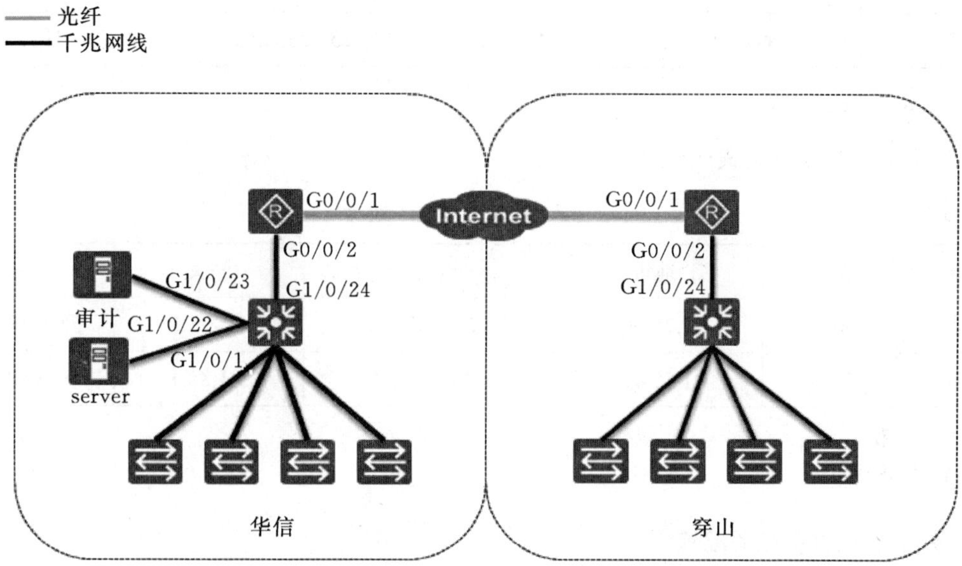

图 5.4　某集团公司网络拓扑结构图

5.3.5　相关信息规划

5.3.5.1　核心交换机

核心交换机信息规划见表 5.2。

表 5.2　核心交换机信息规划表

序号	使用部门	VLAN 号	IP 地址	掩码	备注
1	华信	10	172.16.10.1	255.255.255.0	
2	穿山	20	172.16.20.1	255.255.255.0	
3	服务器	30	172.16.30.1	255.255.255.0	
4	华信管理 VLAN	254	172.16.254.1	255.255.255.0	
5	穿山管理 VLAN	255	172.16.255.1	255.255.255.0	
6	华信互联 1	1000	172.16.250.1	255.255.255.252	与路由器互联
7	穿山互联 2	1001	172.16.251.1	255.255.255.252	与路由器互联

5.3.5.2　出口路由器

出口路由器的信息规划见表 5.3。

表 5.3　　出口路由器信息规划表

序号	接口	IP 地址	掩码	备注
1	G0/0/2	172.16.250.2	255.255.255.252	与核心交换机互联
2	G0/0/2	172.16.251.2	255.255.255.252	与路由器互联

5.3.5.3 ISP 出口规划

ISP 出口信息规划见表 5.4。

表 5.4　　ISP 出口信息规划表

序号	接口	IP 地址	掩码	网关	备注
华信	G0/0/1	1.1.25.36	255.255.255.192	1.1.25.1	静态 IP 地址
穿山	G0/0/1	1.1.35.47	255.255.255.192	1.1.35.1	静态 IP 地址

5.3.6 操作步骤

5.3.6.1 华信出口设备相关配置

（1）外网出口配置 IP 地址，方法如下：

```
<Huawei> system-view
[Huawei] sysname GLL_HX _AR3260
[GLL_HX _AR3260] interface gigabitethernet 0/0/1
[GLL_HX _AR3260-GigabitEthernet0/0/1] ip address 1.1.25.36 255.255.255.192
```

（2）配置上网 ACL，方法如下：

```
[GLL_HX _AR3260] acl 3001
[GLL_HX _AR3260-acl-adv-3001] rule5 deny ip source 172.16.10.00.0.0.255 destination 172.16.20.0 0.0.0.255
[GLL_HX _AR3260-acl-adv-3001] rule10 deny ip source 172.16.30.00.0.0.255 destination 172.16.20.0 0.0.0.255
[GLL_HX _AR3260-acl-adv-3001] rule 15 permit ip source 172.16.10.00.0.255.255 destination  Any
[GLL_HX _AR3260-acl-adv-3001] rule 15 permit ip source 172.16.20.00.0.255.255 destination  Any
```

（3）NAT 地址转换，方法如下：

```
[GLL_HX _AR3260] interface gigabitethernet 0/0/1
[GLL_HX _AR3260-GigabitEthernet0/0/1]nat outbound 3001
[GLL_HX _AR3260-GigabitEthernet0/0/1] quit
```

（4）默认路由，方法如下：

```
[GLL_HX _AR3260] ip route-static 0.0.0.0 0 1.1.25.1
```

（5）配置 IPSec VPN。

1）定义要保护的数据流，方法如下：

```
[GLL_HX _AR3260] acl number 3002
[GLL_HX _AR3260-acl-adv-3002] rule permit ip source 172.16.10.0 0.0.0.255 destination 172.16.20.0 0.0.0.255
[GLL_HX _AR3260-acl-adv-3002] rule permit ip source 172.16.30.0 0.0.0.255 destination 172.16.20.0 0.0.0.255
[GLL_HX _AR3260-acl-adv-3002] rule permit ip source 172.16.255.0 0.0.0.255 destination 172.16.254.0 0.0.0.255
[GLL_HX _AR3260-acl-adv-3002] quit
```

2）配置 IPSec 安全提议，方法如下：

```
[GLL_HX _AR3260] ipsec proposal GLL
[GLL_HX _AR3260-ipsec-proposal-GLL] esp authentication-algorithm sha2-256
[GLL_HX _AR3260-ipsec-proposal-GLL] esp encryption-algorithm aes-128
[GLL_HX _AR3260-ipsec-proposal-GLL] quit
```

3）创建安全策略，方法如下：

```
[GLL_HX _AR3260] ipsec policy GLL 10 manual
[GLL_HX _AR3260-ipsec-policy-manual-GLL-10] security acl 3002
[GLL_HX _AR3260-ipsec-policy-manual-GLL-10] proposal tran1
[GLL_HX _AR3260-ipsec-policy-manual-GLL-10] tunnel remote 1.1.35.47
[GLL_HX _AR3260-ipsec-policy-manual-GLL-10] tunnel local 1.1.25.36
[GLL_HX _AR3260-ipsec-policy-manual-GLL-10] sa spi outbound esp 12345
[GLL_HX _AR3260-ipsec-policy-manual-GLL-10] sa spi inbound esp 54321
[GLL_HX _AR3260-ipsec-policy-manual-GLL-10] sa string-key outbound esp cipher Huawei
[GLL_HX _AR3260-ipsec-policy-manual-GLL-10] sa string-key inbound esp cipher Huawei
[GLL_HX _AR3260-ipsec-policy-manual-GLL-10] quit
```

4）接口上引用安全策略，方法如下：

```
[GLL_HX _AR3260] interface gigabitethernet 0/0/1
[GLL_HX _AR3260-GigabitEthernet1/0/0] ipsec policy GLL
```

5）配置内网互联地址，方法如下：

```
[GLL_HX _AR3260] interface gigabitethernet 0/0/2
[GLL_HX _AR3260-GigabitEthernet0/0/1] ip address 172.16.250.2 255.255.255.252
```

6）配置业务回程路由，方法如下：

```
ip route-static 172.16.10.0 255.255.255.0172.16.250.1
ip route-static 172.16.30.0 255.255.255.0 172.16.250.1
ip route-static 172.16.254.0 255.255.255.0 172.16.250.1
ip route-static 172.16.20.0 255.255.255.0 1.1.35.47
```

5.3.6.2 穿山出口设备相关配置

（1）外网出口配置 IP 地址，方法如下：

```
<Huawei> system-view
[Huawei] sysname GLL_CS_AR3260
[GLL_CS_AR3260] interface gigabitethernet 0/0/1
[GLL_CS_AR3260-GigabitEthernet0/0/1] ip address 1.1.35.47 255.255.255.192
```

（2）配置上网 ACL，方法如下：

```
[GLL_CS_AR3260] acl 3001
[GLL_CS_AR3260-acl-adv-3001] rule5 deny ip source 172.16.20.00.0.255.255 destination 172.16.10.0 0.0.0.255
[GLL_CS_AR3260-acl-adv-3001] rule10 deny ip source 172.16.20.00.0.255.255 destination 172.16.30.0 0.0.0.255
[GLL_CS_AR3260-acl-adv-3001] rule 15 permit ip source 172.16.20.00.0.255.255 destination any
```

（3）NAT 地址转换，方法如下：

```
[GLL_CS _AR3260] interface gigabitethernet 0/0/1
[GLL_CS_AR3260-GigabitEthernet0/0/1]nat outbound 3001
[GLL_CS_AR3260-GigabitEthernet0/0/1] quit
```

（4）默认路由，方法如下：

```
[GLL_CS _AR3260] ip route-static 0.0.0.0 0 1.1.35.1
```

（5）配置 IPSec VPN。

1）定义要保护的数据流，方法如下：

```
[GLL_CS _AR3260] acl number 3002
[GLL_CS _AR3260-acl-adv-3002] rule permit ip source 172.16.20.0 0.0.0.255 destination 172.16.10.0 0.0.0.255
[GLL_CS _AR3260-acl-adv-3002] rule permit ip source 172.16.20.0 0.0.0.255 destination 172.16.30.0 0.0.0.255
[GLL_CS _AR3260-acl-adv-3002] rule permit ip source 172.16.254.0 0.0.0.255 destination 172.16.255.0 0.0.0.255
[GLL_CS _AR3260-acl-adv-3002] quit
```

2）配置 IPSec 安全提议，方法如下：

```
[GLL_CS _AR3260] ipsec proposal GLL
[GLL_CS _AR3260-ipsec-proposal-GLL] esp authentication-algorithm sha2-256
[GLL_CS _AR3260-ipsec-proposal-GLL] esp encryption-algorithm aes-128
[GLL_CS _AR3260-ipsec-proposal-GLL] quit
```

3）创建安全策略，方法如下：

```
[GLL_CS _AR3260] ipsec policy GLL 10 manual
[GLL_CS _AR3260-ipsec-policy-manual-GLL-10] security acl 3002
[GLL_CS _AR3260-ipsec-policy-manual-GLL-10] proposal tran1
[GLL_CS _AR3260-ipsec-policy-manual-GLL-10] tunnel remote 1.1.35.47
[GLL_CS _AR3260-ipsec-policy-manual-GLL-10] tunnel local 1.1.25.36
[GLL_CS _AR3260-ipsec-policy-manual-GLL-10] sa spi outbound esp 12345
[GLL_CS _AR3260-ipsec-policy-manual-GLL-10] sa spi inbound esp 54321
[GLL_CS _AR3260-ipsec-policy-manual-GLL-10] sa string-key outbound esp cipher Huawei
[GLL_CS _AR3260-ipsec-policy-manual-GLL-10] sa string-key inbound esp cipher Huawei
[GLL_CS _AR3260-ipsec-policy-manual-GLL-10] quit
```

4）接口上引用安全策略，方法如下：

```
[GLL_CS _AR3260] interface gigabitethernet 0/0/1
[GLL_CS _AR3260-GigabitEthernet1/0/0] ipsec policy GLL
```

5）配置内网互联地址，方法如下：

```
[GLL_HX _AR3260] interface gigabitethernet 0/0/2
[GLL_HX _AR3260-GigabitEthernet0/0/1] ip address 172.16.251.2 255.255.255.252
```

6）配置业务回程路由，方法如下：

```
ip route-static 172.16.20.0 255.255.255.0172.16.251.1
ip route-static 172.16.255.0  255.255.255.0 172.16.251.1
ip route-static 172.16.10.0  255.255.255.01.1.25.1
ip route-static 172.16.30.0  255.255.255.0 1.1.25.1
```

5.3.6.3　华信核心交换机配置

（1）配置设备间的网络互连，方法如下：

```
<Huawei>system-view
[Huawei] sysname GLL_SW_HX
[GLL_SW_HX] vlan batch 10 30 254 1000
[GLL_SW_HX] interface gigabitethernet1/0/24
[GLL_SW_HX - GigabitEthernet1/0/24] port link-type access
[GLL_SW_HX - GigabitEthernet1/0/24] port default vlan 1000
```

（2）启用 DHCP，配置 VLAN IP 地址，方法如下：

```
[GLL_SW_HX] dhcp enable
[GLL_SW_HX] interface vlanif 10
[GLL_SW_HX -Vlanif1000] ip address 172.16.10.1 24
[GLL_SW_HX -Vlanif1000] dhcp select interface
[GLL_SW_HX -Vlanif1000] quit
[GLL_SW_HX] interface vlanif 30
[GLL_SW_HX -Vlanif1000] ip address 172.16.30.1 24
[GLL_SW_HX -Vlanif1000] quit
[GLL_SW_HX] interface vlanif 254
[GLL_SW_HX - Vlanif254] ip address 172.16.254.1 24
[GLL_SW_HX  - Vlanif254] quit
[GLL_SW_HX] interface vlanif 1000
[GLL_SW_HX -Vlanif1000] ip address 172.16.250.1 30
[GLL_SW_HX -Vlanif1000] quit
```

（3）配置接口加入 VLAN(列举)，方法如下：

```
[GLL_SW_HX] interface gigabitethernet1/0/1
[GLL_SW_HX - GigabitEthernet1/0/1] port link-type access
[GLL_SW_HX - GigabitEthernet1/0/1] port port default vlan 10
[GLL_SW_HX - GigabitEthernet1/0/1]quit
[GLL_SW_HX] interface gigabitethernet1/0/2
[GLL_SW_HX - GigabitEthernet1/0/2] port link-type access
[GLL_SW_HX - GigabitEthernet1/0/2] port port default vlan 10
[GLL_SW_HX - GigabitEthernet1/0/2]quit
[GLL_SW_HX] interface gigabitethernet1/0/22
[GLL_SW_HX - GigabitEthernet1/0/22] port link-type access
[GLL_SW_HX - GigabitEthernet1/0/22] port port default vlan 30
[GLL_SW_HX] interface gigabitethernet1/0/23
[GLL_SW_HX - GigabitEthernet1/0/23] port link-type access
[GLL_SW_HX - GigabitEthernet1/0/23] port port default vlan 30
[GLL_SW_HX - GigabitEthernet1/0/23]quit
```

（4）配置默认路由，方法如下：

```
[GLL_SW_HX]ip route-static 0.0.0.0 0.0.0.0 192.168.250.2
```

（5）配置观察端口，方法如下：

```
[GLL_SW_HX] observe-port interface gigabitethernet 1/0/23
```

（6）配置镜像端口，方法如下：

```
[GLL_SW_HX] interface gigabitethernet 1/0/22
[GLL_SW_HX -GigabitEthernet1/0/22] mirror to observe-port inbound
[GLL_SW_HX -GigabitEthernet1/0/22] return
```

5.3.6.4　华信接入交换机配置 DHCP Snooping，方法如下：

```
<Huawei> system-view
[Huawei] sysname YLRXGA _SW_S5700
[GLL_SW_HX _S5700] dhcp enable
[GLL_SW_HX] dhcp snooping enable
[GLL_SW_HX] port-group port-group group-member G0/0/1 TO G0/0/24
[GLL_SW_HX -port-group] dhcp snooping enable
[GLL_SW_HX -port-group] quit
[GLL_SW_HX] interface gigabitethernet 0/0/24
[GLL_SW_HX -GigabitEthernet0/0/24] undo dhcp snooping enable
[GLL_SW_HX -GigabitEthernet0/0/24] dhcp snooping trusted
```

5.3.6.5　穿山核心交换机配置

（1）配置设备间的网络互连，方法如下：

```
<Huawei>system-view
[Huawei] sysname GLL_SW_CS
[GLL_SW_CS] vlan batch 20 255 1001
[GLL_SW_CS] interface gigabitethernet1/0/24
[GLL_SW_CS -GigabitEthernet1/0/24] port link-type access
[GLL_SW_CS -GigabitEthernet1/0/24] port default vlan 1001
```

（2）启用 DHCP，配置 VLAN IP 地址，方法如下：

```
[GLL_SW_CS] dhcp enable
[GLL_SW_CS] interface vlanif 20
[GLL_SW_CS -Vlanif20] ip address 172.16.20.1 24
[GLL_SW_CS -Vlanif20] dhcp select interface
[GLL_SW_CS -Vlanif20] quit
[GLL_SW_CS] interface vlanif 255
[GLL_SW_CS -Vlanif255] ip address 172.16.255.1 24
[GLL_SW_CS -Vlanif255] quit
[GLL_SW_CS] interface vlanif 1001
```

```
[GLL_SW_CS -Vlanif1001] ip address 172.16.251.1 30
[GLL_SW_CS -Vlanif1001] quit
```

（3）配置接口加入 VLAN(列举)，方法如下：

```
[GLL_SW_CS] interface gigabitethernet1/0/1
[GLL_SW_CS-GigabitEthernet1/0/1] port link-type access
[GLL_SW_CS-GigabitEthernet1/0/1] port port default vlan 20
[GLL_SW_CS-GigabitEthernet1/0/1]quit
[GLL_SW_CS] interface gigabitethernet1/0/2
[GLL_SW_CS-GigabitEthernet1/0/2] port link-type access
[GLL_SW_CS-GigabitEthernet1/0/2] port port default vlan 20
[GLL_SW_CS-GigabitEthernet1/0/2]quit
```

（4）配置默认路由，方法如下：

```
[GLL_SW_CS]ip route-static 0.0.0.0 0.0.0.0 192.168.251.2
```

5.3.6.6　穿山接入交换机配置 DHCP Snooping

```
<Huawei> system-view
[Huawei] sysname YLRXGA _SW_S5700
[GLL_SW_CS_S5700] dhcp enable
[GLL_SW_CS] dhcp snooping enable
[GLL_SW_CS] port-group port-group group-member G0/0/1 TO G0/0/24
[GLL_SW_CS-port-group] dhcp snooping enable
[GLL_SW_CS-port-group] quit
[GLL_SW_CS] interface gigabitethernet 0/0/24
[GLL_SW_CS -GigabitEthernet0/0/24] undo dhcp snooping enable
[GLL_SW_CS -GigabitEthernet0/0/24] dhcp snooping trusted
```

5.4　思政融入

现代网络工程项目日趋复杂，随着学习的深入，涉及的网络工程项目规模越来越大，涉及内容多，复杂性高。在工程设计和实施过程中，要通过提高辩证思维能力，科学总结ICT 项目实践经验，将马克思主义科学实践观立场、观点、方法贯穿始终，指导项目的开展。在网络工程项目实施应用过程中的三个核心问题：网络图绘制、网络计划优化和工程信息化管理上，利用马克思主义科学实践观进行指导，能快速认清问题本质，抓住问题关键，提高网络计划技术的应用水平。

5.5　思考与训练

1. IPSec 是一个框架性架构，具体由哪两类协议组成?
2. IPSec VPN 的应用场景有哪些?
3. DCHP 监听的作用是什么?
4. 启用了 DHCP 监听的交换机，将交换机端口划分为哪两类?
5. 什么是端口镜像?

Project 5

Project 5 Network project of a group company

5.1 Project import

With the popularization and deepening of the application of communication technology, computer technology and network technology, the office of many employees is no longer limited to the same physical location. Even in offices, branches, business trips and at home, they can work together like working in the headquarters of the company. In particular, natural factors (such as typhoon) lead to the need for employees to work together remotely, which requires the establishment of a safe, fast, economic and convenient information exchange platform to transmit the information exchange between remote office employees and the company.

As a well-known domestic enterprise, the sensitivity of information determines that they have always been an important concern of all kinds of people with ulterior motives, and even the content of great interest to internal employees, which reminds us that we should pay more attention to the construction of network security.

5.2 Relevant knowledge

5.2.1 IPSec VPN

5.2.1.1 Summary

Because VPN (virtual private network) is cheaper, safer and more flexible than leased line, more and more companies are using VPN to connect employees who work at home and on business trips, as well as to replace the standard WAN connecting branches and partners. VPN is built on the public network architecture of the Internet. It generally encrypts data at the sending end and decrypts data at the receiving end through encryption protocol to ensure the privacy of data.

IPSec (IP Security) refers to a VPN technology that uses IPSec protocol to realize remote access. IPSec is fully known as Internet Protocol Security. It is a three-layer tunnel encryption protocol formulated by IETF. It provides high-quality, interoperable and cryptography based security guarantee for data transmitted on the Internet. It is a traditional security technology to realize three-layer VPN (virtual private network), Between specific communicators, the privacy, integrity and authenticity of data packets transmitted on the Internet are guaranteed through IP layer encryption and data source verification. IPSec VPN is a technology with very high click through rate in current VPN technology. It provides VPN and information encryption at the same time.

IPSec protects IP packets or upper layer protocols through two security protocols. And this implementation will not affect the user's host or other Internet components. Users can also implement different encryption algorithms without affecting the implementation of other parts.

5.2.1.2 IPSec protocol type

IPSec is a framework architecture, which is composed of two types of protocols: the first is AH protocol (authentication header, less used): it can provide security features such as data integrity confirmation, data source confirmation and anti replay at the same time; AH commonly used summary algorithm (one-way hash function) MD5 and SHA1 realize this feature. Since AH cannot provide data encryption, all data are transmitted in plaintext, while ESP provides data encryption; Moreover, AH cannot cross NAT because it provides data source confirmation (AH verification fails once the source IP address is changed), so AH protocol is rarely used in practice.

The second type is ESP protocol (widely used): it can provide security features such as data integrity confirmation, data encryption and anti replay at the same time; ESP usually uses DES, 3DES, AES and other encryption algorithms to realize data encryption, and MD5 or SHA1 to realize data integrity.

5.2.1.3 IPSec working mode

There are two main working modes of IPSec:

(1) Tunnel mode: the entire IP packet of user is used to calculate the AH or ESP header, and the AH or ESP header and ESP encrypted user data are encapsulated in a new IP packet. Typically, tunnel mode is applied to the communication between two security gateways.

(2) Transport mode: only the transport layer data is used to calculate the AH or ESP header, and the AH or ESP header and ESP encrypted user data are placed behind the original IP packet header. Typically, the transmission mode is applied to the communication between two hosts, or between a host and a security gateway.

5.2.1.4 IPSec VPN working principle

IPSec provides three different forms to protect private data transmitted through public or private IP networks.

(1) Authentication: it can be determined that the received data is consistent with the sent data, and it can be determined that the applicant sender is actually the real sender, not the disguised sender.

(2) Data integrity: ensure that there is no undetectable data loss and change during the transmission of data from the origin to the destination.

(3) Confidentiality: enable the corresponding receiver to obtain the real content of the sent data, while the receiver who does not intend to obtain the data cannot obtain the real content of the data. In IPSec, three basic elements provide the above three forms of protection: authentication protocol header (AH), secure load encapsulation (ESP) and Internet key management protocol (IKMP). The authentication protocol header and security loading encapsulation can be used separately or in combination to achieve the desired level of protection.

5.2.1.5 Application scenarios of IPSec VPN

Application scenario of IPSec VPN as a VPN technology, IPSec's most remarkable feature is that it can provide encryption and data source verification services for load data to protect the confidentiality and integrity of data. At the same time, the key can be updated and maintained regularly through IKE to enhance the security of the system. In view of these characteristics, IPSec is widely used in VPN networks that transmit sensitive data.

Site-to-Site (Site to Site or Gateway to Gateway): For example, three organisations in an enterprise are distributed in three different places on the Internet, and each of them uses a gateway to establish VPN tunnel, and the data between the intranet (a number of PCs) of the enterprise is interconnected through the IPSec tunnels established by these gateways.

End-to-End (End to End or PC to PC): the communication between two PCs is protected by the IPSec session between two PCs, not the gateway.

End-to-Site (End-to-Site or PC-to-Gateway): communication between the two PCs is protected by IPSec between the gateway and the off-site PC.

5.2.2 DHCP Snooping

5.2.2.1 Role of DHCP Snooping

The function of DHCP Snooping is to set up a DHCP server, which can automatically assign IP address, mask, default gateway, DNS server and other network parameters for clients, simplify network configuration and improve management efficiency. However, there are some problems in the management of DHCP service, such as the impersonation of DHCP server, the exhaustion attack of DHCP server, some users arbitrarily specify IP addresses, resulting in IP address conflicts, etc. DHCP Snooping is a DHCP security feature. Switches support enabling DHCP listening on the basis of each VLAN. The second layer of DHCP can intercept all packets in the domain.

DHCP Snooping also plays a very important role in creating a DHCP Snooping binding table. Once a client connected to the untrusted port obtains a legal DHCP offer, the switch will automatically add a binding entry in the DHCP listening binding table, including the client IP address, MAC address, port number, VLAN number, lease term and other information of the untrusted port.

5.2.2.2 listening port

For switches with DHCP listening enabled, switch ports are divided into two categories:

(1) Untrusted port: usually the port connecting terminal equipment, such as PC, network printer, etc.

(2) Trust port: the port connecting the legal DHCP server or the uplink port connecting the aggregation switch.

By turning on the DHCP monitoring feature, the switch restricts the user port (untrusted port) to only send DHCP requests and discards all other DHCP messages from the user port, such as DHCP offer messages. Moreover, not all DHCP requests from the user port are allowed to pass through. The switch will also compare the source MAC address (in the message header) of the DHCP request message with the hardware address (i.e. CHADDR field) of the DHCP client (in the message content). Only the two same request messages will be forwarded, otherwise they will be discarded. This prevents DHCP exhaustion attacks.

The trusted port can receive all DHCP messages. By setting only the ports on the switch that are connected to a legitimate DHCP server as trusted ports and the other ports as untrusted ports, users can be prevented from forging the DHCP server to attack the network. The DHCP monitoring feature can also speed limit the DHCP message of the port. By limiting the speed under each untrusted ports, the broadcast attack of legal DHCP request message can be prevented.

5.2.2.3 Basic configuration code

```
[Huawei]dhcp enable  # Enable DHCP function
[Huawei]dhcp snooping enable  # Enable DHCP Snooping function
Other untrusted ports:
[Huawei-GigabitEthernet0/0/3]dhcp snooping enable
Trust Port:
[Huawei]interface GigabitEthernet 0/0/1
[Huawei-GigabitEthernet0/0/3]dhcp snooping trusted # Set port  G0/0/1 as trusted port
[Huawei-GigabitEthernet0/0/3]arp dhcp-Snooping-detect enable
```

```
#Configure the linkage function between ARP and DHCP Snooping
[Huawei]dhcp snooping user-offline remove mac-address
#Configure the function of clearing the corresponding MAC entries in time after the user goes offline
[Huawei]dhcp snooping check dhcp-giaddr enable vlan 2
#Configure to discard DHCP messages with non-zero giaddr field
[Huawei]dhcp server detect #Enable DHCP server probing
[Huawei]dhcp snooping  check dhcp-rate enable
#Prevent DHCP message flooding attack
```

5.2.3 Port mirror

The port mirroring function sends a copy of the traffic of the mirrored port on the switch or router to the observation port for the traffic analysis equipment (software) connected to the observation port to analyze the traffic of the copied mirrored port. Using the image function in the enterprise can well monitor and manage the internal network data of the enterprise, and can quickly locate the fault in case of network failure.

5.2.3.1 Classification of images

(1) Port mirror. Port image is a port based image, which is divided into local port image, layer-2 remote port image and layer-3 remote port image. The mirrored traffic can be inbound or outbound.

(2) Stream mirroring. Flow mirroring is a flow based mirroring, which is based on the flow matching the flow policy configured by the user. It only supports the in direction (of the mirror port) and does not support the out direction. Stream image is divided into local stream image, layer-2 remote stream image and layer-3 remote stream image.

(3) VLAN mirroring. VLAN image is a VLAN based image, which copies the incoming traffic of all active interfaces in the specified VLAN to the observation port, and does not support outgoing traffic. VLAN image is divided into local VLAN image and layer-2 remote VLAN image.

(4) Address mirroring. Based on the image of the MAC address, the traffic matching the incoming direction of the source or destination MAC address is copied to the observation gateway, and the outgoing direction is not supported. MAC address mirroring supports local MAC address mirroring and layer-2 remote MAC address mirroring.

5.2.3.2 Port mirror configuration

(1) Local port mirroring configuration, as shown in Figure 5.1.

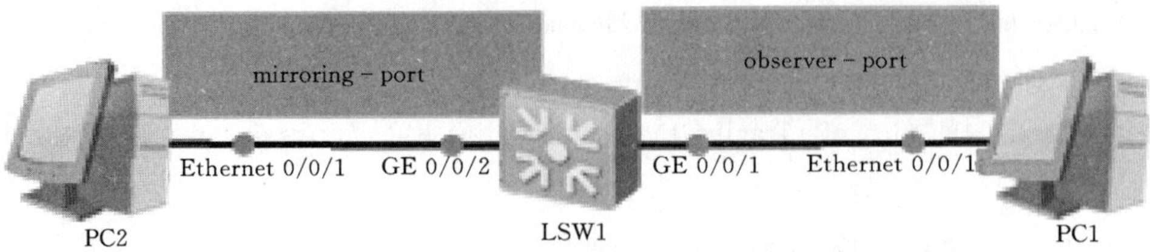

Figure 5.1 local port mirroring configuration

```
[Huawei]observe-port 1 interface g0/0/1
#Configure an observation port g0/0/1 with serial number 1
[Huawei]interface g0/0/2
#Configure mirror port
[Huawei-GigabitEthernet0/0/2]port-mirroring to observe-port 1 both
#Configure a mirror port to copy bidirectional traffic to the observation port with serial number 1
[Huawei]display observe-port  # View port configuration status
---------------------------------------------------------------
  Index    : 1
  Interface: GigabitEthernet0/0/1
  Used     : 2
---------------------------------------------------------------

[Huawei]display port-mirroring #View the port image
Port-mirror:
---------------------------------------------------------------
Mirror-port        Direction       Observe-port
---------------------------------------------------------------
GigabitEthernet0/0/2        Both        GigabitEthernet0/0/1
---------------------------------------------------------------
```

(2) The layer-2 remote port image is shown in Figure 5.2. The principle of layer-2 remote image of port is to create a VLAN, copy the traffic of the image port to the observation port, the observation port broadcasts in the VLAN, and send the copied traffic to the port connected to the monitoring device through the VLAN broadcast for monitoring. It is worth noting that the image gateway and the observation port must be on the same device, and the image port cannot be in the VLAN (in fact, the image port can also be found in the VLAN), while the observation port must be in the VLAN because it needs to collect the traffic of the image port. In the topology shown in Figure 5.2, the mirror port on LSW 1 copies the traffic to the observation port, and the

observation port sends the traffic copied from the mirror port to the GE 0/0/2 interface connected to the monitoring server Server 1 through broadcasting in VLAN 2 for Server 1 to analyze the traffic.

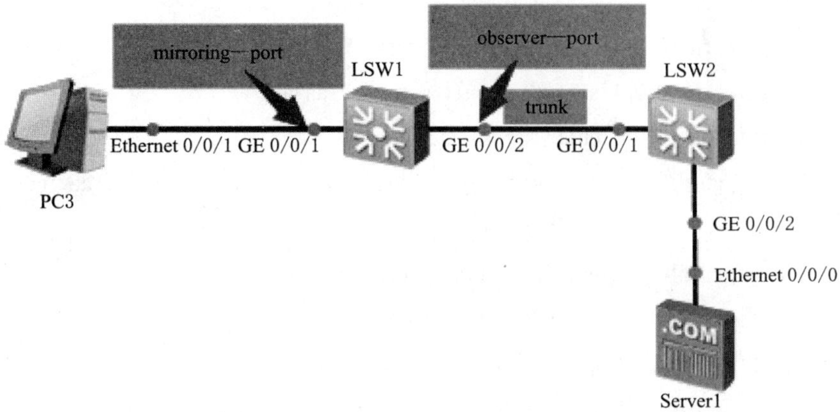

Figure 5.2 Remote port image on the Layer-2

Configuration: first, create a VLAN for broadcasting image traffic, create VLAN 2 on LSW 1 and LSW 2 respectively, and divide the interface GE0 / 0 / 2 connected by Server 1 into VLAN 2 for receiving image traffic for replication. By default, PC 3 does not belong to VLAN 2 in VLAN 1. There is no need to make other configuration. VLAN 2 traffic is allowed to pass between LSWs. Note that the MAC address learning function must be turned off in VLANs, because this function conflicts with the image function.

```
[Huawei]vlan 2
[Huawei-vlan2]mac-address learning disable
#The MAC address learning function must be turned off in the observation VLAN
[Huawei-vlan2]quit
[Huawei]interface g0/0/2
[Huawei-GigabitEthernet0/0/2]port link-type trunk
[Huawei-GigabitEthernet0/0/2]port trunk allow-pass vlan all
[Huawei-GigabitEthernet0/0/2]q
[Huawei]observe-port 1 interface g0/0/2 vlan 2
#Specify the observation port and broadcast the replicated traffic in vlan 2
[Huawei]interface g0/0/1
[Huawei-GigabitEthernet0/0/1]port-mirroring to observe-port 1 both
#Specify the mirror port and copy the traffic to the observation port with serial number 1
```

```
[Huawei-GigabitEthernet0/0/1]q
[Huawei]vlan 2
[Huawei-vlan2]quit
[Huawei]interface g0/0/1
[Huawei-GigabitEthernet0/0/1]port link-type trunk
[Huawei-GigabitEthernet0/0/1]port trunk allow-pass vlan all
[Huawei-GigabitEthernet0/0/1]quit
[Huawei]interface g0/0/2      # Add the port connected to the monitoring device to vlan 2 and collect the traffic copied from the mirror port
[Huawei-GigabitEthernet0/0/2]port link-type access
[Huawei-GigabitEthernet0/0/2]port default vlan 2
[Huawei-GigabitEthernet0/0/2]quit
```

(3) Layer–3 remote mirror port configuration, as shown in Figure 5.3.The principle of the three-layer remote image of the port is to copy the traffic of the image port to the observation port by establishing a GRE Tunnel in the IP layer. The observation port sends the traffic to the end port where the monitoring equipment is located through the GRE tunnel for traffic analysis. In the topology shown in Figure 5.3, the image port on huawei copies the traffic to the observation port, and the observation port sends it to the E0/0/2 interface of huawei connected to server 2 monitoring server through GRE tunnel for server 2 to analyze the traffic of the image port.

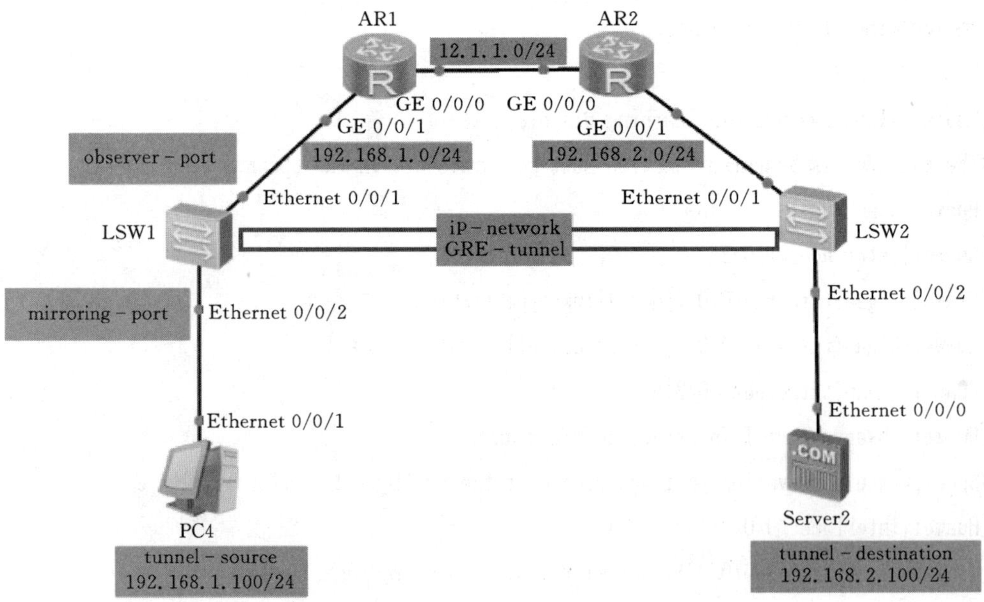

Figure 5.3 Layer-3 remote mirror port

The configuration is as follows: first, configure the routing network segment and static routing on AR1 and AR 2 respectively to ensure the layer-3 interoperability, and then configure the GRE tunnel of the mirror port and observation port on the switch LSW 1.

The routing part is omitted, and the configuration on the switch LSW 1 is as follows.

```
[Huawei]observe-port 1 interface e0/0/1 destination-ip 192.168.2.100 source-ip 192.168.1.100    # Create observation ports and tunnels
[Huawei]interface e0/0/2
[Huawei-Ethernet0/0/1]port-mirroring to observe-port 1 both
#Specify the mirror port and serial number 1
[Huawei-Ethernet0/0/1]quit
```

5.3 Project implementation

5.3.1 Project requirements

(1) The whole network uses static route interaction information.

(2) The same intranet is built between two branches through IPSec VPN.

(3) The server traffic needs to be mirrored to the audit device.

(4) Some people internally use small routers. It is necessary to prevent the access of small routers from affecting the normal DHCP function.

5.3.2 Equipment list

The network equipment of this project mainly uses access layer switches, core layer switches, routers, etc. The specific models and quantities are shown in Table 5.1.

Table 5.1 List of network equipment

Serial No.	Equipment model	Quantity	Function	Remarks
1	AR3260	2	Exit router	1 set for each branch
2	S7703	2	Core switch	1 set for each branch
3	S5700–52P–LI–AC	20	access switch	10 set for each branch

5.3.3 Access switch project topology

The network topology of the project is shown in Figure 5.4.

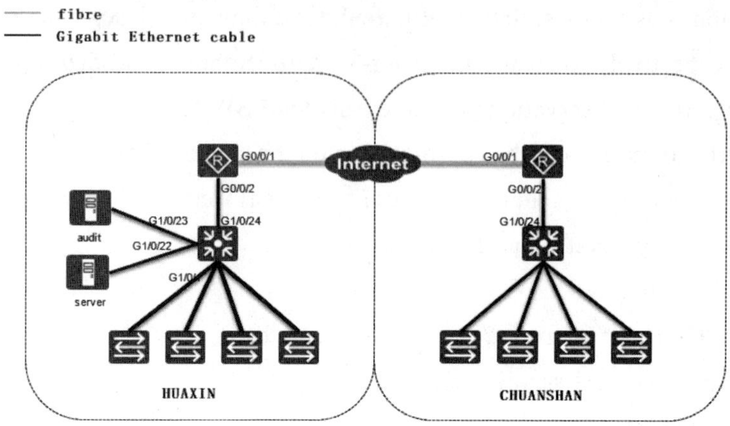

Figure 5.4 Network topology of a group company

5.3.4 Configuration ideas

The following ideas are adopted for configuration:

(1) The exit router is configured with IPS ec VPN to establish a VPN tunnel.

(2) Two branch internal addresses cannot conflict with each other.

(3) Enable Snooping on the access switch to prevent private routers from sending DHCP messages.

(4) Enable port mirroring on the core switch.

5.3.5 Relevant information planning

5.3.5.1 Core switch

The information planning of the core switch is shown in Table 5.2.

Table 5.2 Core switch information planning table

Serial No.	User department	VLAN No.	IP address	Mask	Remarks
1	Huaxin	10	172.16.10.1	255.255.255.0	
2	Chuanshan	20	172.16.20.1	255.255.255.0	
3	The server	30	172.16.30.1	255.255.255.0	
4	Huaxin management VLAN	254	172.16.254.1	255.255.255.0	
5	Chuanshan management VLAN	255	172.16.255.1	255.255.255.0	
6	Huaxin internet 1	1000	172.16.250.1	255.255.255.252	Interconnect with router
7	Chuanshan internet 2	1001	172.16.251.1	255.255.255.252	Interconnect with router

5.3.5.2 Exit router

The information planning of the export router is shown in Table 5.3.

Table 5.3 Export router information planning

Serial No.	Interface	IP address	Mask	Remarks
1	G0/0/2	172.16.250.2	255.255.255.252	Interconnection with core switch
2	G0/0/2	172.16.251.2	255.255.255.252	Interconnect with router

5.3.5.3 ISP export planning

ISP export information planning is shown in Table 5.4.

Table 5.4 ISP Export information planning

Equipment	Interface	IP address	Mask	Gateway	Remarks
Huaxin	G0/0/1	1.1.25.36	255.255.255.192	1.1.25.1	Static IP address
Chuanshan	G0/0/1	1.1.35.47	255.255.255.192	1.1.35.1	Static IP address

5.3.6 Operating steps

5.3.6.1 Relevant configuration of Huaxin export equipment

(1) Address configured at the Internet outlet.

```
<Huawei> system-view
[Huawei] sysname GLL_HX _AR3260
[GLL_HX _AR3260] interface gigabitethernet 0/0/1
[GLL_HX _AR3260-GigabitEthernet0/0/1] ip address 1.1.25.36 255.255.255.192
```

(2) Configure Internet ACL.

```
[GLL_HX _AR3260] acl 3001
[GLL_HX _AR3260-acl-adv-3001] rule5 deny ip source 172.16.10.00.0.0.255 destination 172.16.20.0 0.0.0.255
[GLL_HX _AR3260-acl-adv-3001] rule10 deny ip source 172.16.30.00.0.0.255 destination 172.16.20.0 0.0.0.255
[GLL_HX _AR3260-acl-adv-3001] rule 15 permit ip source 172.16.10.00.0.255.255 destination Any
[GLL_HX _AR3260-acl-adv-3001] rule 15 permit ip source 172.16.20.00.0.255.255 destination Any
```

(3) NAT address translation.

```
[GLL_HX _AR3260] interface gigabitethernet 0/0/1
```

```
[GLL_HX _AR3260-GigabitEthernet0/0/1]nat outbound 3001
[GLL_HX _AR3260-GigabitEthernet0/0/1] quit
```

(4) Default route.

```
[GLL_HX _AR3260] ip route-static 0.0.0.0 0 1.1.25.1
```

(5) Configure IPSec VPN.

1) Define the data flow to be protected.

```
[GLL_HX _AR3260] acl number 3002
[GLL_HX _AR3260-acl-adv-3002] rule permit ip source 172.16.10.0 0.0.0.255 destination 172.16.20.0 0.0.0.255
[GLL_HX _AR3260-acl-adv-3002] rule permit ip source 172.16.30.0 0.0.0.255 destination 172.16.20.0 0.0.0.255
[GLL_HX _AR3260-acl-adv-3002] rule permit ip source 172.16.255.0 0.0.0.255 destination 172.16.254.0 0.0.0.255
[GLL_HX _AR3260-acl-adv-3002] quit
```

2) Configure IPSec security proposal.

```
[GLL_HX _AR3260] ipsec proposal GLL
[GLL_HX _AR3260-ipsec-proposal-GLL] esp authentication-algorithm sha2-256
[GLL_HX _AR3260-ipsec-proposal-GLL] esp encryption-algorithm aes-128
[GLL_HX _AR3260-ipsec-proposal-GLL] quit
```

3) Create security policy.

```
[GLL_HX _AR3260] ipsec policy GLL 10 manual
[GLL_HX _AR3260-ipsec-policy-manual-GLL-10] security acl 3002
[GLL_HX _AR3260-ipsec-policy-manual-GLL-10] proposal tran1
[GLL_HX _AR3260-ipsec-policy-manual-GLL-10] tunnel remote 1.1.35.47
[GLL_HX _AR3260-ipsec-policy-manual-GLL-10] tunnel local 1.1.25.36
[GLL_HX _AR3260-ipsec-policy-manual-GLL-10] sa spi outbound esp 12345
[GLL_HX _AR3260-ipsec-policy-manual-GLL-10] sa spi inbound esp 54321
[GLL_HX _AR3260-ipsec-policy-manual-GLL-10] sa string-key outbound esp cipher Huawei
[GLL_HX _AR3260-ipsec-policy-manual-GLL-10] sa string-key inbound esp cipher Huawei
[GLL_HX _AR3260-ipsec-policy-manual-GLL-10] quit
```

4) Reference security policy on interface.

```
[GLL_HX _AR3260] interface gigabitethernet 0/0/1
[GLL_HX _AR3260-GigabitEthernet1/0/0] ipsec policy GLL
```

5) Configure intranet interconnection address.

```
[GLL_HX _AR3260] interface gigabitethernet 0/0/2
[GLL_HX _AR3260-GigabitEthernet0/0/1] ip address 172.16.250.2 255.255.255.252
```

6) Configure service backhaul routing.

```
ip route-static 172.16.10.0 255.255.255.0172.16.250.1
ip route-static 172.16.30.0 255.255.255.0 172.16.250.1
ip route-static 172.16.254.0 255.255.255.0 172.16.250.1
ip route-static 172.16.20.0 255.255.255.0 1.1.35.47
```

5.3.6.2 Relevant configuration of Chuanshan export equipment

(1) IP address configured at the Internet outlet.

```
<Huawei> system-view
[Huawei] sysname GLL_CS_AR3260
[GLL_CS_AR3260] interface gigabitethernet 0/0/1
[GLL_CS_AR3260-GigabitEthernet0/0/1] ip address 1.1.35.47 255.255.255.192
```

(2) Configure Internet ACL.

```
[GLL_CS_AR3260] acl 3001
[GLL_CS_AR3260-acl-adv-3001] rule5 deny ip source 172.16.20.00.0.255.255 destination 172.16.10.0 0.0.0.255
[GLL_CS_AR3260-acl-adv-3001] rule10 deny ip source 172.16.20.00.0.255.255 destination 172.16.30.0 0.0.0.255
[GLL_CS_AR3260-acl-adv-3001] rule 15 permit ip source 172.16.20.00.0.255.255 destination any
```

(3) NAT address translation.

```
[GLL_CS _AR3260] interface gigabitethernet 0/0/1
[GLL_CS_AR3260-GigabitEthernet0/0/1]nat outbound 3001
[GLL_CS_AR3260-GigabitEthernet0/0/1] quit
```

(4) Default route.

```
[GLL_CS _AR3260] ip route-static 0.0.0.0 0 1.1.35.1
```

(5) Configure IPSec VPN.

1) Define the data flow to be protected.

```
[GLL_CS _AR3260] acl number 3002
[GLL_CS _AR3260-acl-adv-3002] rule permit ip source 172.16.20.0 0.0.0.255 destination 172.16.10.0 0.0.0.255
[GLL_CS _AR3260-acl-adv-3002] rule permit ip source 172.16.20.0 0.0.0.255 destination 172.16.30.0 0.0.0.255
[GLL_CS _AR3260-acl-adv-3002] rule permit ip source 172.16.254.0 0.0.0.255 destination
```

```
172.16.255.0 0.0.0.255
[GLL_CS _AR3260-acl-adv-3002] quit
```

2) Configure IPSec security proposal.

```
[GLL_CS _AR3260] ipsec proposal GLL
[GLL_CS _AR3260-ipsec-proposal-GLL] esp authentication-algorithm sha2-256
[GLL_CS _AR3260-ipsec-proposal-GLL] esp encryption-algorithm aes-128
[GLL_CS _AR3260-ipsec-proposal-GLL] quit
```

3) Create security policy.

```
[GLL_CS _AR3260] ipsec policy GLL 10 manual
[GLL_CS _AR3260-ipsec-policy-manual-GLL-10] security acl 3002
[GLL_CS _AR3260-ipsec-policy-manual-GLL-10] proposal tran1
[GLL_CS _AR3260-ipsec-policy-manual-GLL-10] tunnel remote 1.1.35.47
[GLL_CS _AR3260-ipsec-policy-manual-GLL-10] tunnel local 1.1.25.36
[GLL_CS _AR3260-ipsec-policy-manual-GLL-10] sa spi outbound esp 12345
[GLL_CS _AR3260-ipsec-policy-manual-GLL-10] sa spi inbound esp 54321
[GLL_CS _AR3260-ipsec-policy-manual-GLL-10] sa string-key outbound esp cipher Huawei
[GLL_CS _AR3260-ipsec-policy-manual-GLL-10] sa string-key inbound esp cipher Huawei
[GLL_CS _AR3260-ipsec-policy-manual-GLL-10] quit
```

4) Reference security policy on interface.

```
[GLL_CS _AR3260] interface gigabitethernet 0/0/1
[GLL_CS _AR3260-GigabitEthernet1/0/0] ipsec policy GLL
```

5) Configure intranet interconnection address.

```
[GLL_HX _AR3260] interface gigabitethernet 0/0/2
[GLL_HX _AR3260-GigabitEthernet0/0/1] ip address 172.16.251.2 255.255.255.252
```

6) Configure service backhaul routing.

```
ip route-static 172.16.20.0 255.255.255.0172.16.251.1
ip route-static 172.16.255.0  255.255.255.0 172.16.251.1
ip route-static 172.16.10.0  255.255.255.01.1.25.1
ip route-static 172.16.30.0  255.255.255.0 1.1.25.1
```

5.3.6.3 Huaxin core switch configuration.

(1) Configure network interconnection between devices.

```
<Huawei>system-view
[Huawei] sysname GLL_SW_HX
[GLL_SW_HX] vlan batch 10 30 254 1000
```

```
[GLL_SW_HX] interface gigabitethernet1/0/24
[GLL_SW_HX - GigabitEthernet1/0/24] port link-type access
[GLL_SW_HX - GigabitEthernet1/0/24] port default vlan 1000
```

（2）Enable DHCP and configure VLAN IP address.

```
[GLL_SW_HX] dhcp enable
[GLL_SW_HX] interface vlanif 10
[GLL_SW_HX -Vlanif1000] ip address 172.16.10.1 24
[GLL_SW_HX -Vlanif1000] dhcp select interface
[GLL_SW_HX -Vlanif1000] quit
[GLL_SW_HX] interface vlanif 30
[GLL_SW_HX -Vlanif1000] ip address 172.16.30.1 24
[GLL_SW_HX -Vlanif1000] quit
[GLL_SW_HX] interface vlanif 254
[GLL_SW_HX - Vlanif254] ip address 172.16.254.1 24
[GLL_SW_HX  - Vlanif254] quit
[GLL_SW_HX] interface vlanif 1000
[GLL_SW_HX -Vlanif1000] ip address 172.16.250.1 30
[GLL_SW_HX -Vlanif1000] quit
```

（3）Configure interface to join VLAN (list).

```
[GLL_SW_HX] interface gigabitethernet1/0/1
[GLL_SW_HX - GigabitEthernet1/0/1] port link-type access
[GLL_SW_HX - GigabitEthernet1/0/1] port port default vlan 10
[GLL_SW_HX - GigabitEthernet1/0/1]quit
[GLL_SW_HX] interface gigabitethernet1/0/2
[GLL_SW_HX - GigabitEthernet1/0/2] port link-type access
[GLL_SW_HX - GigabitEthernet1/0/2] port port default vlan 10
[GLL_SW_HX - GigabitEthernet1/0/2]quit
[GLL_SW_HX] interface gigabitethernet1/0/22
[GLL_SW_HX - GigabitEthernet1/0/22] port link-type access
[GLL_SW_HX - GigabitEthernet1/0/22] port port default vlan 30
[GLL_SW_HX] interface gigabitethernet1/0/23
[GLL_SW_HX - GigabitEthernet1/0/23] port link-type access
[GLL_SW_HX - GigabitEthernet1/0/23] port port default vlan 30
[GLL_SW_HX - GigabitEthernet1/0/23]quit
```

(4) Configure default routes.

```
[GLL_SW_HX]ip route-static 0.0.0.0 0.0.0.0 192.168.250.2
```

(5) Configure observation port.

```
[GLL_SW_HX] observe-port interface gigabitethernet 1/0/23
```

(6) Configuring the mirror port.

```
[GLL_SW_HX] interface gigabitethernet 1/0/22
[GLL_SW_HX -GigabitEthernet1/0/22] mirror to observe-port inbound
[GLL_SW_HX -GigabitEthernet1/0/22] return
```

5.3.6.4 Configure DHCP Snooping for Huaxin access switch

```
<Huawei> system-view
[Huawei] sysname YLRXGA _SW_S5700
[GLL_SW_HX _S5700] dhcp enable
[GLL_SW_HX] dhcp snooping enable
[GLL_SW_HX] port-group port-group group-member G0/0/1 TO G0/0/24
[GLL_SW_HX -port-group] dhcp snooping enable
[GLL_SW_HX -port-group] quit
[GLL_SW_HX] interface gigabitethernet 0/0/24
[GLL_SW_HX -GigabitEthernet0/0/24] undo dhcp snooping enable
[GLL_SW_HX -GigabitEthernet0/0/24] dhcp snooping trusted
```

5.3.6.5 Chuanshan core switch configuration

(1) Configure network interconnection between devices.

```
<Huawei>system-view
[Huawei] sysname GLL_SW_CS
[GLL_SW_CS] vlan batch 20 255 1001
[GLL_SW_CS] interface gigabitethernet1/0/24
[GLL_SW_CS -GigabitEthernet1/0/24] port link-type access
[GLL_SW_CS -GigabitEthernet1/0/24] port default vlan 1001
```

(2) Enable DHCP and configure VLAN IP address.

```
[GLL_SW_CS] dhcp enable
[GLL_SW_CS] interface vlanif 20
[GLL_SW_CS -Vlanif20] ip address 172.16.20.1 24
[GLL_SW_CS -Vlanif20] dhcp select interface
[GLL_SW_CS -Vlanif20] quit
[GLL_SW_CS] interface vlanif 255
```

```
[GLL_SW_CS -Vlanif255] ip address 172.16.255.1 24
[GLL_SW_CS -Vlanif255] quit
[GLL_SW_CS] interface vlanif 1001
[GLL_SW_CS -Vlanif1001] ip address 172.16.251.1 30
[GLL_SW_CS -Vlanif1001] quit
```

(3) Configure interface to join VLAN (list).

```
[GLL_SW_CS] interface gigabitethernet1/0/1
[GLL_SW_CS - GigabitEthernet1/0/1] port link-type access
[GLL_SW_CS - GigabitEthernet1/0/1] port port default vlan 20
[GLL_SW_CS - GigabitEthernet1/0/1]quit
[GLL_SW_CS] interface gigabitethernet1/0/2
[GLL_SW_CS - GigabitEthernet1/0/2] port link-type access
[GLL_SW_CS - GigabitEthernet1/0/2] port port default vlan 20
[GLL_SW_CS - GigabitEthernet1/0/2]quit
```

(4) Configure default routes.

```
[GLL_SW_CS]ip route-static 0.0.0.0 0.0.0.0 192.168.251.2
```

5.3.6.6 Configure DHCP Snooping for Chuanshan access switch

```
<Huawei> system-view
[Huawei] sysname YLRXGA _SW_S5700
[GLL_SW_CS_S5700] dhcp enable
[GLL_SW_CS] dhcp snooping enable
[GLL_SW_CS] port-group port-group group-member G0/0/1 TO G0/0/24
[GLL_SW_CS-port-group] dhcp snooping enable
[GLL_SW_CS-port-group] quit
[GLL_SW_CS] interface gigabitethernet 0/0/24
[GLL_SW_CS-GigabitEthernet0/0/24] undo dhcp snooping enable
[GLL_SW_CS-GigabitEthernet0/0/24] dhcp snooping trusted
```

5.4 Ideological and political integration

Modern network engineering projects are becoming more and more complex, and with the deep-going of learning, the scale of network engineering projects involved is getting bigger and bigger, involving more content and high complexity.In the process of engineering design and implementation, it is necessary to scientifically summarize the practical experience of ICT projects by improving the ability of dialectical thinking, and guide the development of the project by running the position, viewpoint and method of the Marxist scientific concept of practice throughout.In the three core issues in the implementation and application of network engineering projects: network diagram drawing, network plan optimization and engineering information management, the use of Marxist scientific practice concept for guidance can quickly identify the essence of the problem, grasp the key to the problem, and improve the application level of network planning technology.

5.5 Thinking and training

1. IPSec is a framework architecture that consists of what two specific types of protocols?
2. What are the application scenarios of IPSec VPN?
3. What does DCHP monitoring do?
4. For switches with DHCP Snooping enabled, what are the two types of switch ports?
5. What is port mirroring?

项目 6

项目 6　某县人民医院网络项目

6.1　项目导入

大多数医院信息化建设发展经历了从早期的单机单用户应用阶段，到部门级和全院级管理信息系统应用；从以财务、药品和管理为中心，开始向以病人信息为中心的临床业务支持和电子病历应用；从局限在医院内部应用，发展到区域医疗信息化应用。

近几年，随着以“以病人为中心，以提高医疗服务质量为主题”的医院管理年活动，各地医院纷纷加快信息化建设步伐。根据权威机构对医院信息化现状调查显示，以费用和管理为中心的全院网络化系统应用已经超过了 80%。住院医生工作站系统，电子病历、全院 PACS、无线查房、腕带技术、RFID、万兆网络、服务器集群、数据虚拟容灾等先进的系统和网络技术已经开始应用。

医院通过信息化系统建设，优化就诊流程，减少患者排队挂号等候时间，实行挂号、检验、交费、取药等一站式、无胶片、无纸化服务，简化看病流程，杜绝“三长一短”现象，有效解决了群众“看病难”问题。本项目讲述了医院新建大楼的内网部署，使之达到如今医疗系统对网络的普遍要求，网络整体高速稳定地运行，易于管理且安全可靠、实现了海量数据的高速传输以及可扩展性等。

6.2　相关知识点

6.2.1　QoS

6.2.1.1　QoS 概述

一般来说，基于存储转发机制的 Internet(IPv4 标准) 只为用户提供了“尽力而为 (Best-Effort)”的服务，不能保证数据包传输的实时性、完整性以及到达的顺序性，不能保证服务的质量，所以主要应用在文件传送和电子邮件服务。随着 Internet 的飞速发展，人们对于在 Internet 上传输分布式多媒体应用的需求越来越大。一般说来，用户对不同的分布式多媒体应用有着不同的服务质量要求，这就要求网络应能根据用户的要求分配和调度资源，因此，传统的所采用的“尽力而为”转发机制，已经不能满足用户的要求。

QoS（Quality of Service，服务质量）指一个网络能够利用各种基础技术，为指定的网络通信提供更好的服务能力。QoS 是网络的一种安全机制，是用来解决网络延迟和阻塞等问题的一种技术。在网络中可以通过保证传输的带宽、降低传送的时延、降低数据的丢包率以及时延抖动等措施来提高服务质量。当网络发生拥塞的时候，所有的数据流都有可能被丢弃；为满足用户对不同应用不同服务质量的要求，就需要网络能根据用户的要求分配和调度资源，对不同的数据流提供不同的服务质量：对实时性强且重要的数据报文优先处理；对于实时性不强的普通数据报文，提供较低的处理优先级，网络拥塞时甚至丢弃。

6.2.1.2　QoS 服务模型

通常 QoS 提供三种服务模型：Best–Effort Service（尽力而为服务模型）、Integrated Service（综合服务模型，简称 Int–Service）、Differentiated Service（区分服务模型，简称 Diff–Serv）。

（1）Best–Effort Service。Best–Effort Service 是一个单一的服务模型，也是最简单的服务模型。对 Best–Effort 服务模型，网络尽最大的可能性来发送报文。但对时延、可靠性等性能不提供任何保证。Best–Effort 服务模型是网络的缺省服务模型，通过 FIFO 队列来实现。它适用于绝大多数网络应用，如 FTP、E–mail 等。

（2）Integrated Service。Inter–Serv 是一个综合服务模型，它可以满足多种 QoS 需求。该模型使用资源预留协议（RSVP），RSVP 运行在从源端到目的端的每个设备上，可以监视每个流，以防止其消耗资源过多。这种体系能够明确区分并保证每一个业务流的服务质量，为网络提供最细粒度化的服务质量区分。例如用 VOIP，需要 12k 的带宽和 100ms 以内的延迟，集成服务模型就会将其归到事先设定的一种服务等级中。但是，Inter–Serv 模型对设备的要求很高，当网络中的数据流数量很大时，设备的存储和处理能力会遇到很大的压力。Inter–Serv 模型可扩展性很差，难以在 Internet 核心网络实施。这种为单一数据流进行带宽预留的解决思路在 Internet 上想要实现很难，所以该模型在 1994 年推出以后就没有使用过。

（3）Differentiated Service。Diff–Serv 是一个多服务模型，由一系列技术组成，它可以满足不同的 QoS 需求。与 Inter-Serv 不同，它不需要通知网络为每个业务预留资源。

区分服务实现简单，扩展性较好。可以用不同的方法来指定报文的 QoS，如 IP 包的优先级 /Precedence、报文的源地址和目的地址等。网络通过这些信息来进行报文的分类、流量整形、流量监管和排队。

6.2.1.3　QoS 技术综述

（1）流分类：采用一定的规则识别符合某类特征的报文，它是对网络业务进行区分服务的前提和基础。

（2）流量监管：对进入或流出设备的特定流量进行监管。当流量超出设定值时，可以采取限制或惩罚措施，以保护网络资源不受损害。可以作用在接口入方向和接口出方向。

（3）流量整形：一种主动调整流的输出速率的流量控制措施，用来使流量适配下游设备可供给的网络资源，避免不必要的报文丢弃和延迟，通常作用在接口出方向。

（4）拥塞管理：当拥塞发生时制定一个资源的调度策略，以决定报文转发的处理次序，通常作用在接口出方向。

（5）拥塞避免：监督网络资源的使用情况，当发现拥塞有加剧的趋势时采取主动丢弃报文的策略，通过调整队列长度来解除网络的过载，通常作用在接口出方向。

在这些 QoS 技术中，流量分类和标记是基础，是有区别地实施服务的前提；而其他 QoS 技术则从不同方面对网络流量及其分配的资源实施控制，有区别地提供服务。

QoS 技术在网络设备中的处理顺序：通过流分类对各种业务进行识别和区分，它是后续各种动作的基础；通过各种动作对特性的业务进行处理。这些动作需要和流分类关联起来才有意义。具体采取何种动作，与所处的阶段以及网络当前的负载状况有关。例如，当报文进入网络时进行流量监管；流出节点之前进行流量整形；拥塞时对队列进行拥塞管理；拥塞加剧时采取拥塞避免措施等。

6.2.1.4　流量分类和 QoS 标记

流量分类是将数据报文划分为多个优先级或多个服务类。网络管理者可以设置流量分类的策略，这个策略除可以包括 IP 报文的 IP 优先级或 DSCP 值、802.1p 的 CoS 值等带内信令，还可以包括输入接口、源 IP 地址、目的 IP 地址、MAC 地址、IP 协议或应用程序的端口号等。分类的结果是没有范围限制的，它可以是一个由五元组（源 IP 地址、源端口号、协议号、目的 IP 地址、目的端口号）确定的流这样狭小的范围，也可以是到某某网段的所有报文。

标记在网络边界处进行，目的在于将区分数据，表明其之间的不同，这样在网络内部队列技术就可以依据这个标记将数据划分到相应的队列，进行不同的处理。在 IP 报文中有专门的字段进行 QoS 的标记，在 IPv4 中为 TOS，IPv6 中为 Traffic Class。TOS 字段用前 6bit 来标记 DSCP，如果只用前 3bit 就为 IP 优先级。DSCP 和 IP 优先级都是标记的标准。IP 优先级提供 0~7 共 8 种服务质量，6 和 7 都保留，所以常用的是 0~5，每个数字都对应一个名称，比如 0 对应 Routine，这样在更改数据包优先级等配置时，既可以用数字也可以用名称。IP 优先级和 DSCP 不能同时设置，如果同时设置的话只有 DSCP 生效，那么标记了 DSCP 的数据包到了只会识别 IP 优先级的路由器，就只会看前 3bit，而且不管是 IP 优先级还是 DSCP 都是用自己的前 3bit 和二层的 CoS 值形成映射。在二层用 CoS 字段进行标记，正常的以太网帧是没有标记的，但是在 ISL 的报头和 802.1Q 的 Tag 中都有 3bit 用来定义服务级别，0~7 中只有 0~5 可用，6 和 7 都保留。

6.2.1.5　几种常用的队列调度机制

（1）FIFO（先进先出队列）。传统的 Best-Effort 服务策略，默认应用在带宽大于 2.048M 的接口上，只适用于对带宽和延迟不敏感的流量，像是 WWW、FTP、E-mail 等。FIFO 不对报文进行分类，当报文进入接口的速率大于接口能发送速率时，FIFO 按报文到达接口的先后顺序让报文进入队列，同时在队列的出口让报文按进队的顺序出队。每个队列内部报文的发送（次序）关系缺省是 FIFO 先进先出队列。

（2）PQ（优先队列）。PQ 队列是针对关键业务应用设计的。关键业务有一个重要的特点，即在拥塞发生时要求优先获得服务以减小响应的延迟。PQ 可以根据网络协议（如 IP、IPX）、数据流入接口、报文长短、源地址 / 目的地址等灵活地指定优先次序。

4 类报文分别对应 4 个队列：高优先队列、中优先队列、正常优先队列和低优先队列。高优先级队列的报文都发送完了才能发送下一个优先级的报文。这样的机制虽然能保证关键数据总是得到优先处理，但是低优先级的队列很可能因此十分拥塞。缺省情况下，数据流进入正常优先队列。

（3）CQ（自定义队列）。CQ 按照一定的规则将分组分成 16 类（对应于 16 个队列），分组根据自己的类别按照先进先出的策略进入相应的 CQ 队列。CQ 的 1～16 号队列是用户队列。用户可以配置流分类的规则，指定 16 个用户队列占用接口或 PVC 带宽的比例关系。在队列调度时，系统队列中的分组被优先发送。直到系统队列为空，再采用轮询的方式按照预先配置的带宽比例依次从 1～16 号用户队列中取出一定数量的分组发送出去。这样，就可以使不同业务的分组获得不同的带宽，既可以保证关键业务能获得较多的带宽，又不至于使非关键业务得不到带宽。缺省情况下，数据流进入 1 号队列。定制队列的另一个优点是：可根据业务的繁忙程度分配带宽，适用于对带宽有特殊需求的应用。虽然 16 个用户队列的调度是轮询进行的，但对每个队列不是固定地分配服务时间片——如果某个队列为空，那么马上换到下一个队列调度。因此，当没有某些类别的报文时，CQ 调度机制能自动增加现存类别的报文可占的带宽。

（4）WFQ（加权公平队列）。WFQ 是为了公平地分享网络资源，尽可能使所有流的延迟和抖动达到最优而推出的。它照顾了各方面的利益，主要表现在：不同的队列获得公平的调度机会，从总体上均衡各个流的延迟。短报文和长报文获得公平的调度：如果不同队列间同时存在多个长报文和短报文等待发送，应当顾及短报文的利益，让短报文优先获得调度，从而在总体上减少各个流的报文间的抖动。

WFQ 按数据流的会话信息自动进行流分类（相同源 IP 地址、目的 IP 地址、原端口号、目的端口号、协议号、IP 优先级的报文同属一个流），并且尽可能多地划分出 N 个队列，以将每个流均匀地放入不同队列中，从而在总体上均衡各个流的延迟。

在出队的时候，WFQ 按流的优先级（Precedence 或 DSCP）来分配每个流应占有出口的带宽。优先级的数值越小，所得的带宽越少。优先级的数值越大，所得的带宽越多。最后，轮询各个队列，按照带宽比例从队列中取出相应数量的报文进行发送。WFQ 是传输速率在 2.048M 以下的接口默认的队列机制。

举例来说，接口中当前共有 5 个流，它们的优先级分别为 0、1、2、3、4，则带宽总配额为所有（流的优先级 +1）的和，即 1+2+3+4+5=15，每个流可得的带宽分别为 1/15、

2/15、3/15、4/15、5/15。

WFQ 的限制：WFQ 不支持隧道或者采用了加密技术的接口，因为这些技术要修改数据包中 WFQ 用于分类的信息。WFQ 提供的带宽控制的精度不如 CBQ，因为是基于流的分类，基于队列的带宽分配，每个队列可能会有多个流，这样无法再针对具体的数据类型指定带宽。

（5）CBQ（基于类的公平队列）。CBQ 是对 WFQ 功能的扩展，为用户提供了定义类的支持。CBQ 为每个用户定义的类分配一个单独的 FIFO 预留队列，用来缓冲同一类的数据。在网络拥塞时，CBQ 对报文根据用户定义的类规则进行匹配，并使其进入相应的队列，在入队列之前必须进行拥塞避免机制和带宽限制的检查。在报文出队列时，加权公平调度每个类对应的队列中的报文。

CBQ 提供一个紧急队列，紧急报文入该队列，该队列采用 FIFO 调度，没有带宽限制。这样，如果 CBQ 加权公平对待所有类的队列，语音报文这类对延迟敏感的数据流就可能得不到及时发送。为此将 PQ 特性引入 CBQ，称其为 LLQ（Low Latency Queuing，低延迟队列），为语音报文这样的对延迟敏感的数据流提供严格优先发送服务。

LLQ 将严格优先队列机制与 CBQ 结合起来使用，用户在定义类时可以指定其享受严格优先服务，这样的类称作优先类。所有优先类的报文将进入同一个优先队列，在入队列之前需对各类报文进行带宽限制的检查。报文出队列时，将首先发送优先队列中的报文，直到发送完后才发送其他类对应的队列的报文。在发送其他队列报文时将仍然按照加权公平的方式调度。

为了不让其他队列中的报文延迟时间过长，在使用 LLQ 时将会为每个优先类指定可用最大带宽，该带宽值用于拥塞发生时监管流量。如果拥塞未发生，优先类允许使用超过分配的带宽。如果拥塞发生，优先类超过分配带宽的数据包将被丢弃。LLQ 还可以指定 Burst-size。

CBQ 对 WFQ 做的一些改进：在 WFQ 中 weight 用来指明队列优先级，而在 CBWFQ 中 weight 用来指明某类流量的优先级。数据包根据 weight 排在相应类的队列中。CBQ 一个队列一种数据，所以可以为某类流量指定相应的带宽，而 WFQ 无法实现，因为基于流种类很多，最后可能每个队列里都有好几种流量。CBQ 分类数据时除了根据 IP 地址和端口号，还可以通过 ACL 或数据输入接口，WFQ 无法实现。PQ 和 CQ 都需要手动配置，在命令中可以看出，并不能依据 IP 优先级或 DSCP 来划分队列，而且配置起来比较麻烦，命令烦琐。在 WFQ 中只有一条命令，执行基本是自动化的，但这样不好控制流量，而且要事先进行 QoS 标记。在 CBQ 中 QoS 的事先标记不是必须的，因为引入了 MQC 的概念，通过结构化的命令行匹配特定的数据流（如果匹配的是 IP 优先级或 DSCP，则需要事先的 QoS 标记）再制定细化的处理策略，不过归队列还是算法自动完成，CBQ 是目前推荐使用的模式。

6.2.2　MSTP+VRRP

MSTP 是多生成树协议，与生成树和快速生成树相比，MSTP 引入了“实例 (INSTANCE)”

的概念。生成树和快速生成树都是基于交换机端口的技术，在进行生成树计算的时候，所有VLAN都共享相同的生成树；而MSTP则是基于实例的。所谓“实例”是指多个VLAN对应的一个集合,MSTP把一台设备的一个或多个VLAN划分为一个实例，有着相同实例配置的设备就组成了一个MST域，运行独立的生成树；这个MST就组成了一个大的设备整体，与其他MST域在进行生成树算法，得出一个整体的生成树。所有VLAN默认映射到实例0，其他实例则称为多生成树实例。

在局域网内，主机发往其他网段的报文都由网关进行转发。当网关发生故障时，本网段内所有发往网关的数据将中断。为了避免网络中断，可以通过在主机上设置多个网关，但是一个主机只允许设置一个默认网关，因此需要管理员手工添加和修改，这样大大增加了网络管理的复杂度。因此我们通常使用VRRP（虚拟路由器冗余）协议，保证用户快速、不间断、透明地切换到另一个网关。

VRRP协议首先采用竞选的方法选择主路由器（Master），主路由器负责提供实际的数据转发服务，主路由器选出后，其他设备作为备份路由器（Backup），并通过主路由设备定时发出的VRRP报文监测主路由设备的状态。如果组内的备份路由设备在规定的时间内没有收到来自主路由设备的报文，则将自己状态转为Master，由于切换非常迅速而且用户终端不需要改变默认网关的IP地址和MAC地址，所以对用户而言是透明的。

主路由负责转发数据，备份路由不负责转发数据，因此在主路由转发数据的同时，备份路由却一直处于空闲状态，这样势必造成了网络带宽资源的浪费。我们通过在VRRP中使用负载均衡技术，创建不同的VRRP组，使得路由器在不同的VRRP组中担任不同的角色。

VRRP与MSTP配合使用的组网方式，不仅可以为网关设备提供冗余备份，还可以为下行的二层链路提供冗余备份，并使用MSTP技术阻塞网络中的冗余链路以消除二层环路，极大地提高了网络的可靠性。

6.3 项目实施

6.3.1 项目需求

（1）全网使用静态路由交互信息。

（2）核心采用VRRP技术进行网关设备的备份，提高网关的可靠性。当一台网关设备出现故障时，局域网内的主机仍然可以通过另一台网关设备访问外部网络。

（3）在网关设备工作正常时，区域A用户通过网关设备Switch A进行数据转发；区域B用户通过网关设备Switch B进行数据转发，实现流量的负载分担。

（4）当网关设备的上行链路出现故障时，降低该网关设备的优先级，以避免该网关设备成为Master，导致流量转发中断。

（5）局域网内进行二层链路的冗余备份，保证网关设备下行链路故障时，流量转发不会中断。使用 MSTP 技术避免二层网络中出现环路。

（6）网关设备通过出口设备与 Internet 连接，出口路由器上进行 QoS 的流量限制。

6.3.2　设备清单

本项目主要用到的设备包括核心交换机、接入层交换机和路由器等，具体型号数量见表 6.1。

表 6.1　　网络设备清单

序号	设备型号	数量	功能	备注
1	AR3260	1	出口路由器	
2	S7706	2	核心交换机	
3	S5700-52P-LI-AC	30	接入交换机	

6.3.3　项目拓扑

本项目网络拓扑结构如图 6.1 所示。

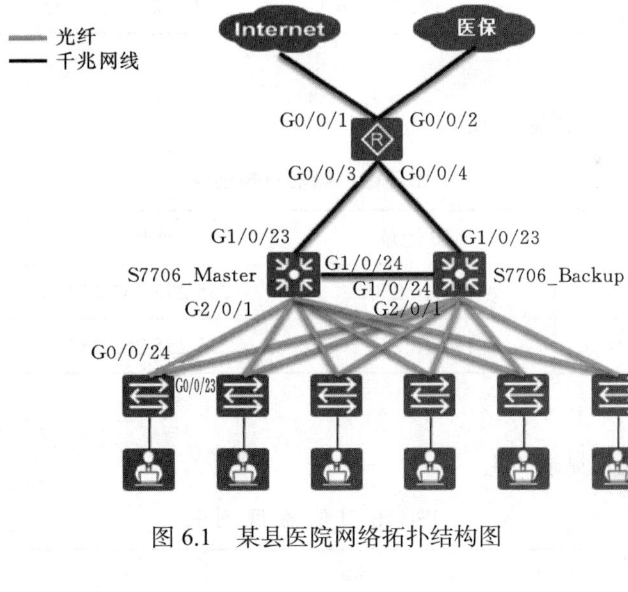

图 6.1　某县医院网络拓扑结构图

6.3.4　配置思路

采用如下的思路配置：

（1）两台核心交换组建 VRRP。

（2）采用 MSTP 进行破环处理。

（3）出口路由器上进行 QoS 的流量限制。

（4）配置 NQA 与静态路由联动，实现路由快速切换。

6.3.5　相关信息规划

6.3.5.1　核心交换机

核心交换机的信息规划见表 6.2。

表 6.2　核心交换机信息规划表

序号	使用部门	VLAN 号	IP 地址	掩　码	备注
1	门诊收费	10	172.16.10.254	255.255.255.0	
2	门诊急诊科	20	172.16.20.254	255.255.255.0	
3	药房	30	172.16.30.254	255.255.255.0	
4	管理	255	172.16.255.1	255.255.255.0	
5	互联 1	1000	172.16.250.1	255.255.255.252	与路由器互联
6	互联 2	1001	172.16.251.1	255.255.255.252	与路由器互联
7	互联 2	1002	172.16.254.1	255.255.255.252	核心交换机互联

6.3.5.2　出口路由器

出口路由器的信息规划见表 6.3。

表 6.3　出口路由器信息规划表

序号	接口	IP 地址	掩码	备注
1	G0/0/3	172.16.250.2	255.255.255.252	与连接 Master
2	G0/0/4	172.16.251.2	255.255.255.252	与连接 Backup

6.3.5.3　ISP 出口规划

ISP 出口的信息规划见表 6.4。

表 6.4　ISP 出口信息规划表

序号	接口	IP 地址	掩　码	网关	备注
1	G0/0/1	1.1.25.67	255.255.255.192	1.1.25.1	静态 IP 地址，连接 Internet
2	G0/0/2	1.1.125.2	255.255.255.0	1.1.125.1	连接医保，对端医保业务地址为 1.1.125.0/24

6.3.6 操作步骤

6.3.6.1 出口设备相关配置

（1）外网出口配置 IP 地址，方法如下：

```
<Huawei> system-view
[Huawei] sysname WZCWXRMYY
[WZCWXRMYY_AR3260] interface gigabitethernet 0/0/1
[WZCWXRMYY_AR3260-GigabitEthernet0/0/1] ip address 1.1.25.67  255.255.255.192
[WZCWXRMYY_AR3260] interface gigabitethernet 0/0/2
[WZCWXRMYY_AR3260-GigabitEthernet0/0/2] ip address 1.1.125.2 255.255.255.0
```

（2）配置上网 ACL，方法如下：

```
[WZCWXRMYY _AR3260] acl 3001
[WZCWXRMYY _AR3260-acl-adv-3001] rule 10 permit ip source 172.16.0.00.0.255.255
```

（3）NAT 地址转换，方法如下：

```
[WZCWXRMYY _AR3260] interface gigabitethernet 0/0/1
[WZCWXRMYY _AR3260-GigabitEthernet0/0/1] nat outbound 3001
[WZCWXRMYY _AR3260-GigabitEthernet0/0/1] quite

[WZCWXRMYY _AR3260] interface gigabitethernet 0/0/2
[WZCWXRMYY _AR3260-GigabitEthernet0/0/2] nat outbound 3001
[WZCWXRMYY _AR3260-GigabitEthernet0/0/2] quite
```

（4）配置内网互联地址，方法如下：

```
[WZCWXRMYY _AR3260] interface gigabitethernet 0/0/3
[WZCWXRMYY _AR3260-GigabitEthernet0/0/3] ip address 172.16.250.1 255.255.255.252
[WZCWXRMYY _AR3260] interface gigabitethernet 0/0/4
[WZCWXRMYY _AR3260-GigabitEthernet0/0/4] ip address 172.16.251.1 255.255.255.252
```

（5）配置 NQA，方法如下：

```
[WZCWXRMYY _AR3260] nqa test-instance aa bb
[WZCWXRMYY _AR3260-nqa-aa-bb] test-type icmp
[WZCWXRMYY _AR3260-nqa-aa-bb] destination-address ipv4 172.16.250.1
[WZCWXRMYY _AR3260-nqa-aa-bb] frequency 5
[WZCWXRMYY _AR3260-nqa-aa-bb] probe-count 1
[WZCWXRMYY _AR3260-nqa-aa-bb] start now
[WZCWXRMYY _AR3260-nqa-aa-bb] quit
```

（6）默认路由，方法如下：

```
[WZCWXRMYY _AR3260] ip route-static 0.0.0.0 0.0.0.0 1.1.25.1
[WZCWXRMYY _AR3260] ip route-static 59.0.0.0 255.255.255.0  1.1.125.1
[WZCWXRMYY _AR3260] ip route-static 172.16.0.0 255.255.0.0 172.16.250.1 track nqa aa bb
[WZCWXRMYY _AR3260] ip route-static 172.16.0.0 255.255.0.0 172.16.251.1 preference 100
```

（7）限制下载速度，方法如下：

```
[WZCWXRMYY _AR3260] interface gigabitethernet 0/0/3
[WZCWXRMYY _AR3260-GigabitEthernet0/0/3]qos car inbound source-ip-address range 172.16.10.2. to 172.16.10.252 per-address cir 4096 cbs 6016 pbs 10016 green pass yellow pass red discard
[WZCWXRMYY _AR3260-GigabitEthernet0/0/3]qos car inbound source-ip-address range 172.16.20.2. to 172.16.20.252 per-address cir 4096 cbs 6016 pbs 10016 green pass yellow pass red discard
[WZCWXRMYY _AR3260-GigabitEthernet0/0/3]qos car inbound source-ip-address range 172.16.30.2. to 172.16.30.252 per-address cir 4096 cbs 6016 pbs 10016 green pass yellow pass red discard
```

6.3.6.2　核心交换机（Master）配置

（1）配置设备间的网络互联，方法如下：

```
<Huawei>system-view
[Huawei] sysname WZCWXRMYY_Master
[WZCWXRMYY_Master] vlan batch 10 20 30  255 1000 1001
[WZCWXRMYY_Master] interface gigabitethernet 1/0/23
[WZCWXRMYY_Master -GigabitEthernet1/0/23] port link-type access
[WZCWXRMYY_Master -GigabitEthernet1/0/23] port default vlan 1000
```

（2）配置地址池，方法如下：

```
[WZCWXRMYY_Master]ip pool vlan10
[WZCWXRMYY_Master-ip-pool-vlan10]gateway-list 172.16.10.254
[WZCWXRMYY_Master-ip-pool-vlan10]network 172.16.10.0 mask 255.255.255.0
[WZCWXRMYY_Master-ip-pool-vlan10]excluded-ip-address 172.16.10.252 172.16.10.253    #排除地址
[WZCWXRMYY_Master]ip pool vlan20
[WZCWXRMYY_Master-ip-pool-vlan20]gateway-list 172.16.20.254
[WZCWXRMYY_Master-ip-pool-vlan20]network 172.16.20.0 mask 255.255.255.0
[WZCWXRMYY_Master-ip-pool-vlan20]excluded-ip-address 172.16.20.252 172.16.20.253

[WZCWXRMYY_Master]ip pool vlan30
[WZCWXRMYY_Master-ip-pool-vlan30]gateway-list 172.16.30.254
[WZCWXRMYY_Master-ip-pool-vlan30]network 172.16.30.0 mask 255.255.255.0
```

```
[WZCWXRMYY_Master-ip-pool-vlan30]excluded-ip-address 172.16.30.252 172.16.30.253
```

（3）配置互联 IP，方法如下：

```
[WZCWXRMYY_Master] interface vlanif 1000
[WZCWXRMYY_Master-Vlanif1000] ip address 172.16.250.1 255.255.255.252
[WZCWXRMYY_Master-Vlanif1000] quit

[WZCWXRMYY_Master] interface vlanif 254
[WZCWXRMYY_Master-Vlanif1000] ip address 172.16.254.1  255.255.255.0
[WZCWXRMYY_Master-Vlanif1000] quit
```

（4）配置 VRRP，方法如下：

```
[WZCWXRMYY_Master] interface vlanif 10
[WZCWXRMYY_Master-Vlanif10] ip address 172.16.10.253
[WZCWXRMYY_Master-Vlanif10]vrrp vrid 10 virtual-ip 172.16.10.254
[WZCWXRMYY_Master-Vlanif10]vrrp vrid 10 priority 200
[WZCWXRMYY_Master-Vlanif10]dhcp select global
[WZCWXRMYY_Master-Vlanif10] quit

[WZCWXRMYY_Master] interface vlanif 20
[WZCWXRMYY_Master-Vlanif20] ip address 172.16.20.253
[WZCWXRMYY_Master-Vlanif20]vrrp vrid 20 virtual-ip 172.16.20.254
[WZCWXRMYY_Master-Vlanif20]vrrp vrid 20 priority 200
[WZCWXRMYY_Master-Vlanif20]dhcp select global
[WZCWXRMYY_Master-Vlanif20] quit

[WZCWXRMYY_Master] interface vlanif 30
[WZCWXRMYY_Master-Vlanif30] ip address 172.16.30.253
[WZCWXRMYY_Master-Vlanif30]vrrp vrid 30 virtual-ip 172.16.30.254
[WZCWXRMYY_Master-Vlanif30]vrrp vrid 30 priority 200
[WZCWXRMYY_Master-Vlanif30]dhcp select global
[WZCWXRMYY_Master-Vlanif30] quit
[WZCWXRMYY_Master] interface vlanif 255
[WZCWXRMYY_Master-Vlanif255] ip address 172.16.255.253
[WZCWXRMYY_Master-Vlanif255]vrrp vrid 255 virtual-ip 172.16.30.254
[WZCWXRMYY_Master-Vlanif255]vrrp vrid 255 priority 200
```

```
[WZCWXRMYY_Master-Vlanif255] quit
```

（5）配置 NQA，方法如下：

```
[WZCWXRMYY_Master] nqa test-instance aa bb
[WZCWXRMYY_Master-nqa-aa-bb] test-type icmp
[WZCWXRMYY_Master-nqa-aa-bb] destination-address ipv4 172.16.250.2
[WZCWXRMYY_Master-nqa-aa-bb] frequency 5
[WZCWXRMYY_Master-nqa-aa-bb] probe-count 1
[WZCWXRMYY_Master-nqa-aa-bb] start now
[WZCWXRMYY_Master-nqa-aa-bb] quit
```

（6）配置默认路由，方法如下：

```
[WZCWXRMYY_Master]ip route-static 0.0.0.0 0.0.0.0 192.168.250.2 rack nqa aa bb
[WZCWXRMYY_Master]ip route-static 0.0.0.0 0.0.0.0 192.168.254.2 preference 100
```

（7）配置 Trunk，方法如下：

```
[WZCWXRMYY_Master] interface gigabitethernet 2/0/1
[WZCWXRMYY_Master-GigabitEthernet2/0/1]port link-type trunk
[WZCWXRMYY_Master-GigabitEthernet2/0/1]port trunk allow-pass all

[WZCWXRMYY_Master] interface gigabitethernet 1/0/24
[WZCWXRMYY_Master-GigabitEthernet1/0/24]port link-type trunk
[WZCWXRMYY_Master-GigabitEthernet1/0/24]port trunk allow-pass all
```

（8）配置 MSTP，方法如下：

```
[WZCWXRMYY_Master] stp region-configuration
[WZCWXRMYY_Master-mst-region] region-name RG1
[WZCWXRMYY_Master-mst-region] instance 1 vlan 10
[WZCWXRMYY_Master-mst-region] instance 1 vlan 20
[WZCWXRMYY_Master-mst-region] instance 1 vlan 30
[WZCWXRMYY_Master-mst-region] instance 1 vlan 255
[WZCWXRMYY_Master-mst-region] active region-configuration
[WZCWXRMYY_Master-mst-region] quit
[WZCWXRMYY_Master] stp instance 1 root primary
[WZCWXRMYY_Master] stp enable
```

6.3.6.3　核心交换机（Backup）配置

（1）配置设备间的网络互联，方法如下：

```
<Huawei> system-view
```

```
[Huawei] sysname WZCWXRMYY_Backup
[WZCWXRMYY_Backup] vlan batch 10 20 30 254 255 1000 1001
[WZCWXRMYY_Backup] interface gigabitethernet 1/0/23
[WZCWXRMYY_Backup-GigabitEthernet1/0/23] port link-type access
[WZCWXRMYY_Backup-GigabitEthernet1/0/23] port default vlan 1001
```

（2）配置地址池，方法如下：

```
[WZCWXRMYY_Backup]ip pool vlan10
[WZCWXRMYY_Backup-ip-pool-vlan10]gateway-list 172.16.10.254
[WZCWXRMYY_Backup-ip-pool-vlan10]network 172.16.10.0 mask 255.255.255.0
[WZCWXRMYY_Backup-ip-pool-vlan10]excluded-ip-address 172.16.10.252 172.16.10.253

[WZCWXRMYY_Backup]ip pool vlan20
[WZCWXRMYY_Backup-ip-pool-vlan20]gateway-list 172.16.20.254
[WZCWXRMYY_Backup-ip-pool-vlan20]network 172.16.20.0 mask 255.255.255.0
[WZCWXRMYY_Backup-ip-pool-vlan20]excluded-ip-address 172.16.20.252 172.16.20.253

[WZCWXRMYY_Backup]ip pool vlan30
[WZCWXRMYY_Backup-ip-pool-vlan30]gateway-list 172.16.30.254
[WZCWXRMYY_Backup-ip-pool-vlan30]network 172.16.30.0 mask 255.255.255.0
[WZCWXRMYY_Backup-ip-pool-vlan30]excluded-ip-address 172.16.30.252 172.16.30.253
```

（3）配置互联 IP，方法如下：

```
[WZCWXRMYY_Backup] interface vlanif 1001
[WZCWXRMYY_Backup-Vlanif1001] ip address 172.16.251.1 255.255.255.252
[WZCWXRMYY_Backup-Vlanif1001] quit

[WZCWXRMYY_Backup] interface vlanif 254
[WZCWXRMYY_Backup-Vlanif254] ip address 172.16.254.2  255.255.255.0
[WZCWXRMYY_Backup-Vlanif254] quit
```

（4）配置 VRRP，方法如下：

```
[WZCWXRMYY_Backup] interface vlanif 10
[WZCWXRMYY_Backup-Vlanif10] ip address 172.16.10.252
[WZCWXRMYY_Backup-Vlanif10]vrrp vrid 10 virtual-ip 172.16.10.254
[WZCWXRMYY_Backup-Vlanif10]dhcp select global
[WZCWXRMYY_Backup-Vlanif10] quit
```

```
[WZCWXRMYY_Backup] interface vlanif 20
[WZCWXRMYY_Backup-Vlanif20] ip address 172.16.20.252
[WZCWXRMYY_Backup-Vlanif20]vrrp vrid 20 virtual-ip 172.16.20.254
[WZCWXRMYY_Backup-Vlanif20]dhcp select global
[WZCWXRMYY_Backup-Vlanif20] quit

[WZCWXRMYY_Backup] interface vlanif 30
[WZCWXRMYY_Backup-Vlanif30] ip address 172.16.30.252
[WZCWXRMYY_Backup-Vlanif30]vrrp vrid 30 virtual-ip 172.16.30.254
[WZCWXRMYY_Backup-Vlanif30]dhcp select global
[WZCWXRMYY_Backup-Vlanif30] quit

[WZCWXRMYY_Backup] interface vlanif 255
[WZCWXRMYY_Backup-Vlanif255] ip address 172.16.255.252
[WZCWXRMYY_Backup-Vlanif255]vrrp vrid 30 virtual-ip 172.16.255.254
[WZCWXRMYY_Backup-Vlanif255] quit
```

（5）配置 NQA，方法如下：

```
[WZCWXRMYY_Backup] nqa test-instance aa bb
[WZCWXRMYY_Backup-nqa-aa-bb] test-type icmp
[WZCWXRMYY_Backup-nqa-aa-bb] destination-address ipv4 172.16.254.2
[WZCWXRMYY_Backup-nqa-aa-bb] frequency 5
[WZCWXRMYY_Backup-nqa-aa-bb] probe-count 1
[WZCWXRMYY_Backup-nqa-aa-bb] start now
[WZCWXRMYY_Backup-nqa-aa-bb] quit
```

（6）配置默认路由，方法如下：

```
[WZCWXRMYY_Backup]ip route-static 0.0.0.0 0.0.0.0 192.168.251.2 rack nqa aa bb
[WZCWXRMYY_Backup]ip route-static 0.0.0.0 0.0.0.0 192.168.254.1 preference 100
```

（7）配置 Trunk，方法如下：

```
[WZCWXRMYY_Backup] interface gigabitethernet 2/0/1
[WZCWXRMYY_Backup-GigabitEthernet2/0/1]port link-type trunk
[WZCWXRMYY_Backup-GigabitEthernet2/0/1]port trunk allow-pass all

[WZCWXRMYY_Backup] interface gigabitethernet 1/0/24
```

```
[WZCWXRMYY_Backup - GigabitEthernet1/0/24]port link-type trunk
[WZCWXRMYY_Backup - GigabitEthernet1/0/24]port trunk allow-pass all
```

（8）配置 MSTP，方法如下：

```
[WZCWXRMYY_Backup] stp region-configuration
[WZCWXRMYY_Backup-mst-region] region-name RG1
[WZCWXRMYY_Backup-mst-region] instance 1 vlan 10
[WZCWXRMYY_Backup-mst-region] instance 1 vlan 20
[WZCWXRMYY_Backup-mst-region] instance 1 vlan 30
[WZCWXRMYY_Backup-mst-region] instance 1 vlan 255
[WZCWXRMYY_Backup-mst-region] active region-configuration
[WZCWXRMYY_Backup-mst-region] quit

[WZCWXRMYY_Master] stp instance 1 root secondary
[WZCWXRMYY_Master] stp enable
```

6.3.6.4　接入交换机(列举)

（1）配置 VLAN，方法如下：

```
[WZCWXRMYY_MZYF] vlan batch 10 20 30 255
[WZCWXRMYY_MZYF] interface gigabitethernet 0/0/1
[WZCWXRMYY_MZYF -GigabitEthernet0/0/1] port link-type access
[WZCWXRMYY_MZYF -GigabitEthernet0/0/1] port default vlan 10
```

（2）配置 Trunk，方法如下：

```
[WZCWXRMYY_MZYF] interface gigabitethernet 1/0/24
[WZCWXRMYY_MZYF-GigabitEthernet1/0/24]port link-type trunk
[WZCWXRMYY_MZYF-GigabitEthernet1/0/24]port trunk allow-pass all

[WZCWXRMYY_MZYF] interface gigabitethernet 1/0/23
[WZCWXRMYY_MZYF-GigabitEthernet1/0/23]port link-type trunk
[WZCWXRMYY_MZYF-GigabitEthernet1/0/23]port trunk allow-pass all
[WZCWXRMYY_MZYF-GigabitEthernet1/0/23] stp instance 1 cost 20000
```

（3）配置 MSTP，方法如下：

```
[WZCWXRMYY_MZYF] stp region-configuration
[WZCWXRMYY_MZYF -mst-region] region-name RG1
[WZCWXRMYY_MZYF -mst-region] instance 1 vlan 10
[WZCWXRMYY_MZYF -mst-region] instance 1 vlan 20
```

```
[WZCWXRMYY_MZYF -mst-region] instance 1 vlan 30
[WZCWXRMYY_MZYF -mst-region] active region-configuration
[WZCWXRMYY_MZYF -mst-region] quit
[WZCWXRMYY_MZYF ] stp enable

[GLL_SW_HX] interface vlanif 255
[GLL_SW_HX -Vlanif1000] ip address 172.16.255.3
[GLL_SW_HX -Vlanif1000] quit
```

6.4 思政融入

随着医院信息化进程的深入，医院信息平台的运行将越来越依赖基础网络的建设。网络成为医院各种关键数据的信息进行交互和传递的重要途径。一旦网络数据丢失，将会给医院和病人带来巨大的灾难和难以弥补的损失。因此，方法是否客观，过程是否准确，操作是否严谨，数据是否真实，任何一个环节的缺失都将导致后果的严重性。大学生应该培养严谨求实的优良学风，树立以事实为依据、实事求是、严谨认真的科学精神和工匠精神。同时，医院要保持干净、整洁、亮丽的良好就医环境，在项目实施过程中，注重手脑并用，树立劳动意识，强化整理、整顿、清扫、清洁、素养等5S管理规范，以劳树德、以劳增智、以劳强体。

6.5 思考与训练

1. QoS是什么？其作用是什么？
2. QoS的三种服务模型是什么？
3. VRRP与MSTP配合使用的组网方式有哪些优点？有哪些可提升的地方？
4. 局域网内进行二层链路的冗余备份，保证网关设备下行链路故障时，流量转发不会中断。使用MSTP技术避免二层网络中出现环路，请写出配置的基本步骤。
5. 当网关设备的上行链路出现故障时，降低该网关设备的优先级，以避免该网关设备成为Master，导致流量转发中断，请写出配置命令。

Project 6　Network project of a county people's hospital

6.1　Project import

The development of most hospital informatization construction has gone through the early stage of single machine and single user application to the application of department level and hospital level management information system; From the center of finance, medicine and management to the application of clinical business support and electronic medical record centered on patient information; From being limited to the internal application of hospitals, to has developed into an attempt to apply regional medical informatization.

In recent years, with the hospital management year activity of "taking patients as the center and improving the quality of medical services as the theme", hospitals everywhere have strengthened the pace of information construction. According to the survey on the current situation of hospital informatization by authoritative institutions, the application of hospital wide networked system centered on cost and management has exceeded 80%. Resident workstation system, electronic medical record, PACS of the whole hospital, wireless ward round, wristband technology, RFID, 10 Gigabit network, server cluster, data virtual disaster recovery and other advanced systems and network technologies have been applied.

Through the construction of information system, the hospital optimized the treatment process, reduced the waiting time for patients to queue up for registration, implemented one-stop, film-free and paperless services such as registration, inspection, payment and medicine collection, simplified the medical treatment process, put an end to the phenomenon of "three longs and one short", and effectively solved the problem of "difficult medical treatment" for the masses. This project describes the internal network deployment of the new building of the hospital, so as to meet the general requirements of today's medical system for the network. The overall high-speed and stable operation of the network is easy to manage, safe and reliable, and the high-speed transmission and scalability of massive data are realized.

6.2 Relevant knowledge

6.2.1 QoS

6.2.1.1 QoS overview

Generally speaking, the Internet (IPv4 standard) based on store and forward mechanism only provides users with "best effort" service, which can not guarantee the real-time, integrity and arrival sequence of data packet transmission, and the quality of service. Therefore, it is mainly used in file transmission and e-mail service. With the rapid development of the Internet, people's demand for transmitting distributed multimedia applications on the Internet is increasing. Generally speaking, users have different quality of service requirements for different distributed multimedia applications, which requires that the network should be able to allocate and schedule resources according to the requirements of users. Therefore, the traditional "best effort" forwarding mechanism can no longer meet the requirements of users.

The full English name of QoS is "Quality of Service", which means that a network can use various basic technologies to provide better service capability for specified network communication. QoS is a security mechanism of the network and a technology used to solve the problems of network delay and congestion. In the network, the quality of service can be improved by ensuring the transmission bandwidth, reducing the transmission delay, reducing the packet loss rate and delay jitter. When the network is congested, all data streams may be discarded; In order to meet the requirements of users for different applications and different service qualities, the network needs to be able to allocate and schedule resources according to the requirements of users, and provide different service qualities for different data streams: give priority to the processing of real-time and important data packets; For ordinary data packets with weak real-time performance, it provides lower processing priority and even discards them in case of network congestion.

6.2.1.2 QoS service model

Generally, QoS provides three service models: Best-Effort Service model, Integrated Service model (Int-Service) and Differentiated Service model (Diff Serv)

(1) Best-Effort Service. Best-Effort Service is a single service model and the simplest service model. For the Best-Effort Service model, the network sends messages as much as possible. However, it does not provide any guarantee for delay, reliability and other performance. The Best-Effort Service model is the default service model of the network, which is implemented through FIFO queue. It is suitable for most network applications, such as FTP, E-mail and so on.

(2) Integrated Service. Inter Serv is an integrated service model, which can meet a variety

of QoS requirements. The model uses resource reservation protocol (RSVP), which runs on each device from the source end to the destination end, and can monitor each flow to prevent it from consuming too much resources. This system can clearly distinguish and ensure the service quality of each business flow, and provide the most fine-grained service quality differentiation for the network. For example, when using VoIP, 12K bandwidth and delay within 100ms are needed. The integrated service model will classify it into a preset service level. However, the Inter-Serv model has high requirements for devices. When the number of data streams in the network is large, the storage and processing capacity of devices will encounter great pressure. Inter-Serv model has poor scalability and is difficult to implement in Internet core network. This solution of bandwidth reservation for a single data stream is difficult to implement on the Internet, so the model has not been used since it was launched in 1994.

(3) Differentiated Service. Diff-Serv is a multi service model composed of a series of technologies, which can meet different QoS requirements. Unlike Inter-Serv, it does not need to notify the network to reserve resources for each service.

Differentiated services have simple implementation and good scalability. Different methods can be used to specify the QoS of the message, such as the priority of the IP packet, the source address and destination address of the message, etc. The network uses this information to classify messages, traffic shaping, traffic supervision and queuing.

6.2.1.3 Overview of QoS technology

(1) Flow classification: Adopting certain rules to identify messages that conform to certain types of characteristics, it is the prerequisite and basis of differentiated services for network services.

(2) Flow supervision: supervise the specific flow into or out of the equipment. When the traffic exceeds the set value, restrictive or punitive measures can be taken to protect network resources from damage. It can act on the input direction and output direction of the interface.

(3) Traffic shaping: a flow control measure that actively adjusts the output rate of the flow. It is used to adapt the flow to the network resources that can be supplied by the downstream equipment and avoid unnecessary message discarding and delay. It usually acts on the outgoing direction of the interface.

(4) Congestion management: when congestion occurs, to formulate a resource scheduling strategy to determine the processing order of message forwarding, which usually acts on the outgoing direction of the interface.

(5) Congestion avoidance: monitor the use of network resources. When congestion is found to be increasing, take the strategy of actively discarding packets, and relieve the overload of the network by adjusting the queue length, which usually acts on the outgoing direction of the interface.

Among these QoS technologies, traffic classification and marking are the basis and the premise of differentiated service implementation; Other QoS technologies control the network traffic and its allocated resources from different aspects, and provide services differently.

The processing sequence of QoS technology in network equipment: Identify and distinguish various services through flow classification, which is the basis of subsequent actions; Process the characteristic business through various actions. These actions need to be associated with flow classification to be meaningful. The specific action is related to the stage and the current load condition of the network. For example, when the message enters the network, conduct traffic supervision; Flow shaping before exiting the node; Congestion management of queue in case of congestion; Take congestion avoidance measures when congestion intensifies.

6.2.1.4 Traffic classification and QoS marking

Traffic classification is to divide data packets into multiple priorities or multiple service classes. The network manager can set the policy of traffic classification. This policy can include not only the IP priority or DSCP value of IP message, the CoS value of 802.1p and other in band signaling, but also the input interface, source IP address, destination IP address, MAC address, IP protocol or application port number. The result of classification has no range limit. It can be a narrow range of streams determined by five tuples (source IP address, source port number, protocol number, destination IP address and destination port number), or all packets to a certain network segment.

The purpose of marking at the network boundary is to distinguish the data and indicate the differences between them. In this way, within the network, the queue technology can divide the data into corresponding queues according to this mark for different processing. There is a special field in the IP message to mark the QoS, which is TOS in IPv4 and Traffic Class in IPv6. The TOS field marks DSCP with the first 6 bits. If only the first 3 bits are used, it is the IP priority. DSCP and IP priority are the standards of marking. IP priority provides 8 kinds of quality of service from 0 to 7, and 6 and 7 are reserved. Therefore, 0 to 5 is commonly used. Each number corresponds to a name, for example, 0 corresponds to route. In this way, when changing packet priority and other configurations, you can use both numbers and names. IP priority and DSCP cannot be set at the same time. If they are set at the same time, only DSCP takes effect. When the packet marked with DSCP reaches the router that can only identify IP priority, it will only look at the first 3 bits, and both IP priority and DSCP form a mapping with their first 3 bits and COS values of layer-2. In layer-2, the cos field is used for marking. The normal Ethernet frame is unmarked, but there are 3 bits in the header of ISL and the tag of 802.1Q to define the service level, from 0 to 7, but only 0 to 5 is available, and 6 and 7 are reserved.

6.2.1.5 Several common queue scheduling mechanisms

(1) FIFO (First In First Out Queuing). The traditional best effort service strategy is applied to the interface with bandwidth greater than 2.048M by default. It is only applicable to the traffic insensitive to bandwidth and delay, such as WWW, FTP, E-mail, etc. FIFO does not classify messages. When the rate of messages entering the interface is greater than the sending rate of the interface, FIFO allows messages to enter the queue according to the order of messages arriving at the interface, and allows messages to leave the queue according to the order of entering the queue at the exit of the queue. The sending (order) relationship of messages in each queue defaults to FIFO first in first out queue.

(2) PQ (Priority Queuing). PQ queue is designed for key business applications. There is a key feature to reduce service delay when congestion occurs. PQ can specify priority according to network protocol (such as IP, IPX), data inflow interface, message length, source address/destination address, etc. flexibly.

The four types of messages correspond to four queues: high priority queue, medium priority queue, normal priority queue and low priority queue. The message of the next priority can only be sent after all the messages in the high priority queue are sent. Although such a mechanism can ensure that key data is always processed first, low priority queues are likely to be very congested. By default, data flows enter the normal priority queue.

(3) CQ (Custom Queuing). CQ divides the group into 16 categories (corresponding to 16 queues) according to certain rules, and the group enters the corresponding CQ queue according to its own category and the first in first out strategy. Queues 1 to 16 of CQ are user queues. Users can configure the rules of flow classification and specify the proportion of interface or PVC bandwidth occupied by 16 user queues. In queue scheduling, packets in the system queue are sent first. Until the system queue is empty, a certain number of packets are taken out from user queues 1 to 16 by polling according to the pre configured bandwidth ratio. In this way, the packets of different services can obtain different bandwidth, which can not only ensure that key services can obtain more bandwidth, but also prevent non key services from obtaining bandwidth. By default, the data flow enters queue 1. Another advantage of customized queue is that it can allocate bandwidth according to the busy degree of business, which is suitable for applications with special requirements for bandwidth. Although the scheduling of 16 user queues is polling, the service time slice is not allocated to each queue - if one queue is empty, it will immediately switch to the next queue scheduling. When there is no CQ scheduling mechanism for some types of messages, it can automatically increase the bandwidth of some existing types of messages.

(4) WFQ (Weighted Fair Queuing). WFQ is introduced to share network resources fairly and optimize the delay and jitter of all streams as much as possible. It takes care of the interests of all aspects, which is mainly reflected in: different queues get fair scheduling opportunities and balance the delay of each flow as a whole. Fair scheduling of short messages and long messages: if there are multiple long messages and short messages waiting to be sent between different queues at the same time, the interests of short messages should be taken into account and short messages should be scheduled first, so as to reduce the jitter between messages of each flow as a whole.

WFQ automatically classifies the streams according to the session information of the data stream (the same source IP address, destination IP address, origin slogan, destination port number, protocol number and IP priority messages belong to the same stream), and divides *N* queues as much as possible to put each stream evenly into different queues, so as to balance the delay of each stream as a whole.

When leaving the queue, WFQ allocates the bandwidth that each flow should occupy at the exit according to the priority of the flow (priority or DSCP). The smaller the priority value, the less bandwidth. The greater the bandwidth, the higher the priority. Finally, each queue is polled and a corresponding number of messages are taken from the queue according to the bandwidth ratio for transmission. WFQ is the default queue mechanism for interfaces with a transmission rate below 2.048M.

For example,there are currently five streams in the interface,and their priorities are 0,1,2,3 and4 respectively,so the total bandwidth quota is the sum of all (stream priority+1).That is,1+2+3+4+5=15.The available bandwidth of each stream is 1/15, 2/15, 3/15, 4/15 and 5/15 respectively.

Limitations of WFQ: WFQ does not support tunnels or interfaces with encryption technology, because these technologies need to modify the information used by WFQ for classification in the packet. The accuracy of bandwidth control provided by WFQ is not as good as that provided by CBQ, because it is based on stream classification and queue based bandwidth allocation. Each queue may have multiple streams, so the bandwidth cannot be specified for specific data types.

(5) CBQ (Class-Based Weighted Fair Queuing). CBQ is an extension of WFQ function,which provides users with the support of defining classes.CBQ allocates a separate FIFO reserved queue for each user-defined class to buffer the data of the same class.In case of network congestion, CBQ matches the message according to the user-defined class rules and makes it enter the corresponding queue.Before entering the queue,it must check the congestion avoidance

mechanism and bandwidth limit.When the message goes out of the queue,the message in the queue corresponding to each class is dispatched in a weighted and fair manner.

CBQ provides an emergency queue into which emergency messages are sent.The queue adopts FIFO scheduling and has no bandwidth limit.In this way,if CBQ weights and treats all kinds of queues fairly,delay sensitive data streams such as voice messages may not be sent in time. Therefore,PQ feature is introduced into CBQ,called LLQ (low latency queuing),which provides strict priority sending service for delay sensitive data streams such as voice messages.

LLQ combines the strict priority queue mechanism with CBQ.Users can specify that they enjoy strict priority services when defining classes.Such classes are called priority classes. All priority messages will enter the same priority queue. Before entering the queue,check the bandwidth limit of various messages.When the message leaves the queue,the message in the priority queue will be sent first, and the messages in the priority queue are not sent until the messages in the priority queue have been sent. When sending other queue messages,it will still be scheduled in a weighted fair manner.

In order not to delay messages in other queues too long,when using LLQ,the maximum available bandwidth will be specified for each priority class,which is used to supervise traffic when congestion occurs.If congestion does not occur, Priority classes are allowed to use more than their allocated bandwidth. If congestion occurs, packets whose priority class exceeds the allocated bandwidth will be discarded.LLQ can also specify Burst-size.

CBQ makes some improvements to WFQ:in WFQ, weight is used to indicate the queue priority,while in CBWFQ, weight is used to indicate the priority of a certain type of traffic. Data packets are queued in corresponding classes according to weight.CBQ is a kind of data in a queue,so the corresponding bandwidth can be specified for a certain type of traffic, but WFQ cannot be realized because it is based on the flow. There are many kinds of traffic. Finally, there may be several kinds of traffic in each queue. In addition to IP address and port number, CBQ can also classify data through ACL or data input interface, which cannot be realized by WFQ. PQ and CQ need to be configured manually.It can be seen from the command that the queue cannot be divided according to IP priority or DSCP,and the configuration and the command are cumbersome. There is only one command in WFQ,and the execution is basically automatic,but it is not easy to control the traffic,and QoS marking should be carried out in advance.In CBQ, the pre-marking of QoS is not necessary, because the concept of MQC is introduced.The specific data flow is matched through the structured command line (if the matching is IP priority or DSCP, the pre-marking of QoS is required) and then the detailed processing strategy is formulated.However,the back-up queue is completed automatically.CBQ is the recommended mode at present.

6.2.2 MSTP+VRRP

MSTP is a multi spanning tree protocol.Compared with spanning tree and fast spanning tree,MSTP introduces the concept of "instance".Spanning tree and fast spanning tree are based on switch port technology. When calculating spanning tree, all VLANs share the same spanning tree; MSTP is instance based. The so-called "instance" refers to a set corresponding to multiple VLANs. MSTP divides one or more VLANs of a device into an instance. Devices with the same instance configuration form an MST domain and run an independent spanning tree; This MST constitutes a large device as a whole, and carries out spanning tree algorithm with other MST domains to obtain an overall spanning tree. All VLANs are mapped to instance 0 by default, and other instances are called multi spanning tree instances.

In the LAN, the messages sent by the host to other network segments are forwarded by the gateway. When the gateway fails, all data sent to the gateway in this network segment will be interrupted. In order to avoid network interruption, multiple gateways can be set on the host, but only one default gateway can be set on a host. Therefore, it needs to be added and modified manually by the administrator, which greatly increases the complexity of network management. Therefore, we usually use VRRP (virtual router redundancy) protocol to ensure that users can switch to another gateway quickly, continuously and transparently.

VRRP protocol first selects the master router by election. The master router is responsible for providing the actual data forwarding service. After the master router is selected, other devices are used as the backup router, and the status of the master routing device is monitored through the VRRP message sent by the master routing device regularly. If the backup routing device in the group does not receive the message from the main routing device within the specified time, it will change its status to master. Because the switching is very fast and the user terminal does not need to change the IP address and MAC address of the default gateway, it is transparent to the user.

The primary route is responsible for forwarding data, while the backup route is not responsible for forwarding data. Therefore, while the primary route forwards data, the backup route is always idle, which is bound to cause a waste of network bandwidth resources. By using load balancing technology in VRRP, we create different VRRP groups, so that routers play different roles in different VRRP groups.

The networking mode of VRRP combined with MSTP can not only provide redundant backup for gateway equipment, but also provide redundant backup for downlink layer-2 links, and use MSTP technology to block redundant links in the network to eliminate layer-2 loops. The reliability of the network is greatly improved.

6.3 Project implementation

6.3.1 Project requirements

(1) The entire network uses static routes to exchange information.

(2) The core adopts VRRP technology to back up gateway devices and improve the reliability of gateways.When one gateway device fails, hosts within the LAN can still access the external network through another gateway device.

(3) When the gateway device is working normally, the zone A users forward data through the gateway device Switch A.Zone B users forward data through Switch B to achieve traffic load sharing.

(4) When the uplink of a gateway device fails, lower the priority of the gateway device to prevent the gateway device from becoming the master and causing traffic forwarding to be interrupted.

(5) Redundant backup of Layer-2 links is performed in the LAN to ensure that traffic forwarding will not be interrupted when the downlink of the gateway device fails.Use MSTP technology to avoid loops in Layer-2 networks.

(6) The gateway device is connected to the Internet through the egress device, and the traffic restriction of QoS on the egress router is carried out.

6.3.2 Device inventory

The main equipment used in this project includes core switches, access layer switches, routers, etc., and the specific number of models is shown in Table 6.1.

Table 6.1 List of network devices

Serial No.	Device model	Quantity	Function	Remark
1	AR3260	1	Egress router	
2	S7706	2	Core switch	
3	S5700–52P–LI–AC	30	Access switch	

6.3.3 Project topology

The network topology of this project is shown in Figure 6.1.

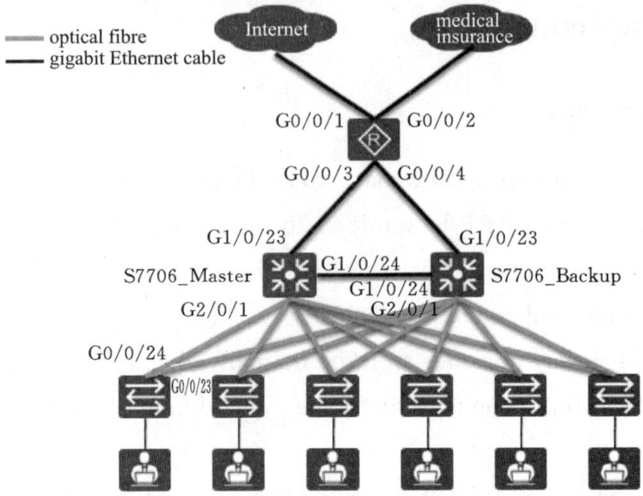

Figure 6.1 Topology diagram of hospital network in a county

6.3.4 Configuration ideas

The configuration is as follows:

(1) The two cores are exchanged to form VRRP.

(2) Loop destruction using MSTP.

(3) Traffic restriction on QoS on the egress router.

(4) Configure NQA to link with static routes to achieve fast route switchover.

6.3.5 Planning of relevant information

6.3.5.1 Core switch

Table 6.2 shows the information planning of the core switch.

Table 6.2 Core switch information planning table

Serial No.	Use departments	VLAN No.	IP address	Mask	Remark
1	Outpatient charges	10	172.16.10.254	255.255.255.0	
2	Outpatient emergency department	20	172.16.20.254	255.255.255.0	
3	Pharmacy	30	172.16.30.254	255.255.255.0	
4	Management	255	172.16.255.1	255.255.255.0	

continue

Serial No.	Use departments	VLAN No.	IP address	Mask	Remark
5	Interconnection1	1000	172.16.250.1	255.255.255.252	Interconnect with the router
6	Interconnection2	1001	172.16.251.1	255.255.255.252	Interconnect with the router
7	Interconnection2	1002	172.16.254.1	255.255.255.252	Interconnect with the Core switch

6.3.5.2 Egress router

The information planning of the egress router is shown in Table 6.3.

Table 6.3 Egress router information planning table

Serial No.	Interface	IP address	Mask	Remark
1	G0/0/3	172.16.250.2	255.255.255.252	Connect with Master
2	G0/0/4	172.16.251.2	255.255.255.252	Connect with Backup

6.3.5.3 ISP export planning

Table 6.4 shows the information planning of ISP exports.

Table 6.4 ISP export information planning table

Serial No.	Interface	IP address	Mask	Gateway	Remark
1	G0/0/1	1.1.25.67	255.255.255.192	1.1.25.1	Static IP address Connect to the Internet
2	G0/0/2	1.1.125.2	255.255.255.0	1.1.125.1	Connect to medical insurance, and the address of the peer medical insurance business is 1.1.125.0/24

6.3.6 Procedure

6.3.6.1 Egress device configuration

(1) Configure the IP address for the egress of the Internet.

```
<Huawei> system-view
```

```
[Huawei] sysname WZCWXRMYY
[WZCWXRMYY_AR3260] interface gigabitethernet 0/0/1
[WZCWXRMYY_AR3260-GigabitEthernet0/0/1] ip address 1.1.25.67  255.255.255.192
[WZCWXRMYY_AR3260] interface gigabitethernet 0/0/2
[WZCWXRMYY_AR3260-GigabitEthernet0/0/2] ip address 1.1.125.2 255.255.255.0
```

(2) Configure Internet ACLs.

```
[WZCWXRMYY _AR3260] acl 3001
[WZCWXRMYY _AR3260-acl-adv-3001] rule 10 permit ip source 172.16.0.00.0.255.255
```

(3) NAT address translation.

```
[WZCWXRMYY _AR3260] interface gigabitethernet 0/0/1
[WZCWXRMYY _AR3260-GigabitEthernet0/0/1] nat outbound 3001
[WZCWXRMYY _AR3260-GigabitEthernet0/0/1] quite

[WZCWXRMYY _AR3260] interface gigabitethernet 0/0/2
[WZCWXRMYY _AR3260-GigabitEthernet0/0/2] nat outbound 3001
[WZCWXRMYY _AR3260-GigabitEthernet0/0/2] quite
```

(4) Configure the intranet interconnect address.

```
[WZCWXRMYY _AR3260] interface gigabitethernet 0/0/3
[WZCWXRMYY _AR3260-GigabitEthernet0/0/3] ip address 172.16.250.1 255.255.255.252
[WZCWXRMYY _AR3260] interface gigabitethernet 0/0/4
[WZCWXRMYY _AR3260-GigabitEthernet0/0/4] ip address 172.16.251.1 255.255.255.252
```

(5) Configure NQA.

```
[WZCWXRMYY _AR3260] nqa test-instance aa bb
[WZCWXRMYY _AR3260-nqa-aa-bb] test-type icmp
[WZCWXRMYY _AR3260-nqa-aa-bb] destination-address ipv4 172.16.250.1
[WZCWXRMYY _AR3260-nqa-aa-bb] frequency 5
[WZCWXRMYY _AR3260-nqa-aa-bb] probe-count 1
[WZCWXRMYY _AR3260-nqa-aa-bb] start now
[WZCWXRMYY _AR3260-nqa-aa-bb] quit
```

(6) Default route.

```
[WZCWXRMYY _AR3260] ip route-static 0.0.0.0 0.0.0.0 1.1.25.1
[WZCWXRMYY _AR3260] ip route-static 59.0.0.0 255.255.255.0  1.1.125.1
[WZCWXRMYY _AR3260] ip route-static 172.16.0.0 255.255.0.0 172.16.250.1 track nqa aa bb
[WZCWXRMYY _AR3260] ip route-static 172.16.0.0 255.255.0.0 172.16.251.1 preference 100
```

(7) Limit download speed.

```
[WZCWXRMYY _AR3260] interface gigabitethernet 0/0/3
[WZCWXRMYY _AR3260-GigabitEthernet0/0/3]qos car inbound source-ip-address range 172.16.10.2. to 172.16.10.252 per-address cir 4096 cbs 6016 pbs 10016 green pass yellow pass red discard
[WZCWXRMYY _AR3260-GigabitEthernet0/0/3]qos car inbound source-ip-address range 172.16.20.2. to 172.16.20.252 per-address cir 4096 cbs 6016 pbs 10016 green pass yellow pass red discard
[WZCWXRMYY _AR3260-GigabitEthernet0/0/3]qos car inbound source-ip-address range 172.16.30.2. to 172.16.30.252 per-address cir 4096 cbs 6016 pbs 10016 green pass yellow pass red discard
```

6.3.6.2 Core switch (Master) configuration

(1) Configure network interconnection between devices.

```
<Huawei>system-view
[Huawei] sysname WZCWXRMYY_Master
[WZCWXRMYY_Master] vlan batch 10 20 30  255 1000 1001
[WZCWXRMYY_Master] interface gigabitethernet 1/0/23
[WZCWXRMYY_Master-GigabitEthernet1/0/23] port link-type access
[WZCWXRMYY_Master-GigabitEthernet1/0/23] port default vlan 1000
```

(2) Configure address pools.

```
[WZCWXRMYY_Master]ip pool vlan10
[WZCWXRMYY_Master-ip-pool-vlan10]gateway-list 172.16.10.254
[WZCWXRMYY_Master-ip-pool-vlan10]network 172.16.10.0 mask 255.255.255.0
[WZCWXRMYY_Master-ip-pool-vlan10]excluded-ip-address 172.16.10.252 172.16.10.253    #Exclude addresses
[WZCWXRMYY_Master]ip pool vlan20
[WZCWXRMYY_Master-ip-pool-vlan20]gateway-list 172.16.20.254
[WZCWXRMYY_Master-ip-pool-vlan20]network 172.16.20.0 mask 255.255.255.0
[WZCWXRMYY_Master-ip-pool-vlan20]excluded-ip-address 172.16.20.252 172.16.20.253

[WZCWXRMYY_Master]ip pool vlan30
[WZCWXRMYY_Master-ip-pool-vlan30]gateway-list 172.16.30.254
[WZCWXRMYY_Master-ip-pool-vlan30]network 172.16.30.0 mask 255.255.255.0
[WZCWXRMYY_Master-ip-pool-vlan30]excluded-ip-address 172.16.30.252 172.16.30.253
```

(3) Configure the interconnect IP.

```
[WZCWXRMYY_Master] interface vlanif 1000
[WZCWXRMYY_Master -Vlanif1000] ip address 172.16.250.1 255.255.255.252
```

```
[WZCWXRMYY_Master-Vlanif1000] quit

[WZCWXRMYY_Master] interface vlanif 254
[WZCWXRMYY_Master-Vlanif1000] ip address 172.16.254.1  255.255.255.0
[WZCWXRMYY_Master-Vlanif1000] quit
```

(4) Configure VRRP.

```
[WZCWXRMYY_Master] interface vlanif 10
[WZCWXRMYY_Master-Vlanif10] ip address 172.16.10.253
[WZCWXRMYY_Master-Vlanif10]vrrp vrid 10 virtual-ip 172.16.10.254
[WZCWXRMYY_Master-Vlanif10]vrrp vrid 10 priority 200
[WZCWXRMYY_Master-Vlanif10]dhcp select global
[WZCWXRMYY_Master-Vlanif10] quit

[WZCWXRMYY_Master] interface vlanif 20
[WZCWXRMYY_Master-Vlanif20] ip address 172.16.20.253
[WZCWXRMYY_Master-Vlanif20]vrrp vrid 20 virtual-ip 172.16.20.254
[WZCWXRMYY_Master-Vlanif20]vrrp vrid 20 priority 200
[WZCWXRMYY_Master-Vlanif20]dhcp select global
[WZCWXRMYY_Master-Vlanif20] quit

[WZCWXRMYY_Master] interface vlanif 30
[WZCWXRMYY_Master-Vlanif30] ip address 172.16.30.253
[WZCWXRMYY_Master-Vlanif30]vrrp vrid 30 virtual-ip 172.16.30.254
[WZCWXRMYY_Master-Vlanif30]vrrp vrid 30 priority 200
[WZCWXRMYY_Master-Vlanif30]dhcp select global
[WZCWXRMYY_Master-Vlanif30] quit
[WZCWXRMYY_Master] interface vlanif 255
[WZCWXRMYY_Master-Vlanif255] ip address 172.16.255.253
[WZCWXRMYY_Master-Vlanif255]vrrp vrid 255 virtual-ip 172.16.30.254
[WZCWXRMYY_Master-Vlanif255]vrrp vrid 255 priority 200
[WZCWXRMYY_Master-Vlanif255] quit
```

(5) Configure NQA.

```
[WZCWXRMYY_Master] nqa test-instance aa bb
[WZCWXRMYY_Master-nqa-aa-bb] test-type icmp
```

```
[WZCWXRMYY_Master-nqa-aa-bb] destination-address ipv4 172.16.250.2
[WZCWXRMYY_Master-nqa-aa-bb] frequency 5
[WZCWXRMYY_Master-nqa-aa-bb] probe-count 1
[WZCWXRMYY_Master-nqa-aa-bb] start now
[WZCWXRMYY_Master-nqa-aa-bb] quit
```

(6) Configure the default route.

```
[WZCWXRMYY_Master]ip route-static 0.0.0.0 0.0.0.0 192.168.250.2 rack nqa aa bb
[WZCWXRMYY_Master]ip route-static 0.0.0.0 0.0.0.0 192.168.254.2 preference 100
```

(7) Configure Trunk.

```
[WZCWXRMYY_Master] interface gigabitethernet 2/0/1
[WZCWXRMYY_Master-GigabitEthernet2/0/1]port link-type trunk
[WZCWXRMYY_Master-GigabitEthernet2/0/1]port trunk allow-pass all

[WZCWXRMYY_Master] interface gigabitethernet 1/0/24
[WZCWXRMYY_Master-GigabitEthernet1/0/24]port link-type trunk
[WZCWXRMYY_Master-GigabitEthernet1/0/24]port trunk allow-pass all
```

(8) Configure MSTP.

```
[WZCWXRMYY_Master] stp region-configuration
[WZCWXRMYY_Master-mst-region] region-name RG1
[WZCWXRMYY_Master-mst-region] instance 1 vlan 10
[WZCWXRMYY_Master-mst-region] instance 1 vlan 20
[WZCWXRMYY_Master-mst-region] instance 1 vlan 30
[WZCWXRMYY_Master-mst-region] instance 1 vlan 255
[WZCWXRMYY_Master-mst-region] active region-configuration
[WZCWXRMYY_Master-mst-region] quit
[WZCWXRMYY_Master] stp instance 1 root primary
[WZCWXRMYY_Master] stp enable
```

6.3.6.3 Core switch (Backup) configuration

(1) Configure network interconnection between devices.

```
<Huawei> system-view
[Huawei] sysname WZCWXRMYY_Backup
[WZCWXRMYY_Backup] vlan batch 10 20 30 254 255 1000 1001
[WZCWXRMYY_Backup] interface gigabitethernet 1/0/23
[WZCWXRMYY_Backup-GigabitEthernet1/0/23] port link-type access
```

```
[WZCWXRMYY_Backup-GigabitEthernet1/0/23] port default vlan 1001
```

(2) Configure address pools.

```
[WZCWXRMYY_Backup]ip pool vlan10
[WZCWXRMYY_Backup-ip-pool-vlan10]gateway-list 172.16.10.254
[WZCWXRMYY_Backup-ip-pool-vlan10]network 172.16.10.0 mask 255.255.255.0
[WZCWXRMYY_Backup-ip-pool-vlan10]excluded-ip-address 172.16.10.252 172.16.10.253

[WZCWXRMYY_Backup]ip pool vlan20
[WZCWXRMYY_Backup-ip-pool-vlan20]gateway-list 172.16.20.254
[WZCWXRMYY_Backup-ip-pool-vlan20]network 172.16.20.0 mask 255.255.255.0
[WZCWXRMYY_Backup-ip-pool-vlan20]excluded-ip-address 172.16.20.252 172.16.20.253

[WZCWXRMYY_Backup]ip pool vlan30
[WZCWXRMYY_Backup-ip-pool-vlan30]gateway-list 172.16.30.254
[WZCWXRMYY_Backup-ip-pool-vlan30]network 172.16.30.0 mask 255.255.255.0
[WZCWXRMYY_Backup-ip-pool-vlan30]excluded-ip-address 172.16.30.252 172.16.30.253
```

(3) Configure the interconnect IP.

```
[WZCWXRMYY_Backup] interface vlanif 1001
[WZCWXRMYY_Backup-Vlanif1001] ip address 172.16.251.1 255.255.255.252
[WZCWXRMYY_Backup-Vlanif1001] quit

[WZCWXRMYY_Backup] interface vlanif 254
[WZCWXRMYY_Backup-Vlanif254] ip address 172.16.254.2  255.255.255.0
[WZCWXRMYY_Backup-Vlanif254] quit
```

(4) Configure VRRP.

```
[WZCWXRMYY_Backup] interface vlanif 10
[WZCWXRMYY_Backup-Vlanif10] ip address 172.16.10.252
[WZCWXRMYY_Backup-Vlanif10]vrrp vrid 10 virtual-ip 172.16.10.254
[WZCWXRMYY_Backup-Vlanif10]dhcp select global
[WZCWXRMYY_Backup-Vlanif10] quit

[WZCWXRMYY_Backup] interface vlanif 20
[WZCWXRMYY_Backup-Vlanif20] ip address 172.16.20.252
[WZCWXRMYY_Backup-Vlanif20]vrrp vrid 20 virtual-ip 172.16.20.254
```

```
[WZCWXRMYY_Backup-Vlanif20]dhcp select global
[WZCWXRMYY_Backup-Vlanif20] quit

[WZCWXRMYY_Backup] interface vlanif 30
[WZCWXRMYY_Backup-Vlanif30] ip address 172.16.30.252
[WZCWXRMYY_Backup-Vlanif30]vrrp vrid 30 virtual-ip 172.16.30.254
[WZCWXRMYY_Backup-Vlanif30]dhcp select global
[WZCWXRMYY_Backup-Vlanif30] quit

[WZCWXRMYY_Backup] interface vlanif 255
[WZCWXRMYY_Backup-Vlanif255] ip address 172.16.255.252
[WZCWXRMYY_Backup-Vlanif255]vrrp vrid 30 virtual-ip 172.16.255.254
[WZCWXRMYY_Backup-Vlanif255] quit
```

(5) Configure NQA.

```
[WZCWXRMYY_Backup] nqa test-instance aa bb
[WZCWXRMYY_Backup-nqa-aa-bb] test-type icmp
[WZCWXRMYY_Backup-nqa-aa-bb] destination-address ipv4 172.16.254.2
[WZCWXRMYY_Backup-nqa-aa-bb] frequency 5
[WZCWXRMYY_Backup-nqa-aa-bb] probe-count 1
[WZCWXRMYY_Backup-nqa-aa-bb] start now
[WZCWXRMYY_Backup-nqa-aa-bb] quit
```

(6) Configure the default route.

```
[WZCWXRMYY_Backup]ip route-static 0.0.0.0 0.0.0.0 192.168.251.2 rack nqa aa bb
[WZCWXRMYY_Backup]ip route-static 0.0.0.0 0.0.0.0 192.168.254.1 preference 100
```

(7) Configure Trunk.

```
[WZCWXRMYY_Backup] interface gigabitethernet 2/0/1
[WZCWXRMYY_Backup-GigabitEthernet2/0/1]port link-type trunk
[WZCWXRMYY_Backup-GigabitEthernet2/0/1]port trunk allow-pass all

[WZCWXRMYY_Backup] interface gigabitethernet 1/0/24
[WZCWXRMYY_Backup-GigabitEthernet1/0/24]port link-type trunk
[WZCWXRMYY_Backup-GigabitEthernet1/0/24]port trunk allow-pass all
```

(8) Configure MSTP.

```
[WZCWXRMYY_Backup] stp region-configuration
```

```
[WZCWXRMYY_Backup-mst-region] region-name RG1
[WZCWXRMYY_Backup-mst-region] instance 1 vlan 10
[WZCWXRMYY_Backup-mst-region] instance 1 vlan 20
[WZCWXRMYY_Backup-mst-region] instance 1 vlan 30
[WZCWXRMYY_Backup-mst-region] instance 1 vlan 255
[WZCWXRMYY_Backup-mst-region] active region-configuration
[WZCWXRMYY_Backup-mst-region] quit

[WZCWXRMYY_Master] stp instance 1 root secondary
[WZCWXRMYY_Master] stp enable
```

6.3.6.4 Access switches (listed)

(1) Configure VLANs.

```
[WZCWXRMYY_MZYF] vlan batch 10 20 30 255
[WZCWXRMYY_MZYF] interface gigabitethernet 0/0/1
[WZCWXRMYY_MZYF-GigabitEthernet0/0/1] port link-type access
[WZCWXRMYY_MZYF-GigabitEthernet0/0/1] port default vlan 10
```

(2) Configure Trunks.

```
[WZCWXRMYY_MZYF] interface gigabitethernet 1/0/24
[WZCWXRMYY_MZYF-GigabitEthernet1/0/24]port link-type trunk
[WZCWXRMYY_MZYF-GigabitEthernet1/0/24]port trunk allow-pass all

[WZCWXRMYY_MZYF] interface gigabitethernet 1/0/23
[WZCWXRMYY_MZYF-GigabitEthernet1/0/23]port link-type trunk
[WZCWXRMYY_MZYF-GigabitEthernet1/0/23]port trunk allow-pass all
[WZCWXRMYY_MZYF-GigabitEthernet1/0/23] stp instance 1 cost 20000
```

(3) Configure MSTP.

```
[WZCWXRMYY_MZYF] stp region-configuration
[WZCWXRMYY_MZYF-mst-region] region-name RG1
[WZCWXRMYY_MZYF-mst-region] instance 1 vlan 10
[WZCWXRMYY_MZYF-mst-region] instance 1 vlan 20
[WZCWXRMYY_MZYF-mst-region] instance 1 vlan 30
[WZCWXRMYY_MZYF-mst-region] active region-configuration
[WZCWXRMYY_MZYF-mst-region] quit
[WZCWXRMYY_MZYF] stp enable
```

```
[GLL_SW_HX] interface vlanif 255
[GLL_SW_HX-Vlanif1000] ip address 172.16.255.3
[GLL_SW_HX-Vlanif1000] quit
```

6.4 Ideological and political integration

With the deep going of hospital informatization, the operation of hospital information platform will increasingly rely on the construction of basic networks. The network has become an important way to interact and transmit information about various key data of the hospital. Once the network data is lost, it will bring huge disasters and irreparable losses to hospitals and patients. Therefore, whether the method is objective, whether the process is accurate, whether the operation is rigorous, whether the data is true, the lack of any link will lead to the seriousness of the consequences. College students should cultivate a rigorous and realistic academic style, establish a scientific spirit and craftsman spirit based on facts, seek truth from facts, and be rigorous and serious. At the same time, the hospital should maintain a clean, tidy and beautiful good medical environment, pay attention to the simultaneous use of hands and brains in the process of project implementation, establish labor awareness, strengthen 5S management norms such as Seiri, Seiton, Seiso, Seiketsu and Shitsuke, and strengthen the body with labor.

6.5 Thinking and training

1. What is QoS? What is its role?
2. What are the three service models for QoS?
3. What are the advantages of VRRP and MSTP networking? What can be improved?
4. Redundant backup of Layer-2 links is performed in the LAN to ensure that traffic forwarding will not be interrupted when the downlink of the gateway device fails.To avoid loops in Layer-2 networks using MSTP technology, write out the basic steps of configuration.
5. When the uplink of a gateway device fails, lower the priority of the gateway device to prevent the gateway device from becoming the master and causing traffic forwarding to be interrupted.Write out the configuration command.

项目 7　某电力电容器有限责任公司网络工程项目

7.1　项目导入

某电力电容器有限责任公司前身为全民所有制企业——某电力电容器总厂，创建于 1967 年 6 月 1 日。2006 年 9 月，改制为股权结构多元化的公司制企业。企业通过改制引入投资、扩大资本规模，增强了企业实力，公司经营更具活力。公司原为机械工业部直属企业，是国家大型二类公司；是我国高压输变电设备制造行业的高压、超高压和特高压电容式电压互感器、耦合电容器、无功补偿及其成套设备两大科研和制造基地之一；也是全国 500 家大型电力机械及设备制造企业之一。

随着科技的发展，公司的规模不断扩大，占地面积不断扩大，管理人员的队伍不断扩大。目前的网络结构已经不能满足当前发展的需求，主要表现在有网络不稳定，网络抖动现象频发，采用的网络线路连接存在成本高、周期长、不易管理、不宜实施等难题。因此建立一个设计规范、功能完备、性能优良、安全可靠、有良好的扩展性与可用性并且具备可管理易维护的网络及系统平台，以高效率、高速度、低成本的方式提高公司员工的工作效率与执行效率是十分必要的。所以采用高端的计算机、网络设备、软件、以先进、成熟的网络通信技术进行组网，利用先进的系统集成技术和管理模式，建立一个高效的办公网络体系是该公司迫在眉睫的任务。

7.2　相关知识点

7.2.1　策略路由

7.2.1.1　概述

策略路由（Policy–Based Routing，PBR）是一种依据用户制定的策略进行路由选择的机制。通过配置策略路由，可以用于提高网络的安全性能和负载分担。

策略路由可以使数据包按照用户指定的策略进行转发。对于某些管理目的，如 QoS 需求或 VPN 拓扑结构，要求某些路由必须经过特定的路径，就可以使用策略路由。例如，一个策略可以指定从某个网络发出的数据包只能转发到某个特定的接口。

策略路由主要有两种：一种是根据路由的目的地址来进行的策略路由，称为目的地址路由；另一种是根据路由源地址来进行策略实施，称为源地址路由。随着策略路由的发展现在有了第三种路由方式：智能均衡的策略方式。

7.2.1.2　策略路由（Policy-Based Routing）与路由策略（Routing Policy）的区别

（1）策略路由的操作对象是数据包。

（2）策略路由是在路由表已经产生的情况下，不按照路由表进行转发，而是根据需要，依照某种策略改变数据包转发路径。

（3）路由策略的操作对象是路由信息。

（4）路由策略主要实现了路由过滤和路由属性设置等功能，它通过改变路由属性（包括可达性）来改变网络流量所经过的路径。

（5）传统的路由转发原理是首先根据报文的目的地址查找路由表，然后进行报文转发。但是目前越来越多的用户希望能够在传统路由转发的基础上根据自己定义的策略进行报文转发和选路。

7.2.1.3　策略路由优点

（1）可以根据用户实际需求制定策略进行路由选择，增强路由选择的灵活性和可控性。

（2）可以使不同的数据流通过不同的链路进行发送，提高链路的利用效率。

（3）在满足业务服务质量的前提下，选择费用较低的链路传输业务数据，从而降低企业数据服务的成本。

7.2.1.4　策略路由的类型

华为 AR 系列支持三种策略路由，具体如下：

（1）本地策略路由。本地策略路由是指仅对本地发送的报文有效的策略路由，对转发的报文不起作用。一条本地策略路由可以配置多个策略点，并且这些策略点具有不同的优先级，本机发送的报文优先匹配优先级高的策略点。本地策略路由支持 acl 或报文长度匹配规则。

本地路由策略会根据本地策略路由节点的优先级依次匹配各个节点绑定的匹配规则发送报文，如果没有找到本地策略路由节点，则按照 IP 报文的一般流程根据目的地址查找路由。本地策略路由节点优先级顺序为（高到低）：报文优先级→出接口→下一跳→缺省出接口。

（2）接口策略路由。仅对转发的报文生效，对本地发送的报文不起作用，且只对接口入方向的报文生效。接口策略路由是通过在流行为中配置重定向实现的。

（3）智能策略路由。智能策略路由（Smart Policy Routing SPR）可以主动探测链路质量并匹配业务的需求，从而选择一条最优链路转发业务数据，可以有效地避免网络黑洞、网络震荡等问题。此功能需要购买华为 license 授权。智能策略路由是基于业务需求的策略路由，通过匹配链路质量和网络业务对链路质量的需求，实现智能选路。

7.2.1.5　本地策略路由配置

配置策略路由可以将到达接口的转发报文重定向到指定的下一跳地址。通过配置重定向，设备将符合流分类规则的报文重定向到指定的下一跳地址。包含重定向动作的流策略只能在全局、接口或 VLAN 的入方向上应用。

（1）创建策略路由和策略点，方法如下：

```
[Huawei]policy-based-route 1 deny node 1
```

（2）设置本地策略匹配规则，方法如下：

```
[Huawei-policy-based-route-1-1]if-match?
acl Access control list #根据 IP 报文中的 acl 匹配
packet-length  Match packet length #根据 IP 报文长度匹配
[Huawei-policy-based-route-1-1]if-match packet-length?
INTEGER<0-65535>  Minimum packet length #最短报文长度
[Huawei-policy-based-route-1-1]if-match packet-length 100?
INTEGER<1-65535>  Maximum packet length# 最长报文长度
```

（3）设置报文的出接口，方法如下：

```
[Huawei-policy-based-route-1-1]apply output-interface? #直接设定出接口
[Huawei-policy-based-route-1-1]apply default output-interface ? #缺省出接口
```

#报文的出接口，匹配成功后将从指定接口发出去。接口不是能以太网等广播类型接口（改为 P2P 即可），因为多个下一跳可能导致报文转发不成功

（4）设置报文的下一跳，方法如下：

```
[Huawei-policy-based-route-1-1]apply ip-address?
default Set default information #缺省下一跳，仅对在路由表中未查到的路由报文起作用
next-hop  Next hop address# 直接设定下一跳
```

（5）设置 VPN 转发实例，方法如下：

```
[Huawei-policy-based-route-1-1]apply access-vpn vpn-instance?
STRING<1-31> VPN instance name
```

（6）设置 IP 报文优先级，方法如下：

```
[Huawei-policy-based-route-1-1]apply ip-precedence?
INTEGER<0-7>    IP precedence value
critical        Set packet precedence to critical(5)（关键）
flash           Set packet precedence to flash(3)（闪速）
flash-override  Set packet precedence to flash override(4)（疾速）
immediate       Set packet precedence to immediate(2)（快速）
internet        Set packet precedence to Internet control(6)（网间）
```

```
network          Set packet precedence to network control(7)（网内）
priority         Set packet precedence to priority(1)（优先）
routine          Set packet precedence to routine(0)（普通）
```

（7）设置本地策略路由刷新 LSP 信息时间间隔，方法如下：

```
[Huawei]ip policy-based-route refresh-time?
INTEGER<1000-65535>  Refresh time value(ms)
<cr>
```

（8）应用本地策略路由，方法如下：

```
[Huawei]ip local policy-based-route ?
STRING<1-19>  Policy name
#一台路由器只能使能一个本地策略路由（可以创建多条）
```

7.2.1.6　接口策略路由配置

（1）设置流分类，方法如下：

```
[Huawei]traffic classifier test1 operator ?#逻辑运算符
and  Rule of matching all of the statements #与关系
or   Rule of matching one of the statements #或关系
```

（2）设置分类匹配规则，方法如下：

```
[Huawei-classifier-test1]if-match?
8021p                 Specify vlan 802.1p to match
acl                   Specify ACL to match
any                   Specify any data packet to match
app-protocol          Specify app-protocol to match
cvlan-8021p           Specify inner vlan 802.1p of QinQ packets to match
cvlan-id              Specify inner vlan id of QinQ packets to match  #基于内层VLAN ID分类
匹配规则
destination-mac       Specify destination MAC address to match
dlci                  Specify a DLCI to match
dscp                  Specify DSCP (DiffServ CodePoint) to match
fr-de                 Specify FR DE to match
inbound-interface     Specify an inbound interface to match
ip-precedence         Specify IP precedence to match
ipv6                  Specify ip v6
l2-protocol           Specify layer-2 protocol to match
mpls-exp              Specify MPLS EXP value to match
```

```
protocol              Specify ipv4 or ipv6 packets to match
protocol-group        Specify protocol-group to match
pvc                   Specify a PVC to match
rtp                   Specify RTP port to match
source-mac            Specify source MAC address to match
tcp                   Specify TCP parameters to match
vlan-id               Specify a vlan id to match #基于外出 VLAN ID
```

（3）设置流重定向。将符合流分类规则的报文重定向到指定的下一跳或指定接口。只能在接口的入方向应用。

（4）设置流行为名，方法如下：

```
[Huawei]traffic behavior test2
[Huawei-behavior-test2]
```

（5）重定向符合流分类的报文，方法如下：

```
[Huawei-behavior-test2]redirect?
interface  Status and configuration about the traffic on the interface #重定向到指定接口（3G 或 dialer）
ip-nexthop     Redirect packets to nexthop #重定向到下一跳
ipv6-nexthop   Redirect packets to ipv6 nexthop
-----------
[Huawei-behavior-test2]redirect ip-nexthop 10.1.1.1?
post-nat  Post nat
track     Track nqa #支持与 NQA 联动
<cr>      Please press ENTER to execute command
```

（6）应用流策略，方法如下：

```
[Huawei]traffic policy test3  #设置流策略名
[Huawei-trafficpolicy-test3]classifier test1 behavior test2 #关联流分类和流行为
[Huawei-GigabitEthernet0/0/2]traffic-policy test3 inbound #在接口应用（入方向）
```

7.2.2　端口映射

端口映射就是将外网主机的 IP 地址的一个端口映射到内网中一台机器，提供相应的服务。当用户访问该 IP 的这个端口时，服务器自动将请求映射到对应局域网内部的机器上。内网的一台电脑要上因特网对外开放服务或接收数据，都需要端口映射。

配置代码如下：

```
[Huawei]interface GigabitEthernet 0/0/0
[Huawei-GigabitEthernet0/0/0] nat server protocol tcp global 外网 ip 外网端口 inside 内网 ip 内网端口
```

7.2.3　Eth-Trunk

一台交换机将这多个接口捆绑，形成一个 Eth-Trunk 接口，从而实现了增加带宽和提高可靠性的目的。

如图 7.1 所示，Eth-Trunk 接口连接的链路可以看成是一条点到点的直连链路。

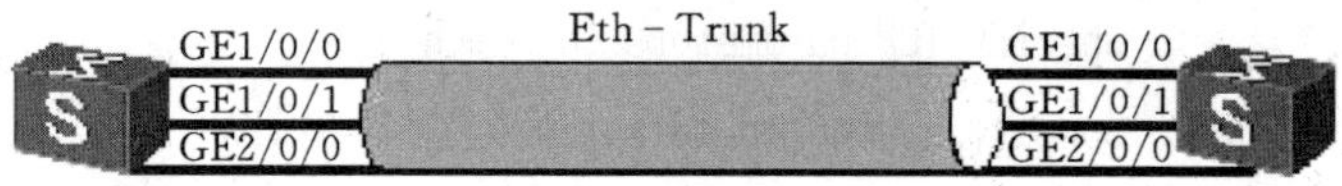

图 7.1　Eth-Trunk 示意图

Eth-Trunk 的优势如下：

（1）负载分担。通过 Trunk 接口可以实现负载分担。在一个 Eth-Trunk 接口内，可以实现流量负载分担。

（2）提高可靠性。当某个成员接口连接的物理链路出现故障时，流量会切换到其他可用的链路上，从而提高整个 Trunk 链路的可靠性。

（3）增加带宽。Eth-Trunk 接口的总带宽是各成员接口带宽之和。

7.2.4　堆叠

7.2.4.1　堆叠简介

堆叠（Intelligent Stack，iStack），是指将多台支持堆叠特性的交换机设备组合在一起，从逻辑上组合成一台交换设备。如图 7.2 所示，SwitchA 与 SwitchB 通过堆叠线缆连接后组成堆叠 iStack，对于上游和下游设备来说，它们就相当于一台交换机 Switch。

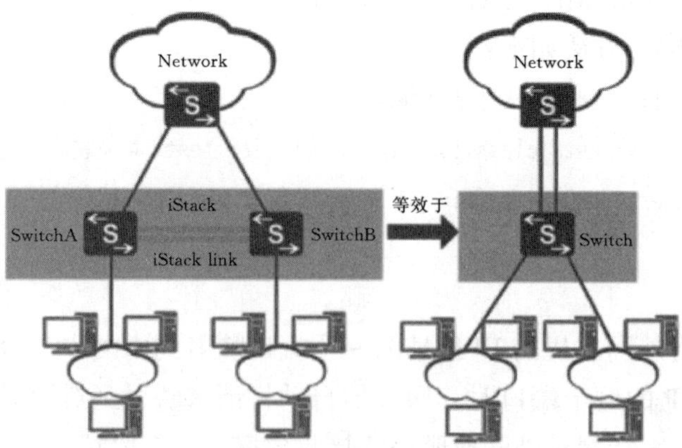

图 7.2　堆叠示意图

通过交换机堆叠，可实现以下效果：

（1）高可靠性。堆叠系统多台成员交换机之间冗余备份；堆叠支持跨设备的链路聚合功能，实现跨设备的链路冗余备份。

（2）强大的网络扩展能力。通过增加成员交换机，可以轻松地扩展堆叠系统的端口数、带宽和处理能力；同时支持成员交换机热插拔，新加入的成员交换机自动同步主交换机的配置文件和系统软件版本。

（3）简化配置和管理。一方面，用户可以通过任何一台成员交换机登录堆叠系统，对堆叠系统所有成员交换机进行统一配置和管理；另一方面，堆叠形成后，不需要配置复杂的二层破环协议和三层保护倒换协议，简化了网络配置。

7.2.4.2　堆叠术语

堆叠涉及以下几个基本术语：

（1）角色。堆叠中所有的单台交换机都称为成员交换机，按照功能不同，可以分为三种角色：

1）主交换机。主交换机（Master）负责管理整个堆叠。堆叠中只有一台主交换机。

2）备交换机。备交换机（Standby）是主交换机的备份交换机。当主交换机故障时，备交换机会接替原主交换机的所有业务。堆叠中只有一台备交换机。

3）从交换机。从交换机（Slave）主要用于业务转发，从交换机数量越多，堆叠系统的转发能力越强。除主交换机和备交换机外，堆叠中其他所有的成员交换机都是从交换机。

（2）堆叠 ID。堆叠 ID，即成员交换机的槽位号（Slot ID），用来标识和管理成员交换机，堆叠中所有成员交换机的堆叠 ID 都是唯一的。

（3）堆叠优先级。堆叠优先级是成员交换机的一个属性，主要用于角色选举过程中确定成员交换机的角色，优先级值越大表示优先级越高，优先级越高当选为主交换机的可能性越大。

7.2.4.3　堆叠建立过程

堆叠建立的过程包括以下 4 个阶段。

1. 物理连接

根据网络需求，选择适当的连接方式和连接拓扑，组建堆叠网络。根据连接介质的不同，堆叠可分为堆叠卡堆叠和业务口堆叠。

如图 7.3 所示，每种连接方式都可组成链形和环形两种连接拓扑。表 7.1 从可靠性、链路带宽利用率和组网布线是否方便的角度对两种连接拓扑进行对比。

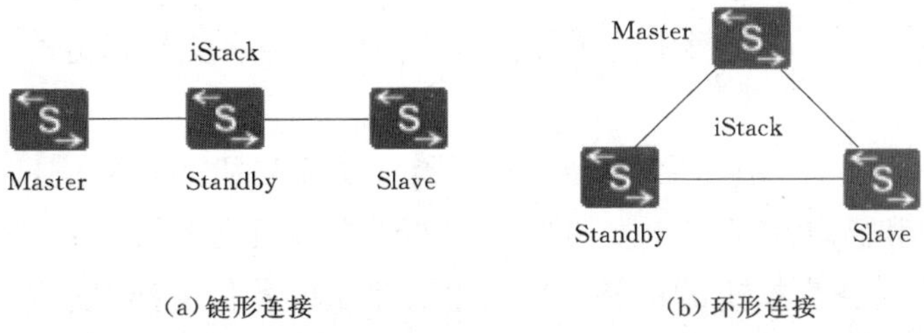

图 7.3　堆叠连接拓扑

表 7.1　　不同连接拓扑优缺点

连接拓扑	优　点	缺　点	适用场景
链形连接	首尾不需要有物理连接，适合长距离堆叠	可靠性低：其中一条堆叠链路出现故障，就会造成堆叠分裂。 堆叠链路带宽利用率低：整个堆叠系统只有一条路径	堆叠成员交换机距离较远时，组建环形连接比较困难，可以使用链形连接
环形连接	可靠性高：其中一条堆叠链路出现故障，环形拓扑变成链形拓扑，不影响堆叠系统正常工作。 堆叠链路带宽利用率高：数据能够按照最短路径转发	首尾需要有物理连接，不适合长距离堆叠	堆叠成员交换机距离较近时，从可靠性和堆叠链路利用率上考虑，建议使用环形连接

2. 选举主交换机

成员交换机之间相互发送堆叠竞争报文，并根据选举原则，选出堆叠系统主交换机。

确定出堆叠的连接方式和连接拓扑，完成成员交换机之间的物理连接之后，所有成员交换机上电。此时，堆叠系统开始进行主交换机的选举。在堆叠系统中每台成员交换机都具有一个确定的角色，其中，主交换机负责管理整个堆叠系统。主交换机选举规则如下（依次从第一条开始判断，直至找到最优的交换机才停止比较）：

（1）运行状态比较。

（2）已经运行的交换机优先处于启动状态的交换机竞争为主交换机。

（3）堆叠主交换机选举超时时间为 20s，堆叠成员交换机上电或重启时，由于不同成员交换机所需的启动时间可能差异比较大，因此不是所有成员交换机都有机会参与主交换机的选举：启动时间与启动最快的成员交换机相比，相差超过 20s 的成员交换机没有机会参与主交换机的选举，只能被动加入堆叠成为非主交换机，加入过程可参见堆叠成员加入与退出。因此，如果希望指定某一成员交换机成为主交换机，则可以先为其上电，待其启动完成后再给其他成员交换机上电。

（4）堆叠优先级高的交换机优先竞争为主交换机。

（5）堆叠优先级相同时，MAC 地址小的交换机优先竞争为主交换机。

3. 拓扑收集和备交换机选举

主交换机收集所有成员交换机的拓扑信息，向所有成员交换机分配堆叠 ID，之后选出堆叠系统备交换机。

主交换机选举完成后，主交换机会收集所有成员交换机的拓扑信息，根据拓扑信息计算出堆叠转发表项和破环点信息下发给堆叠中的所有成员交换机，并向所有成员交换机分配堆叠 ID。之后进行备交换机的选举，作为主交换机的备份交换机。除主交换机外最先完成设备启动的交换机优先被选为备份交换机。当除主交换机外其他交换机同时完成启动时，备交换机的选举规则如下（依次从第一条开始判断，直至找到最优的交换机才停止比较）：

（1）堆叠优先级最高的设备成为备交换机。

（2）堆叠优先级相同时，MAC 地址最小的成为备交换机。

（3）除主交换机和备交换机之外，剩下的其他成员交换机作为从交换机加入堆叠。

4. 稳定运行

主交换机将整个堆叠系统的拓扑信息同步给所有成员交换机，成员交换机同步主交换机的系统软件和配置文件，之后进入稳定运行状态。

角色选举、拓扑收集完成之后，所有成员交换机会自动同步主交换机的系统软件和配置文件：

（1）堆叠具有自动加载系统软件的功能，待组成堆叠的成员交换机不需要具有相同软件版本，只需要版本间兼容即可。当备交换机或从交换机与主交换机的软件版本不一致时，备交换机或从交换机会自动从主交换机下载系统软件，然后使用新系统软件重启，并重新加入堆叠。

（2）堆叠具有配置文件同步机制，备交换机或从交换机会将主交换机的配置文件同步到本设备并执行，以保证堆叠中的多台设备能够像一台设备一样在网络中工作，并且在主交换机出现故障之后，其余交换机仍能够正常执行各项功能。

7.2.4.4　堆叠登录与访问

堆叠建立后，多台成员交换机组成一台虚拟设备存在于网络中，堆叠系统的接口编号规则以及登录与访问的方式都发生了变化。

1. 堆叠接口编号规则

堆叠系统的接口编号采用堆叠 ID 作为标识信息，所有成员交换机的堆叠 ID 都是唯一的。

对于单台没有运行堆叠的设备，接口编号采用：槽位号 / 子卡号 / 端口号（槽位号统一取值为 0）。设备加入堆叠后，接口编号采用：堆叠 ID/ 子卡号 / 端口号。

例如，设备没有运行堆叠时，某个接口的编号为 GigabitEthernet0/0/1；当该设备加入堆叠后，如果堆叠 ID 为 2，则该接口的编号将变为 GigabitEthernet2/0/1。对于管理网口，无论系统是否运行堆叠以及运行堆叠后堆叠 ID 是多少，接口的编号均为 MEth 0/0/1。

子卡号与端口号的编号规则与单机状态下一致。如果设备曾加入过堆叠，在退出堆叠后，仍然会使用组成堆叠时的堆叠 ID 作为自身的槽位号。

2. 堆叠系统的登录

登录堆叠系统的方式如下：

本地登录：通过任意成员交换机的 Console 口登录。

远程登录：通过任意成员交换机的管理网口或其他三层接口登录。只要保证到堆叠系统的路由可达，就可以使用 Telnet、Stelnet、Web 以及 SNMP 等方式进行远程登录。

有管理网口的设备组建堆叠后，系统运行阶段，只有一台成员交换机的管理网口生效，称为主用管理网口。堆叠系统启动后默认选取主成员交换机的管理网口为主用管理网口，若主成员交换机的管理网口异常或不可用，则选取其他成员交换机的管理网口为主用管理网口。如果通过 PC 直连到非主用管理网口，则无法正常登录堆叠系统。

堆叠建立后，竞争为主的交换机的配置文件生效。如果远程登录堆叠，需要主交换机的 IP 地址。

不管通过哪台成员交换机登录到堆叠系统，实际登录的都是主交换机。主交换机负责将用户的配置下发给其他成员交换机，统一管理堆叠系统中所有成员交换机的资源。

3. 堆叠文件系统的访问

文件系统的访问包括对存储器中文件和目录的创建、删除、修改以及文件内容的显示等。设备支持的存储器为 Flash。

通过 drive + path + filename 这种格式，指定到某路径下的文件名。其中，drive 指设备中的存储器；path 指存储器中的目录以及子目录；filename 指文件名。

堆叠环境与单机环境的不同点在于 drive 的命名：flash 指堆叠系统中主交换机 Flash 存储器的根目录；堆叠 ID#flash 指堆叠系统中某成员交换机 Flash 存储器的根目录。例如，slot2#flash 指堆叠 ID 为 2 的成员交换机 Flash 存储器的根目录。

7.2.4.5　堆叠跨设备链路聚合

堆叠支持跨设备链路聚合技术，通过配置跨设备 Eth–Trunk 接口实现。用户可以将不同成员交换机上的物理以太网端口配置成一个聚合端口连接到上游或下游设备上，实现多台设备之间的链路聚合。当其中一条聚合链路故障或堆叠中某台成员交换机故障时，Eth–Trunk 接口能够将流量重新分布到其他聚合链路上，实现了链路间和设备间的备份，保证了数据流量的可靠传输。

如图 7.4 所示，流向网络核心的流量将均匀分布在聚合链路上，当某一条聚合链路失效时，Eth–Trunk 接口将流量通过堆叠线缆重新分布到其他聚合链路上，实现了链路间的备份。如图 7.5 所示，流向网络核心的流量将均匀分布在聚合链路上，当某台成员交换机故障时，Eth–Trunk 接口将流量重新分布到其他聚合链路上，实现了设备间的备份。

7.2.5　边缘端口

边缘端口是指不直接与任何交换机连接，也不通过端口所连接的网络间接与任何交换机相连的端口。可以通过下面两种途径来配置端口为边缘端口或者非边缘端口。用户如果将某个端口指定为边缘端口，那么当该端口由堵塞状态向转发状态迁移时，这个端口可以实现快速迁移，而无需等待延迟时间。用户只能将与终端连接的端口设置为边缘端口。该参数对所有生成树实例有效，也就是说，当端口被配置为边缘端口或非边缘端口时，该端口在所有生成树实例上都被设置为边缘端口或非边缘端口。

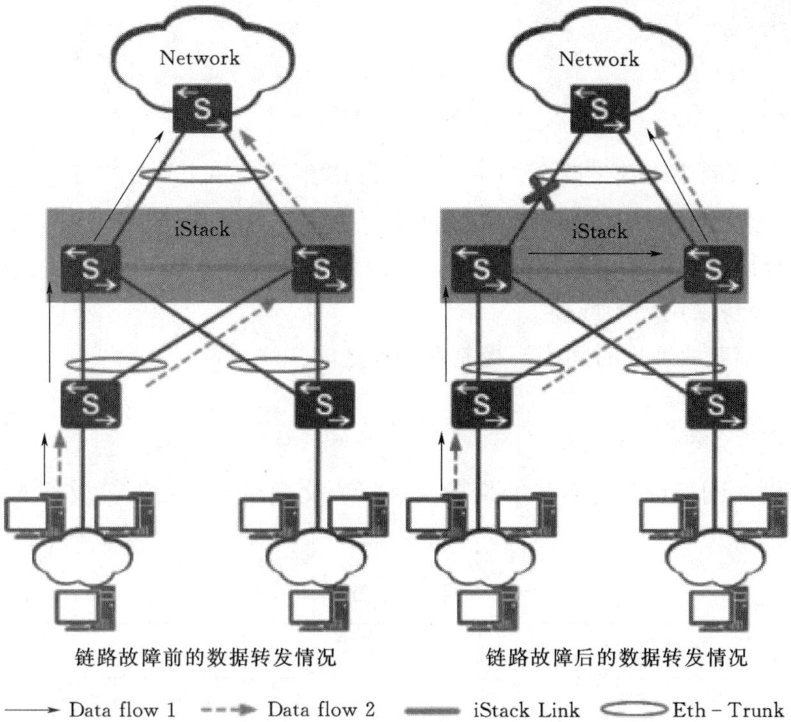

图 7.4　跨设备 Eth-Trunk 接口实现链路间的备份

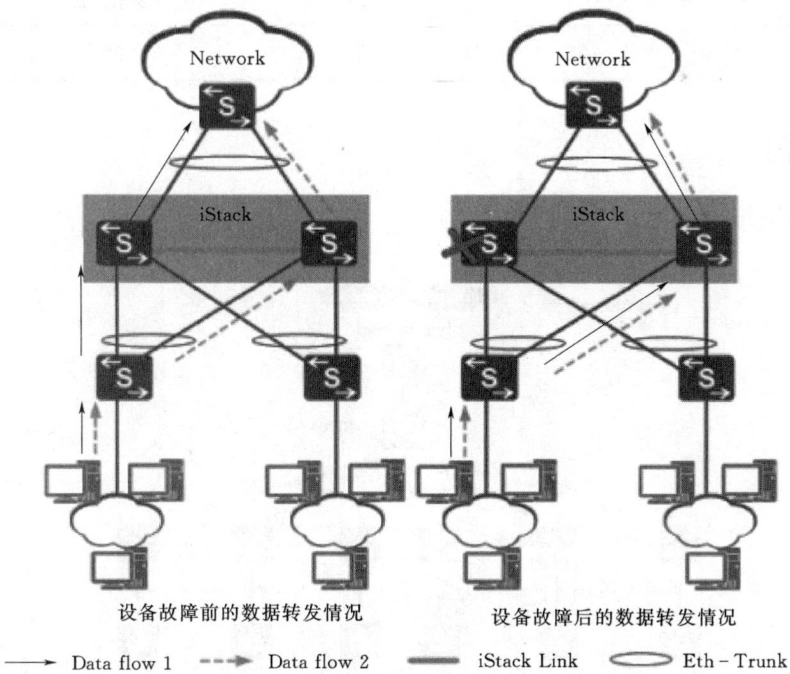

图 7.5　跨设备 Eth-Trunk 接口实现设备间的备份

7.3　项目实施

7.3.1　项目需求

（1）ISP 运营商出口数量两个，一个为静态 IP 形式，一个为 PPPoE 拨号。

（2）要求服务器流量通过静态 IP 地址出口上网，并对相对应的内网服务器进行地址映射，从而使外出人员可以访问内部资源。其他业务网段通过 PPPoE 拨号上网。

（3）接入设备采用环形堆叠。

（4）PC 电脑获取地址时间应当控制在 30s 之内。

7.3.2　设备清单

本项目用到的网络设备包含核心交换机、接入交换机和路由器，设备的型号和数量见表 7.2。

表 7.2　　网络设备清单

序号	设备型号	数量	功能	备注
1	AR3260	1	出口路由器	
2	S7706	1	核心交换机	
3	S5700–28P–LI–AC	30	接入交换机	含堆叠线缆
4	S5700–52P–LI–AC	20	接入交换机	含堆叠线缆

7.3.3　项目拓扑

本项目网络拓扑结构如图 7.6 所示。

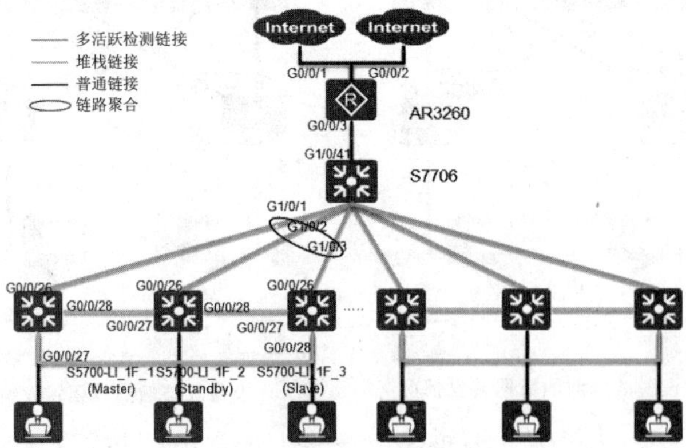

图 7.6　某电力电容器有限责任公司网络拓扑图

7.3.4　配置思路

采用如下的思路配置：

（1）在拨号接口下配置 CHAP 认证，实现设备通过 PPP 认证与 PPPoE Server 建立连接。

（2）配置拨号方式为自动拨号方式，连接断开后，每隔一段时间设备会自动再次尝试建立拨号连接。

（3）分别对多条出接口线路做 NAT 地址转换，使用 ACL 进行源地址的匹配。

（4）路由配置策略路由，实现流量分流。

（5）核心配置 DHCP，实现终端用户自动获取 IP 地址。

（6）通过业务口连接方式组建堆叠时，为了能够在堆叠的成员交换机之间转发数据报文，需要配置逻辑堆叠端口，并添加物理成员端口。

（7）为方便用户管理和识别设备，配置成员交换机的堆叠 ID 和优先级。

（8）将 S5700–LI_1F_1、S5700–LI_1F_2、S5700–LI_1F_2 下电，使用 SFP+ 堆叠电缆连接各物理成员端口后再上电。

（9）为提高可靠性、增加上行链路带宽，配置跨设备 Eth–Trunk。

（10）为有效避免堆叠分裂时网络不可用，配置代理方式多主检测。

（11）开启边缘端口，防止端口 UP&DOWN 触发 STP 的计算。

7.3.5　相关信息规划

7.3.5.1　核心交换机

核心交换机的信息规划见表 7.3。

表 7.3　核心交换机信息规划表

序号	使用部门	VLAN 号	IP 地址	掩码	备注
1	生产部	10	10.10.10.1	255.255.255.0	
2	销售部	11	10.10.20.1	255.255.255.0	
3	财务部	12	10.10.30.1	255.255.255.0	
4	安保部	13	10.10.40.1	255.255.255.0	
5	服务器	100	10.10.100.1	255.255.255.0	
6	管理	255	10.10.255.1	255.255.255.0	
7	互联	1000	10.10.250.1	255.255.255.252	与路由器互联

7.3.5.2　出口路由器

出口路由器的信息规划见表 7.4。

表 7.4　　出口路由器信息规划表

序号	接口	IP 地址	掩码	备　注
1	G0/0/3	10.10.250.2	255.255.255.252	与核心交换机互联

7.3.5.3　ISP 出口规划

出口路由器的信息规划见表 7.5。

表 7.5　　ISP 出 口 信 息 规 划 表

序号	接口	IP 地址	掩码	网关	备注
1	G0/0/1	1.1.1.46	255.255.255.240	1.1.1.33	静态 IP 地址
2	G0/0/2	账号：gxsdxy	密码：123456		PPPoE 拨号

7.3.6　操作步骤

7.3.6.1　出口设备配置

（1）PPPoE 拨号。

1）配置 Dialer 接口，方法如下：

```
<Huawei> system-view
[Huawei] sysname BYDQ_AR3260
[BYDQ_AR3260] interface dialer 0
[BYDQ_AR3260-Dialer1] ppp pap local-user gxsdxy password simple 123456
[BYDQ_AR3260-Dialer1] dialer bundle 1
[BYDQ_AR3260-Dialer1] dialer-group 1
[BYDQ_AR3260-Dialer1] ip address ppp-negotiate
[BYDQ_AR3260-Dialer1] tcp adjust-mss 1220
[BYDQ_AR3260-Dialer1] quit
```

注意：PPPoE 侧接口 MTU 值为 1492 字节，当从这个接口上送的三层转发报文大于 MTU 值且报文被设置为不可分片时就会导致报文无法发送出去，出现用户上网网速慢的情形。此时可以在拨号口下配置 tcp adjust-mss value 命令修改 TCP 协商阶段协商出的 MSS（Max Segment Size）的大小，使报文长度小于拨号口的 MTU 值，保证报文能被正常处理，解决上网网速慢的问题。

2）建立 PPPoE 会话，方法如下：

```
[BYDQ_AR3260] interface gigabitethernet 0/0/2
```

```
[BYDQ_AR3260-GigabitEthernet0/0/2] pppoe-client dial-bundle-number 1
[BYDQ_AR3260-GigabitEthernet0/0/2] quit
```

（2）配置局域网用户通过 NAT 转换将私网地址转换为公网地址，进行拨号上网，方法如下：

```
[BYDQ_AR3260] acl number 3002
[BYDQ_AR3260-acl-adv-3002] rule 5 permit ip source 10.10.10.0 0.0.0.255
[BYDQ_AR3260-acl-adv-3002] rule 10 permit ip source 10.10.20.0 0.0.0.255
[BYDQ_AR3260-acl-adv-3002] rule 15 permit ip source 10.10.30.0 0.0.0.255
[BYDQ_AR3260-acl-adv-3002] rule 20 permit ip source 10.10.40.0 0.0.0.255
[BYDQ_AR3260-acl-adv-3002] quit
[BYDQ_AR3260] interface dialer 1
[BYDQ_AR3260-Dialer1] nat outbound 3002
[BYDQ_AR3260-Dialer1] quit
```

（3）配置到 PPPoE Server 的静态路由，方法如下：

```
[BYDQ_AR3260] ip route-static 0.0.0.0 0 dialer 1
[BYDQ_AR3260] quit
```

（4）配置静态接口，方法如下：

```
[BYDQ_AR3260] interface gigabitethernet 0/0/1
[BYDQ_AR3260-GigabitEthernet0/0/2] ip address 1.1.1.46 255.255.255.240
```

（5）配置局域网用户（服务器地址段）通过 NAT 转换将私网地址转换为公网地址，进行拨号上网，方法如下：

```
[BYDQ_AR3260] acl number 3003
[BYDQ_AR3260-acl-adv-3003] rule 5 permit ip source 10.10.100.0 0.0.0.255
[BYDQ_AR3260] interface gigabitethernet 0/0/1
[BYDQ_AR3260-GigabitEthernet0/0/1]nat outbound 3003
[BYDQ_AR3260- GigabitEthernet0/0/1] quit
```

（6）地址映射，方法如下：

```
[BYDQ_AR3260] interface gigabitethernet 0/0/1
[BYDQ_AR3260-GigabitEthernet0/0/1]nat server protocol tcp global current-interface 8081 inside 10.10.100.100 8081
```

（7）策略路由。由于静态 IP 出口没有配置默认路由，从而导致服务器地址段会走 PPPoE 拨号默认路由出去，所以应采用策略路由来指导服务器流量的转发。

1）配置流分类。在 Router 上创建流分类 VLAN 100 分别匹配目的地址为 10.10.100.0/24 网段的报文，方法如下：

```
[BYDQ_AR3260] traffic classifier vlan 100
[BYDQ_AR3260-classifier-vlan10] if-match acl 3003
[BYDQ_AR3260-classifier-vlan10] quit
```

2）配置流行为，方法如下：

```
[BYDQ_AR3260] traffic behavior vlan 100
[BYDQ_AR3260-behavior-vlan100] redirect ip-nexthop 1.1.1.33
[BYDQ_AR3260-behavior-vlan100] quit
```

3）配置流策略并应用到接口上，方法如下：

```
[BYDQ_AR3260] traffic policy vlan 100
[BYDQ_AR3260-trafficpolicy-vlan100] classifier vlan 10 behavior vlan 100
[BYDQ_AR3260-trafficpolicy-vlan100] quit
```

4）将流策略 VLAN 100 应用到接口 GE0/0/3 入方向（内网接口），方法如下：

```
[BYDQ_AR3260] interface gigabitethernet 0/0/3
[BYDQ_AR3260-GigabitEthernet0/0/3] traffic-policy vlan 100 inbound
[BYDQ_AR3260-GigabitEthernet0/0/3] quit
```

思考：为什么不直接配置两个默认路由而采用策略的方式进行路由选择呢?

（8）配置静态路由。配置静态路由，使得内部流量可以回到核心交换机，方法如下：

```
[BYDQ_AR3260] ip route-static 10.10.0.0 0.0.255.255 10.10.250.2
[BYDQ_AR3260] quit
```

7.3.6.2　核心交换机设备配置

（1）启用使能 DHCP 服务，方法如下：

```
<Huawei>system-view
[Huawei] sysname BYDQ_S7706
[BYDQ_S7706] dhcp enable
```

（2）配置 VLANIF 接口 IP 地址，方法如下：

```
[BYDQ_S7706] interface vlanif 10
[BYDQ_S7706-Vlanif10] ip address 10.10.10.1 24
[BYDQ_S7706-Vlanif10]description to_ShengChanBu
[BYDQ_S7706-Vlanif10] quit
[BYDQ_S7706] interface vlanif 20
[BYDQ_S7706-Vlanif20]description to_XiaoShouBu
[BYDQ_S7706-Vlanif20] ip address 10.10.20.1 24
[BYDQ_S7706-Vlanif20] quit
[BYDQ_S7706] interface vlanif 30
```

```
[BYDQ_S7706-Vlanif30] ip address 10.10.30.1 24
[BYDQ_S7706-Vlanif30]description to_CaiWuBu
[BYDQ_S7706-Vlanif30] quit
[BYDQ_S7706] interface vlanif 40
[BYDQ_S7706-Vlanif40] ip address 10.10.40.1 24
[BYDQ_S7706-Vlanif40]description to_AnBaoBu
[BYDQ_S7706-Vlanif40] quit
[BYDQ_S7706] interface vlanif 100
[BYDQ_S7706-Vlanif100] ip address 10.10.100.1 24
[BYDQ_S7706-Vlanif100]description to_Server
[BYDQ_S7706-Vlanif100] quit
[BYDQ_S7706] interface vlanif 255
[BYDQ_S7706-Vlanif255] ip address 10.10.255.1 24
[BYDQ_S7706-Vlanif255]description to_Manager
[BYDQ_S7706-Vlanif255] quit
[BYDQ_S7706] interface vlanif 1000
[BYDQ_S7706-Vlanif1000] ip address 10.10.252.230
[BYDQ_S7706-Vlanif1000]description to_AR2360
[BYDQ_S7706-Vlanif1000] quit
```

（3）配置接口地址池，方法如下：

```
[BYDQ_S7706] interface vlanif 10
[BYDQ_S7706-Vlanif10] dhcp select interface
[BYDQ_S7706-Vlanif10] dhcp server dns-list 202.103.224.68
[BYDQ_S7706-Vlanif10] quit
[BYDQ_S7706] interface vlanif 20
[BYDQ_S7706-Vlanif20] dhcp select interface
[BYDQ_S7706-Vlanif20] dhcp server dns-list 202.103.224.68
[BYDQ_S7706-Vlanif20] quit
[BYDQ_S7706] interface vlanif 30
[BYDQ_S7706-Vlanif30] dhcp select interface
[BYDQ_S7706-Vlanif30] dhcp server dns-list 202.103.224.68
[BYDQ_S7706-Vlanif30] quit
[BYDQ_S7706] interface vlanif 40
[BYDQ_S7706-Vlanif40] dhcp select interface
```

```
[BYDQ_S7706-Vlanif40] dhcp server dns-list 202.103.224.68
[BYDQ_S7706-Vlanif40] quit
```

（4）配置与出口路由器互联接口，方法如下：

```
[BYDQ_S7706] interface gigabitethernet 1/0/41
[BYDQ_S7706-GigabitEthernet1/0/41] port link-type access
[SW2-GigabitEthernet1/0/41] port default vlan 1000
[SW2-GigabitEthernet1/0/41] quit
```

（5）配置默认路由，方法如下：

```
[BYDQ_S7706] ip route-static 0.0.0.0 0.0.0.0 10.10.250.1
```

（6）配置 Eth-Trunk，方法如下：

```
[BYDQ_S7706] interface eth-trunk 1
[BYDQ_S7706-Eth-Trunk1] trunkport gigabitethernet 1/0/1 to 1/0/3
[BYDQ_S7706-Eth-Trunk1] port trunk allow-pass vlan 10 20 30 40 255
[BYDQ_S7706-Eth-Trunk1] quit
```

（7）配置代理方式多主检测，BYDQ_S7706 做代理设备。

```
# 在代理设备 SwitchD 上，配置 eth-trunk 的代理功能
[BYDQ_S7706] interface eth-trunk 11
[BYDQ_S7706-Eth-Trunk1] mad relay
[BYDQ_S7706-Eth-Trunk1] return
```

7.3.6.3 接入交换机配置

堆叠：

（1）配置逻辑堆叠端口并加入物理成员端口，方法如下：

配置 S5700-LI_1F_1 的业务口 GigabitEthernet0/0/27、GigabitEthernet0/0/28 为物理成员端口，并加入到相应的逻辑堆叠端口

```
<Huawei> system-view
[Huawei] sysname S5700-LI_1F_1
[S5700-LI_1F_1] interface stack-port 0/1
[S5700-LI_1F_1-stack-port0/1] port interface gigabitethernet 0/0/27 enable
Warning: Enabling stack function may cause configuration loss on the interface. Continue? [Y/N]:y
Info: This operation may take a few seconds. Please wait.
[S5700-LI_1F_1-stack-port0/1] quit
[S5700-LI_1F_1] interface stack-port 0/2
[S5700-LI_1F_1-stack-port0/2] port interface gigabitethernet 0/0/28 enable
Warning: Enabling stack function may cause configuration loss on the interface. Continue? [Y/N]:y
```

```
Info: This operation may take a few seconds. Please wait.
[S5700-LI_1F_1-stack-port0/2] quit
```

配置 S5700-LI_1F_2 的业务口 GigabitEthernet0/0/27、GigabitEthernet0/0/28 为物理成员端口，并加入到相应的逻辑堆叠端口

```
<Huawei>system-view
[Huawei] sysname S5700-LI_1F_2
[S5700-LI_1F_2] interface stack-port 0/1
[S5700-LI_1F_2-stack-port0/1] port interface gigabitethernet 0/0/27 enable
Warning: Enabling stack function may cause configuration loss on the interface. Continue? [Y/N]:y
Info: This operation may take a few seconds. Please wait.
[S5700-LI_1F_2-stack-port0/1] quit
[S5700-LI_1F_2] interface stack-port 0/2
[S5700-LI_1F_2-stack-port0/2] port interface gigabitethernet 0/0/28 enable
Warning: Enabling stack function may cause configuration loss on the interface. Continue? [Y/N]:y
Info: This operation may take a few seconds. Please wait.
[S5700-LI_1F_2-stack-port0/2] quit
```

配置 S5700-LI_1F_3 的业务口 GigabitEthernet0/0/27、GigabitEthernet0/0/28 为物理成员端口，并加入到相应的逻辑堆叠端口

```
<Huawei>system-view
[Huawei] sysname S5700-LI_1F_3
[S5700-LI_1F_3] interface stack-port 0/1
[S5700-LI_1F_3-stack-port0/1] port interface gigabitethernet 0/0/27 enable
Warning: Enabling stack function may cause configuration loss on the interface. Continue? [Y/N]:y
Info: This operation may take a few seconds. Please wait.
[S5700-LI_1F_3-stack-port0/1] quit
[S5700-LI_1F_3] interface stack-port 0/2
[S5700-LI_1F_3-stack-port0/2] port interface gigabitethernet 0/0/28 enable
Warning: Enabling stack function may cause configuration loss on the interface. Continue? [Y/N]:y
Info: This operation may take a few seconds. Please wait.
[S5700-LI_1F_3-stack-port0/2] quit
```

（2）配置堆叠 ID 和堆叠优先级，方法如下：

```
[S5700-LI_1F_1] stack slot 0 priority 200# 配置 S5700-LI_1F_1 的堆叠优先级为 200
Warning: Please do not frequently modify Priority because it will make the stack split. Continue? [Y/N]:y
```

```
[S5700-LI_1F_2] stack slot 0 renumber 1# 配置 S5700-LI_1F_2 的堆叠 ID 为 1
Warning: All the configurations related to the slot ID will be lost after the slot ID is modified.
Please do not frequently modify slot ID because it will make the stack split. Continue? [Y/N]:y
Info: Stack configuration has been changed, and the device needs to restart to make the configuration effective.
[S5700-LI_1F_3] stack slot 0 renumber 2# 配置 S5700-LI_1F_3 的堆叠 ID 为 2
Warning: All the configurations related to the slot ID will be lost after the slot ID is modified.
Please do not frequently modify slot ID because it will make the stack split. Continue? [Y/N]:y
Info: Stack configuration has been changed, and the device needs to restart to make the configuration effective.
```

（3）SwitchA、SwitchB、SwitchC 下电，使用 SFP+ 电缆连接后再上电，如图 7.7 所示。

（4）配置跨设备 Eth-Trunk，方法如下：

```
# 在堆叠系统创建 eth-trunk，并将上行物理端口设置为 eth-trunk 成员接口
[S5700-LI_1F] interface eth-trunk 1
[S5700-LI_1F-Eth-Trunk1] trunkport gigabitethernet 0/0/26
[S5700-LI_1F-Eth-Trunk1] trunkport gigabitethernet 1/0/26
[S5700-LI_1F-Eth-Trunk1] trunkport gigabitethernet 2/0/26
[S5700-LI_1F-Eth-Trunk1]port trunk allow-pass vlan 10 20 30 40 255
[S5700-LI_1F-Eth-Trunk1] quit
```

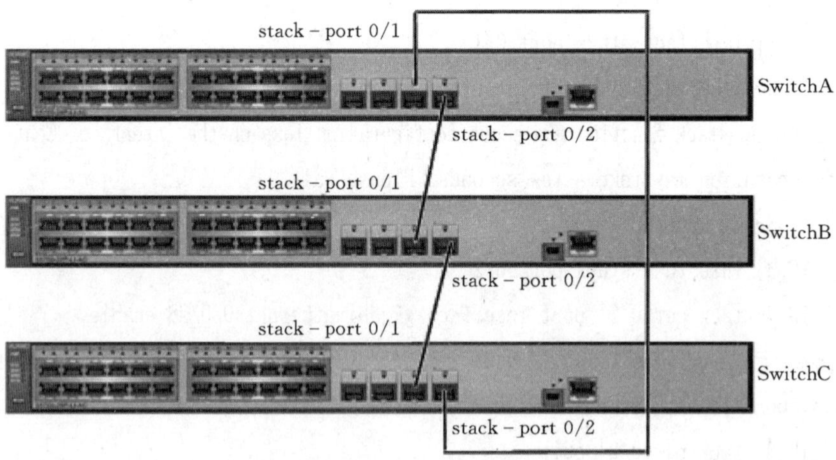

图 7.7　堆叠效果图

（5）配置代理方式多主检测，方法如下：

```
# 在堆叠系统上，配置跨设备 eth-trunk 的代理方式多主检测功能
```

```
[S5700-LI_1F] interface eth-trunk 1
[S5700-LI_1F -Eth-Trunk10] mad detect mode relay
[S5700-LI_1F -Eth-Trunk10] return
```

（6）配置端口 VLAN，并加入到边缘端口，方法如下：

```
[S5700-LI_1F] interface gigabitethernet 0/0/1
[S5700-LI_1F _S7706-GigabitEthernet0/0/1] port link-type access
[S5700-LI_1F -GigabitEthernet0/0/1] port default vlan 10
[S5700-LI_1F -GigabitEthernet0/0/1] stp edged-port enable
[S5700-LI_1F -GigabitEthernet0/0/1] quit
```

（7）配置管理 IP 地址，方法如下：

```
[S5700-LI_1F] interface vlanif 255
[S5700-LI_1F -Vlanif255] ip address 10.10.255.2 24
[S5700-LI_1F -Vlanif255]description to_Manager
[S5700-LI_1F -Vlanif255] quit
```

7.4　思政融入

我国在信息技术领域取得了巨大的进步，与发达国家的差距日渐缩小，计算机网络工程已形成了一项单独的学科，在不同的领域向高精方向进行发展。在网络工程项目建设过程中，要不断追求完美和极致，将一丝不苟、精益求精的工匠精神融入每一个环节。大学生在学习过程中，以“写好一行代码、熔好一根光纤、装好一台设备、做好一个项目”的“四个一”细致要求贯穿技能训练过程，在实践技能和理论知识的螺旋递进过程养成极致、追求卓越的工匠精神意识与职业行为习惯，注重核心职业素养的养成，做出一流的网络工程项目。

7.5　思考与训练

1. 什么是堆叠？堆叠有哪些优点？
2. 通过交换机堆叠，可实现什么效果？
3. 什么是边缘端口？设置边缘端口有什么优点？
4. 策略路由的用途有哪些？
5. 局域网用户通过 NAT 转换将私网地址转换为公网地址进行上网，请写出配置步骤。

Project 7 Network project of a power capacitor limited liability company

7.1 Project import

A power capacitor limited liability company formerly owned by the whole people, a power capacitor factory, was founded on June 1, 1967. In September 2006, it was converted into a company with diversified equity structure. By introducing investment and expanding the scale of capital through restructuring, enterprises have strengthened their strength and made their operations more dynamic. The company was originally a direct subordinate enterprise of the Ministry of Machinery Industry and is a large second-class company of the State. China's high-voltage, ultra-high-voltage and ultra-high-voltage capacitive voltage transformers, coupling capacitors, reactive compensation and their set of equipment are two major research and manufacturing bases in the manufacturing industry of high-voltage transmission and substation equipment. There are 500 large-scale electric machinery and equipment manufacturing enterprises in the country.

With the development of science and technology, the size of the company is expanding, the area is expanding, and the team of managers is expanding. The current network structure can no longer meet the needs of the current development, mainly manifested by unstable network, frequent network jitter, high cost, long cycle, difficult to manage and unsuitable for implementation of the network line connection. Therefore, the establishment of a design specification, complete functionality, excellent performance, safety and reliability, good scalability and usability and have a manageable and easy to maintain the network and system platform to improve the efficiency of the company's staff to improve the efficiency of the work of the company's staff and the implementation of the efficiency of the way of high efficiency, high speed, low cost is very necessary. So using high-end computers, network devices, software, advanced and mature network communication technology to construct networks, and building an efficient office network system with is advanced system integration technology and management mode the company's imminent task.

7.2 Relevant knowledge

7.2.1 Policy-Based Routing

7.2.1.1 Summary

Policy-Based Routing (PBR) is a mechanism for routing based on user-defined policies. Configuring policy routes can be used to improve network security performance and load sharing.

Policy routing enables packets to be forwarded according to user-specified policies. For some administrative purposes, such as QoS requirements or VPN topologies, policy routing can be used if certain routes must go through specific paths. For example, a policy may specify that packets sent from a network can only be forwarded to a particular interface.

There are two main types of policy routing: one is based on the destination address of the route, called destination address routing; The other is based on the route source address for policy implementation, called source address routing. With the development of policy routing, there is now a third routing method: intelligent balanced policy.

7.2.1.2 Differences between Policy-Based Routing and Routing Policy

(1) Policy-Based Routing operates on packets.

(2) Policy-Based Routing refers to changing the packet forwarding path according to a policy instead of following the routing table when the routing table has already been generated.

(3) Routing Policy operates on routing information.

(4) Routing Policy mainly implements such functions as routing filtering and routing attribute settings. It changes the path traveled by network traffic by changing routing attributes, including accessibility.

(5) The traditional routing forwarding principle is to first find the routing table based on the destination address of the message, and then forward the message. However, at present, more and more users want to be able to forward messages and select routes based on their own defined strategies on the basis of traditional routing forwarding.

7.2.1.3 Policy-Based Routing advantages

(1) Routing strategies can be formulated according to users'actual needs to enhance the flexibility and controllability of routing.

(2) Different data streams can be sent through different links to improve link utilization efficiency.

(3) On the premise of meeting the quality of business service, select a link with lower cost to transfer business data, which can reduce the cost of enterprise data services.

7.2.1.4 Types of Policy-Based Routing

The AR Series supports three strategic routes:

(1) Local Policy-Based Routing. Local Policy-Based Routing refers to a Policy-Based Routing that is valid only for locally sent messages and has no effect on forwarded messages. A Local Policy-Based Routing can configure multiple policy points with different priorities, and the messages sent locally match the priority policy points. Local policy routing supports ACL or message length matching rules.

The Local policy-Based Routing sends messages according to the priority of the Local Policy -Based Routing nodes, which in turn matches the matching rules of each node's binding. If the Local Policy-Based Route node is not found, follow the general flow of IP message to find a route based on the destination eddress. Local Policy-Based Routing node priority order is (high to low): message priority → outgoing interface → next hop → default outgoing interface.

(2) Interface Policy Routing. Interface Policy Routing only works for forwarded messages, not for locally sent messages, and only for messages entering the direction of the interface. Interface Policy Routing is achieved by configuring redirection in popularity.

(3) Smart Policy Routing (SPR). Smart Policy Routing (SPR) can proactively detect link quality and match business needs, thus selecting an optimal link to forward business data can effectively avoid network black holes, network oscillations and other issues. This feature requires the purchase of a license authorization. Smart Policy Routing is a policy routing based on business requirements, which achieves intelligent routing by matching link quality and network business requirements for link quality.

7.2.1.5 Local Policy-Based Routing configuration

Configuring policy routing can redirect the forwarding message arriving at the interface to the specified next hop address. By configuring redirection, the device redirects the message conforming to the flow classification rules to the specified next hop address. The flow policy with redirection action can only be applied in the global, interface or VLAN incoming direction.

(1) Create Policy-Based Routing and policy points.

```
[Huawei]policy-based-route 1 deny node 1
```

(2) Set local policy matching rules.

```
[Huawei-policy-based-route-1-1]if-match?
acl Access control list  #Match according to ACL in IP message
packet-length Match packet length  #Match according to IP message length
[Huawei-policy-based-route-1-1]if-match packet-length?
INTEGER<0-65535>  Minimum packet length #Minimum message length
```

[Huawei-policy-based-route-1-1]if-match packet-length 100?

INTEGER<1-65535> Maximum packet length #Maximum message length

(3) Set the outgoing interface of the message.

[Huawei-policy-based-route-1-1]apply output-interface? #Set the interface directly

[Huawei-policy-based-route-1-1]apply default output-interface? #Default out interface

#Serial Serial interface. The outgoing interface of the message will be sent from the specified interface after successful matching. The interface is not a broadcast type interface such as Ethernet (just change to P2P), because multiple next hops may lead to unsuccessful message forwarding.

(4) Set the next hop of the message.

[Huawei-policy-based-route-1-1]apply ip-address?

default Set default information #By default, the next hop only works for routing packets not found in the routing table

next-hop Next hop address #Directly set the next hop

(5) Set VPN forwarding instance.

[Huawei-policy-based-route-1-1]apply access-vpn vpn-instance?

STRING<1-31> VPN instance name

(6) Set IP message priority.

[Huawei-policy-based-route-1-1]apply ip-precedence?

INTEGER<0-7>	IP precedence value
critical	Set packet precedence to critical(5) (crux)
flash	Set packet precedence to flash(3) (flash)
flash-override	Set packet precedence to flash override(4) (speed)
immediate	Set packet precedence to immediate(2) (fast)
internet	Set packet precedence to Internet control(6) (inter network)
network	Set packet precedence to network control(7) (in network)
priority	Set packet precedence to priority(1) (first)
routine	Set packet precedence to routine(0) (ordinary)

(7) Set the time interval for refreshing LSP information of Local Policy-Based Routing.

[Huawei]ip policy-based-route refresh-time?

INTEGER<1000-65535> Refresh time value (ms)

<cr>

(8) Apply Local Policy-Based Routing.

[Huawei]ip local policy-based-route?

STRING<1-19> Policy name

#A router can only enable one local policy route (multiple routes can be created)

7.2.1.6 Interface policy routing configuration

(1) Set flow classification.

```
[Huawei]traffic classifier test1 operator? #Logical operator
and  Rule of matching all of the statements #and operations
or   Rule of matching one of the statements #or operations
```

(2) Set classification matching rules.

```
[Huawei-classifier-test1]if-match?
8021p               Specify vlan 802.1p to match
acl                 Specify ACL to match
any                 Specify any data packet to match
app-protocol        Specify app-protocol to match
cvlan-8021p         Specify inner vlan 802.1p of QinQ packets to match
cvlan-id            Specify inner vlan id of QinQ packets to match  #Set classification matching rules based on inner VLAN ID classification matching rules
destination-mac     Specify destination MAC address to match
dlci                Specify a DLCI to match
dscp                Specify DSCP (DiffServ CodePoint) to match
fr-de               Specify FR DE to match.
inbound-interface   Specify an inbound interface to match
ip-precedence       Specify IP precedence to match
ipv6                Specify ip v6
l2-protocol         Specify layer-2 protocol to match
mpls-exp            Specify MPLS EXP value to match
protocol            Specify ipv4 or ipv6 packets to match
protocol-group      Specify protocol-group to match
pvc                 Specify a PVC to match
rtp                 Specify RTP port to match
source-mac          Specify source MAC address to match
tcp                 Specify TCP parameters to match
vlan-id             Specify a vlan id to match  #Based on outgoing VLAN ID
```

(3) Set stream redirection. Redirect the message conforming to the flow classification rules to the specified next hop or specified interface. It can only be applied in the incoming direction of the interface.

(4) Set traffic behavior.

```
[Huawei]traffic behavior test2
[Huawei-behavior-test2]
```

(5) Redirect message conforming to stream classification.

```
[Huawei-behavior-test2]redirect?
interface            Status and configuration about the traffic on the interface # Redirect to the specified interface (3G or dialer)

ip-nexthop           Redirect packets to nexthop #Redirect to next hop
ipv6-nexthop         Redirect packets to ipv6 nexthop
-----------
[Huawei-behavior-test2]redirect ip-nexthop 10.1.1.1?
post-nat  Post nat
track       Track nqa #Support linkage with NQA
<cr>        Please press ENTER to execute command
```

(6) Application traffic policy.

```
[Huawei]traffic policy test3  #Stream policy settings
[Huawei-trafficpolicy-test3]classifier test1 behavior test2 #Associated flow classification and popularity
[Huawei-GigabitEthernet0/0/2]traffic-policy test3 inbound #Interface (in application direction)
```

7.2.2 Port mapping

Port mapping is to map a port of the IP address of the external network host to a machine in the internal network to provide corresponding services. When the user accesses this port of the IP, the server automatically maps the request to the machine inside the corresponding LAN. If a computer in the intranet wants to access the Internet, open services or receive data, it needs port mapping.

The configuration code is as follows:

```
[Huawei]interface GigabitEthernet 0/0/0
[Huawei-GigabitEthernet0/0/0] nat server protocol tcp global #Nat server protocol TCP global extranet IP extranet port inside intranet IP intranet port
```

7.2.3 Eth-Trunk

One switch bundles these multiple interfaces to form an Eth–Trunk interface, so as to achieve the purpose of increasing bandwidth and improving reliability.

As shown in Figure 7.1, the link connected to the Eth-Trunk interface can be regarded as a point–to–point direct link.

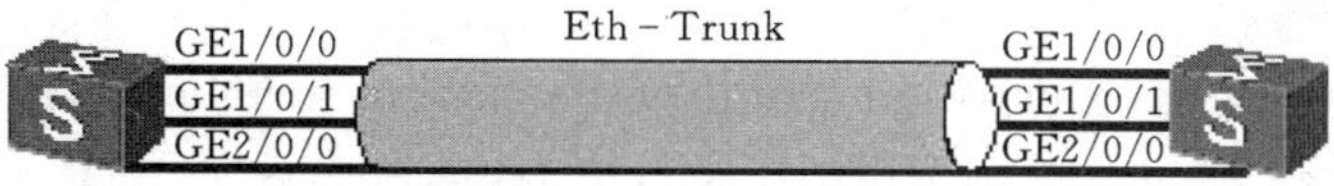

Figure 7.1 Eth-Trunk diagram

The advantages of Eth-Trunk are:

(1) Load sharing: Load sharing can be realized through Trunk interface. Traffic load sharing can be realized in an Eth-Trunk interface.

(2) Improve reliability. When the physical link connected to a member interface fails, the traffic will be switched to other available links, so as to improve the reliability of the whole Trunk link.

(3) Increase bandwidth. The total bandwidth of Eth-Trunk interface is the sum of the bandwidth of each member interface.

7.2.4 Istack stack

7.2.4.1 Stack introduction

Stacking istack (Intelligent stack) refers to the combination of multiple switch devices supporting stacking characteristics to logically combine them into one switch device. As shown in Figure 7.2 above, switchA and switchB are connected through stacked cables to form a stacked istack. For upstream and downstream devices, they are equivalent to a switch.

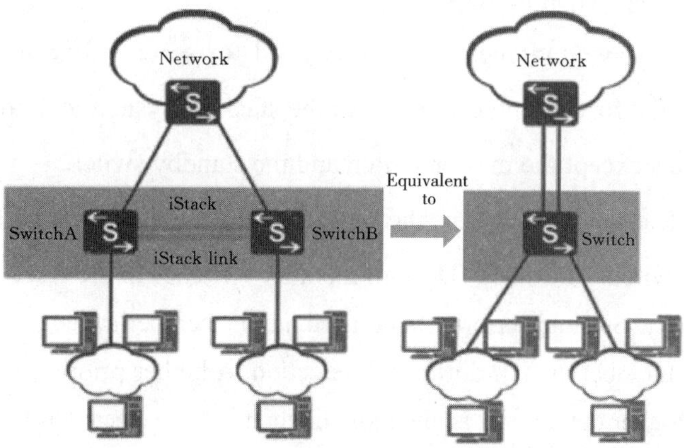

Figure 7.2 Stacking diagram

Through switch stacking, the following effects can be achieved:

(1) High reliability. Redundant backup between multiple member switches of stacking system; Stacking supports the link aggregation function across devices to realize link redundant backup across devices.

(2) Strong network expansion capability. By adding member switches, the number of ports, bandwidth and processing capacity of the stacking system can be easily expanded; At the same time, it supports hot plug of member switches, and the newly added member switches automatically synchronize the configuration file and system software version of the main switch.

(3) Simplify configuration and management. On one hand, users can log in to the stacking system through any member switch to uniformly configure and manage all member switches of the stacking system; On the other hand, after the stack is formed, there is no need to configure complex layer-2 ring breaking protocol and layer-3 protection switching protocol, which simplifies the network configuration.

7.2.4.2 Stacking terms

Stacking involves the following basic terms:

(1) Role. All single switches in the stack are called member switches. According to different functions, they can be divided into three roles:

1) Master switch. The master is responsible for managing the entire stack. There is only one master switch in the stack.

2) Standby switch. The standby switch is the backup switch of the main switch. When the Master switch fails, the Master switch will take over all services of the original Master switch. There is only one standby switch in the stack.

3) Slave switch. Slave switches are mainly used for service forwarding. The more slave switches, the stronger the forwarding capability of the stacking system. All member switches in the stack are slave switches except the master switch and the standby switch.

(2) Stack ID. The stack ID, that is, the Slot ID of the member switch, is used to identify and manage the member switch. The stack ID of all member switches in the stack is unique.

(3) Stack priority. Stacking priority is an attribute of member switch, which is mainly used to determine the role of member switch during role election. A higher priority value indicates a higher loved of priority. The higher the priority is, the more likely it is to be elected as the Master switch.

7.2.4.3 Stack build process

The process of stack establishment includes the following four stages:

1. Physical connection

According to the network requirements, select the appropriate connection mode and connection topology to form a stacked network. According to different connecting media, stacking can be divided into stacking card stacking and service port stacking.

As shown in Figure 7.3 above, each connection mode can form two connection topologies: chain and ring. The following Table 7.1 compares the two connection topologies in terms of reliability, link bandwidth utilization and network cabling convenience.

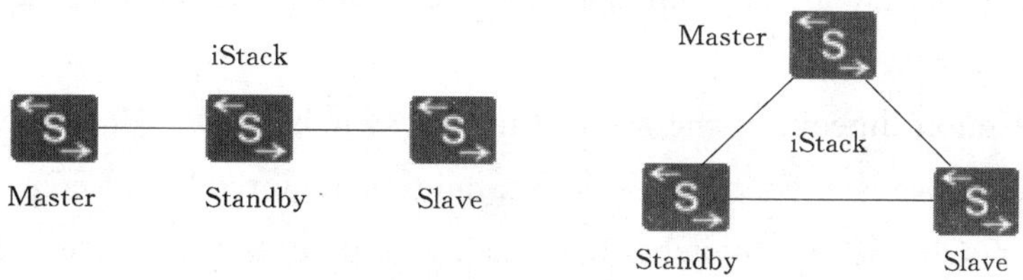

(a) Chain connection (b) Ring connection

Figure 7.3 Stacked connection topology

2. Election master switch

The member switches send stacking competition messages to each other, and select the master switch of the stacking system according to the election principle.

Table 7.1 Advantages and disadvantages of different connection topologies

Connection topology	Advantage	Shortcoming	Applicable scenario
Chain connection	There is no need for physical connection between the head and tail, which is suitable for long-distance stacking	Low reliability: if one of the stacking links fails, the stack will be split. Low utilization of stacking link bandwidth: the whole stacking system has only one path	When the stacked member switches are far away, it is difficult to form a ring connection, and chain connection can be used
Ring connection	High reliability of ring connection: when one of the stacking links fails, the ring topology becomes a chain topology, which does not affect the normal operation of the stacking system. High bandwidth utilization of stacked links: data can be forwarded according to the shortest path	The head and tail need to be physically connected, which is not suitable for long-distance stacking	It is recommended to consider the reliability of the stacking and utilization of the ring switch from the perspective of the short stacking distance of the link members

Determine the connection mode and topology of the stack. After completing the physical connection between member switches, all member switches are powered on. At this point, the stacking system starts the election of the master switch. In the stacking system, each member switch has a certain role, in which the master switch is responsible for managing the whole stacking system. The election rules of the master switch are as follows (judge from the first one in turn until the optimal switch is found):

(1) Operation status comparison.

(2) Switches that are already running will take priority over those in the start-up state and compete for the master switch.

(3) The election timeout of the stacked master switch is 20s. When the stacked member switch is powered on or restarted, because the startup time required by different member switches may vary greatly, not all member switches have the opportunity to participate in the election of the master switch: compared with the fastest member switch, the member switch with a startup time difference of more than 20s has no opportunity to participate in the election of the master switch, You can only passively join the stack to become a non master switch. For the joining process, see joining and exiting stack members. Therefore, if you want to designate a member switch as the master switch, you can power it up first, and then power up other member switches after it is started up.

(4) Switches with high stacking priority compete with the master switch first.

(5) With the same stacking priority, the switch with small MAC address has priority to compete for the master switch.

3. Topology collection and standby switch election

The master switch collects the topology information of all member switches, assigns the stacking ID to all member switches, and then selects the standby switch of the stacking system.

After the election of the master switch is completed, the master switch collects the topology information of all member switches, calculates the stack forwarding table entries and ring breaking point information according to the topology information, sends them to all member switches in the stack, and assigns the stack ID to all member switches. After that, the backup switch is elected as the backup switch of the master switch. In addition to the master switch, the switch that completes the equipment startup first is preferentially selected as the backup switch. When all switches except the master switch are started at the same time, the election rules of the standby switch are as follows (judge from the first one in turn until the optimal switch is found):

(1) The device with the highest stacking priority becomes the standby switch.

(2) With the same stacking priority, the one with the smallest MAC address becomes the

standby switch.

(3) In addition to the master switch and standby switch, the remaining member switches are added to the stack as slave switches.

4. Stable operation

The master switch synchronizes the topology information of the whole stacking system to all member switches. The member switches synchronize the system software and configuration files of the master switch, and then enter the stable operation state.

After the role election and topology collection are completed, all member switches automatically synchronize the system software and configuration files of the master switch:

(1) Stacking has the function of automatically loading system software. The member switches to be stacked do not need to have the same software version, but only need to be compatible between versions. When the software version of the standby switch or slave switch is inconsistent with that of the master switch, the standby switch or slave switch will automatically download the system software from the master switch, then restart with the new system software and rejoin the stack.

(2) The stack has a configuration file synchronization mechanism. The standby switch or slave switch synchronizes the configuration file of the master switch to the device and executes it, so as to ensure that multiple devices in the stack can work in the network like one device, and other switches can still perform various functions normally after the main switch fails.

7.2.4.4 Stack login and access

After the stack is established, a virtual device composed of multiple member switches exists in the network, and the interface numbering rules, login and access methods of the stack system have changed.

1. Stacking interface numbering rules

The interface number of the stacking system uses the stacking ID as the identification information, and the stacking ID of all member switches is unique.

For a single equipment without running stack, the interface number adopts: slot number/sub card number/port number (slot number is uniformly taken as 0). After the equipment is added to the stack, the interface number adopts: stack ID/sub card number/port number.

For example, when the device does not run stacking, the number of an interface is GigabitEthernet 0/0/1; When the device is added to the stack, if the stack ID is 2, the number of the interface will change to GigabitEthernet 2/0/1. For the management network interface, the interface number is MEth 0/0/1 no matter the system running stack or not and what the stacking ID is after stacking.

The numbering rules of sub card number and port number are consistent with those in stand-alone status. If the stacked device is still used as the ID number of the stacked device, it will be used as the ID number of the stacked device.

2. Login of stacking system

Log in to the stacking system as follows:

Local login: log in through the console port of any member switch.

Remote login: log in through the management network port or other layer-3 interfaces of any member switch. As long as the route to the stacking system is reachable, Telnet, Stelnet, Web and SNMP can be used for remote login.

After the equipment with management network interface is stacked, in the system operation stage, only the management network interface of one member switch takes effect, which is called the main management network interface. After the stacking system is started, the management network port of the master member switch is selected as the main management network port by default. If the management network port of the master member switch is abnormal or unavailable, the management network port of other member switches is selected as the main management network port. If you are directly connected to a non master management network port through a PC, you cannot log in to the stacking system normally.

After the stack is established, the configuration file of the contention based switch takes effect. If the remote login stack, the IP address of the master switch is required.

No matter which member switch logs in to the stacking system, the master switch is actually logged in. The master switch is responsible for distributing the user configuration to other member switches and uniformly managing the resources of all member switches in the stacking ystem.

3. Access to stacked file systems

The access of file system includes the creation, deletion and modification of files and directories in memory, as well as the display of file contents. The memory supported by the device is flash.

Specify the file name under a path in the format of drive + path + filename, where: Drive refers to the memory in the device. Path refers to the directory and subdirectories in memory. Filename refers to the file name.

The difference between stacked environment and stand-alone environment lies in the naming of drive: Flash refers to the root directory of the flash memory of the main switch in the stacking system. Stack ID #flash refers to the root directory of the flash memory of a member switch in the stack system. For example: slot2#flash refers to the root directory of the flash memory of the member switch with stack ID 2.

7.2.4.5 Stack cross device link aggregation

Stacking supports cross device link aggregation technology, which is realized by configuring cross device Eth-Trunk interface. Users can configure the physical Ethernet ports on different member switches as an aggregation port to connect upstream or downstream devices to realize link aggregation between multiple devices. When one of the aggregation links fails or a member switch in the stack fails, the Eth-Trunk interface can redistribute the traffic to other aggregation links, realize the backup between links and devices, and ensure the reliable transmission of data traffic.

As shown in Figure 7.4 above, the traffic flowing to the network core will be evenly distributed on the aggregation link. When one aggregation link fails, the Eth-Trunk interface redistributes the traffic to other aggregation links through stacked cables, realizing the backup between links.

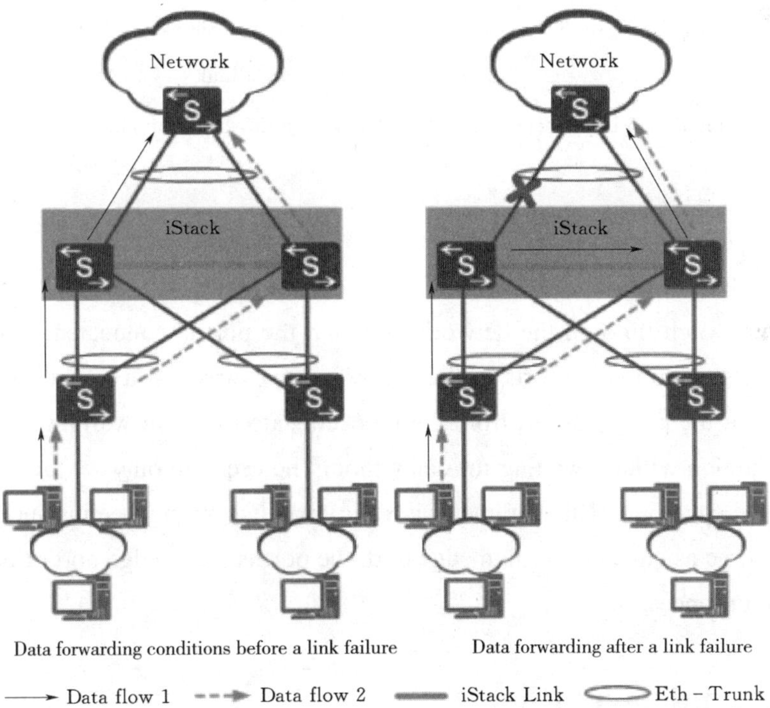

Figure 7.4 Backup between links through cross device Eth–Trunk interface

As shown in Figure 7.5 above, the traffic flowing to the network core will be evenly distributed on the aggregation link. When a member switch fails, the Eth-Trunk interface will redistribute the traffic to other aggregation links to realize the backup between devices.

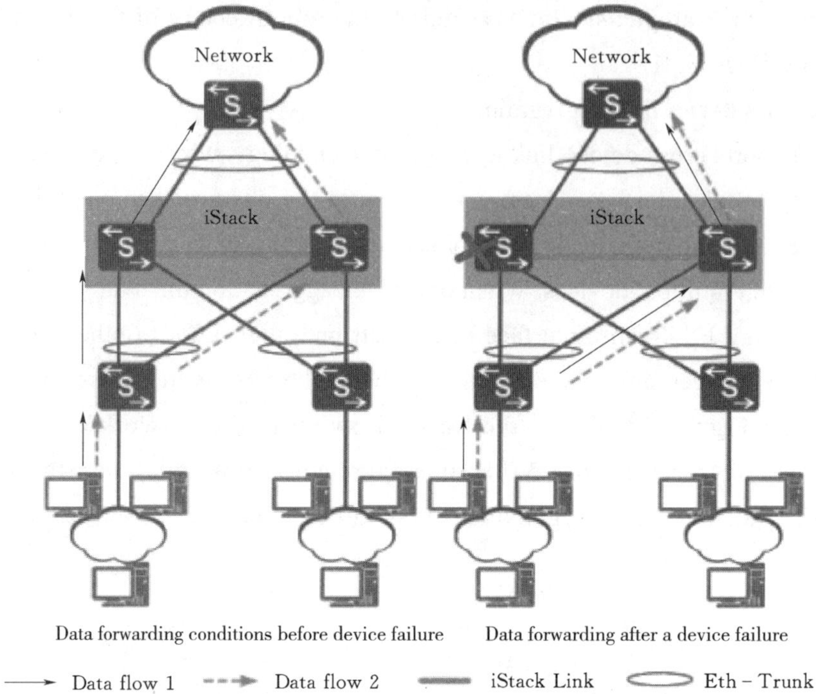

Figure 7.5 Implementation of Eth–Trunk inter device backup interface

7.2.5 Edge port

Edge port refers to a port that is not directly connected to any switch, nor is it indirectly connected to any switch through the network to which the port is connected. You can configure ports as edge ports or non edge ports in the following two ways. If a user designates a port as an edge port, when the port migrates from the blocked state to the forwarding state, the port can realize rapid migration without waiting for delay time. The user can only set the port connected to the terminal as the edge port. This parameter is valid for all spanning tree instances, that is, when the port is configured as edge port or non edge port, the port is set as edge port or non edge port on all spanning tree instances.

7.3 Project implementation

7.3.1 Project requirements

(1) The number of ISP operators is two, one is in the form of static IP and one is PPPoE dial.

(2) Server traffic is required to egress through static IP addresses, and address mapping

is performed on the corresponding intranet server, so that people who go out can access internal resources.Other service network blocks use PPPoE dial–up Internet access.

(3) Access devices are stacked in a ring shape.

(4) The time for the PC to obtain the address should be controlled within 30s.

7.3.2 Device inventory

The network device used in this project includes core switches, access switches, and routers. The models and numbers of devices are shown in Table 7.2.

Table 7.2 List of network devices

Serial No.	Device model	Quantity	Function	Remark
1	AR3260	1	Egress router	
2	S7706	1	Core switch	
3	S5700–28P–LI–AC	30	Access switch	Stacking cable included
4	S5700–52P–LI–AC	20	Access switch	Stacking cable included

7.3.3 Project topology

The network topology of this project is shown in Figure 7.6.

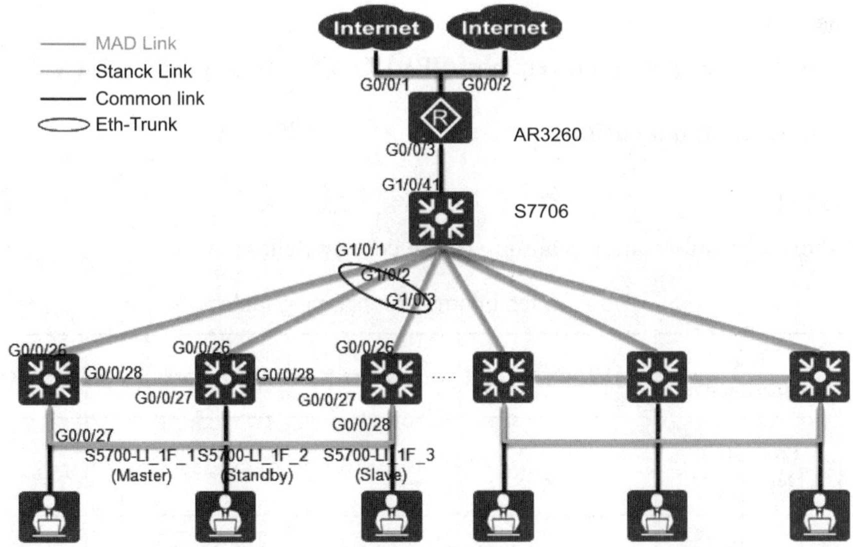

Figure 7.6 Network topology of a power capacitor limited liability company

7.3.4 Configuration ideas

The configuration is as follows:

(1) Configure CHAP authentication under the dialer interface to enable the device to establish a connection with the PPPoE server through PPP authentication.

(2) Configure the dialing method to be autodialed so that every once in a while the device automatically tries to establish a dial-up connection again.

(3) Do NAT address translation on multiple outgoing interface lines separately, using ACLs for source address matching

(4) Routing configures policy routing to implement traffic offloading.

(5) The core configuration is DHCP to automatically obtain IP addresses for end users.

(6) When building a stack through a service port connection, you need to configure logical stack ports and add physical member ports in order to forward data packets between the member switches of the stack.

(7) To facilitate user management and identification of devices, configure the stack ID and priority of the member switches.

(8) Power down the S5700-LI_1F_1, S5700-LI_1F_2, and S5700-LI_1F_2, and use SFP+ stacking cables to connect the physical member ports before powering up.

(9) To improve reliability and increase uplink bandwidth, configure cross-device Eth-Trunk.

(10) To effectively avoid network unavailability when stacking splitting, configure proxy multi-master detection.

(11) Enable the edge port to prevent port UP&DOWN from triggering STP calculation.

7.3.5 Planning of relevant information

7.3.5.1 Core switch

Table 7.3 shows the information planning of the core switches.

Table 7.3 Core switch information planning table

Serial No.	Use departments	VLAN No.	IP address	Mask	Remark
1	Production Department	10	10.10.10.1	255.255.255.0	
2	Sales Department	11	10.10.20.1	255.255.255.0	

Continue

Serial No.	Use departments	VLAN No.	IP address	Mask	Remark
3	Finance Department	12	10.10.30.1	255.255.255.0	
4	Department of Safety and Security	13	10.10.40.1	255.255.255.0	
5	Server	100	10.10.100.1	255.255.255.0	
6	Management	255	10.10.255.1	255.255.255.0	
7	Interconnection	1000	10.10.250.1	255.255.255.252	Interconnect with the router

7.3.5.2 Exit router

The information planning of the exit router is shown in Table 7.4.

Table 7.4 Exit router information planning table

Serial No.	Interface	IP address	Mask	Remark
1	G0/0/3	10.10.250.2	255.255.255.252	Interconnect with core switches

7.3.5.3 ISP export planning

Table 7.5 shows the information planning of ISP exports.

Table 7.5 ISP export information planning table

Serial No.	Interface	IP address	Mask	Gateway	Remark
1	G0/0/1	1.1.1.46	255.255.255.240	1.1.1.33	Static IP address
2	G0/0/2	Account:gxsdxy	Password:123456		PPPoE dialing

7.3.6 Procedure

7.3.6.1 Egress device configuration

(1) PPPoE dialing.

1) Configure the Dialer interface.

```
<Huawei> system-view
[Huawei] sysname BYDQ_AR3260
```

```
[BYDQ_AR3260] interface dialer 0
[BYDQ_AR3260-Dialer1] ppp pap local-user gxsdxy password simple 123456
[BYDQ_AR3260-Dialer1] dialer bundle 1
[BYDQ_AR3260-Dialer1] dialer-group 1
[BYDQ_AR3260-Dialer1] ip address ppp-negotiate
[BYDQ_AR3260-Dialer1] tcp adjust-mss 1220
[BYDQ_AR3260-Dialer1] quit
```

Remark:The MTU value of the PPPoE interface is 1492 bytes, and when the Layer-3 forwarding packets sent from this interface are greater than the MTU value and the packets are set to non-fragmentable, the packets cannot be sent, resulting in slow Internet connection speed. At this time, you can configure the tcp adjust-mss value command under the dial port to modify the size of the MSS (Max Segment Size) negotiated during the TCP negotiation stage, so that the packet length is less than the MTU value of the dial port, ensure that the packet can be processed normally, and solve the problem of slow Internet speed.

2）Establish a PPPoE session.

```
[BYDQ_AR3260] interface gigabitethernet 0/0/2
[BYDQ_AR3260-GigabitEthernet0/0/2] pppoe-client dial-bundle-number 1
[BYDQ_AR3260-GigabitEthernet0/0/2] quit
```

（2）Configure LAN users to use NAT translation to convert private IP addresses to public IP addresses for dial-up Internet access.

```
[BYDQ_AR3260] acl number 3002
[BYDQ_AR3260-acl-adv-3002] rule 5 permit ip source 10.10.10.0 0.0.0.255
[BYDQ_AR3260-acl-adv-3002] rule 10 permit ip source 10.10.20.0 0.0.0.255
[BYDQ_AR3260-acl-adv-3002] rule 15 permit ip source 10.10.30.0 0.0.0.255
[BYDQ_AR3260-acl-adv-3002] rule 20 permit ip source 10.10.40.0 0.0.0.255
[BYDQ_AR3260-acl-adv-3002] quit
[BYDQ_AR3260] interface dialer 1
[BYDQ_AR3260-Dialer1] nat outbound 3002
[BYDQ_AR3260-Dialer1] quit
```

（3）Configure static routes to the PPPoE Server.

```
[BYDQ_AR3260] ip route-static 0.0.0.0 0 dialer 1
[BYDQ_AR3260] quit
```

（4）Configure a static interface.

```
[BYDQ_AR3260] interface gigabitethernet 0/0/1
[BYDQ_AR3260-GigabitEthernet0/0/2] ip address 1.1.1.46 255.255.255.240
```

(5) Configure LAN users (server CIDRs) to convert private IP addresses to public IP addresses through NAT translation for dial-up Internet access.

```
[BYDQ_AR3260] acl number 3003
[BYDQ_AR3260-acl-adv-3003] rule 5 permit ip source 10.10.100.0 0.0.0.255
[BYDQ_AR3260] interface gigabitethernet 0/0/1
[BYDQ_AR3260-GigabitEthernet0/0/1]nat outbound 3003
[BYDQ_AR3260- GigabitEthernet0/0/1] quit
```

(6) Address mapping.

```
[BYDQ_AR3260] interface gigabitethernet 0/0/1
[BYDQ_AR3260-GigabitEthernet0/0/1]nat server protocol tcp global current-interface 8081 inside
10.10.100.100 8081
```

(7) Policy routing. Since the static IP egress is not configured with a default route, the server address segment will go out through the PPPoE dial default route, so we have to use policy routing to guide the forwarding of server traffic.

1) Configure traffic classifier. Create a traffic classifier VLAN 100 on the router to match packets with CIDR blocks with destination CIDR blocks of 10.10.100.0/24.

```
[BYDQ_AR3260] traffic classifier vlan 100
[BYDQ_AR3260-classifier-vlan10] if-match acl 3003
[BYDQ_AR3260-classifier-vlan10] quit
```

2) Configuration traffic behavior.

```
[BYDQ_AR3260] traffic behavior vlan 100
[BYDQ_AR3260-behavior-vlan100] redirect ip-nexthop 1.1.1.33
[BYDQ_AR3260-behavior-vlan100] quit
```

3) Configure and apply the traffic-policy to the interface.

```
[BYDQ_AR3260] traffic policy vlan 100
[BYDQ_AR3260-trafficpolicy-vlan100] classifier vlan 10 behavior vlan 100
[BYDQ_AR3260-trafficpolicy-vlan100] quit
```

4) Apply the traffic-policy VLAN100 to interface GE0/0/3 inbound (intranet interface).

```
[BYDQ_AR3260] interface gigabitethernet 0/0/3
[BYDQ_AR3260-GigabitEthernet0/0/3] traffic-policy vlan 100 inbound
[BYDQ_AR3260-GigabitEthernet0/0/3] quit
```

Question: Why not just configure the two default routes and use policies for routing?

(8) Configure static routes. Configure static routes so that internal traffic can come back to the core switch.

```
[BYDQ_AR3260] ip route-static 10.10.0.0 0.0.255.255 10.10.250.2
[BYDQ_AR3260] quit
```

7.3.6.2 Core switch device configuration

(1) Enable DHCP service.

```
<Huawei>system-view
[Huawei] sysname BYDQ_S7706
[BYDQ_S7706] dhcp enable
```

(2) Configure the VLANIF interface IP address.

```
[BYDQ_S7706] interface vlanif 10
[BYDQ_S7706-Vlanif10] ip address 10.10.10.1 24
[BYDQ_S7706-Vlanif10]description to_ShengChanBu
[BYDQ_S7706-Vlanif10] quit
[BYDQ_S7706] interface vlanif 20
[BYDQ_S7706-Vlanif20]description to_XiaoShouBu
[BYDQ_S7706-Vlanif20] ip address 10.10.20.1 24
[BYDQ_S7706-Vlanif20] quit
[BYDQ_S7706] interface vlanif 30
[BYDQ_S7706-Vlanif30] ip address 10.10.30.1 24
[BYDQ_S7706-Vlanif30]description to_CaiWuBu
[BYDQ_S7706-Vlanif30] quit
[BYDQ_S7706] interface vlanif 40
[BYDQ_S7706-Vlanif40] ip address 10.10.40.1 24
[BYDQ_S7706-Vlanif40]description to_AnBaoBu
[BYDQ_S7706-Vlanif40] quit
[BYDQ_S7706] interface vlanif 100
[BYDQ_S7706-Vlanif100] ip address 10.10.100.1 24
[BYDQ_S7706-Vlanif100]description to_Server
[BYDQ_S7706-Vlanif100] quit
[BYDQ_S7706] interface vlanif 255
[BYDQ_S7706-Vlanif255] ip address 10.10.255.1 24
[BYDQ_S7706-Vlanif255]description to_Manager
[BYDQ_S7706-Vlanif255] quit
```

```
[BYDQ_S7706] interface vlanif 1000
[BYDQ_S7706-Vlanif1000] ip address 10.10.252.230
[BYDQ_S7706-Vlanif1000]description to_AR2360
[BYDQ_S7706-Vlanif1000] quit
```

(3) Configure the interface address pool.

```
[BYDQ_S7706] interface vlanif 10
[BYDQ_S7706-Vlanif10] dhcp select interface
[BYDQ_S7706-Vlanif10] dhcp server dns-list 202.103.224.68
[BYDQ_S7706-Vlanif10] quit
[BYDQ_S7706] interface vlanif 20
[BYDQ_S7706-Vlanif20] dhcp select interface
[BYDQ_S7706-Vlanif20] dhcp server dns-list 202.103.224.68
[BYDQ_S7706-Vlanif20] quit
[BYDQ_S7706] interface vlanif 30
[BYDQ_S7706-Vlanif30] dhcp select interface
[BYDQ_S7706-Vlanif30] dhcp server dns-list 202.103.224.68
[BYDQ_S7706-Vlanif30] quit
[BYDQ_S7706] interface vlanif 40
[BYDQ_S7706-Vlanif40] dhcp select interface
[BYDQ_S7706-Vlanif40] dhcp server dns-list 202.103.224.68
[BYDQ_S7706-Vlanif40] quit
```

(4) Configure the interconnect interface with the exit router.

```
[BYDQ_S7706] interface gigabitethernet 1/0/41
[BYDQ_S7706-GigabitEthernet1/0/41] port link-type access
[SW2-GigabitEthernet1/0/41] port default vlan 1000
[SW2-GigabitEthernet1/0/41] quit
```

(5) Configure the default route.

```
[BYDQ_S7706] ip route-static 0.0.0.0 0.0.0.0 10.10.250.1
```

(6) Configure Eth-Trunk.

```
[BYDQ_S7706] interface eth-trunk 1
[BYDQ_S7706-Eth-Trunk1] trunkport gigabitethernet 1/0/1 to 1/0/3
[BYDQ_S7706-Eth-Trunk1] port trunk allow-pass vlan 10 20 30 40 255
[BYDQ_S7706-Eth-Trunk1] quit
```

(7) Configure the proxy mode multi-master detection and BYDQ_S7706 proxy device.

On the proxy device SwitchD, configure the proxy function of eth-trunk.

```
[BYDQ_S7706] interface eth-trunk 11
[BYDQ_S7706-Eth-Trunk1] mad relay
[BYDQ_S7706-Eth-Trunk1] return
```

7.3.6.3 Access switch configuration

Stack:

(1) Configure logical stacking ports and join physical member ports.

Configure the service ports GigabitEthernet0/0/27 and GigabitEthernet0/0/28 of the S5700-LI_1F_1 as physical member ports and add them to the corresponding logical stack ports

```
<Huawei> system-view
[Huawei] sysname S5700-LI_1F_1
[S5700-LI_1F_1] interface stack-port 0/1
[S5700-LI_1F_1-stack-port0/1] port interface gigabitethernet 0/0/27 enable
Warning: Enabling stack function may cause configuration loss on the interface. Continue? [Y/N]:y
Info: This operation may take a few seconds. Please wait.
[S5700-LI_1F_1-stack-port0/1] quit
[S5700-LI_1F_1] interface stack-port 0/2
[S5700-LI_1F_1-stack-port0/2] port interface gigabitethernet 0/0/28 enable
Warning: Enabling stack function may cause configuration loss on the interface. Continue? [Y/N]:y
Info: This operation may take a few seconds. Please wait.
[S5700-LI_1F_1-stack-port0/2] quit
```

Configure the service ports GigabitEthernet0/0/27 and GigabitEthernet 0/0/28 of the S5700-LI_1F_2 as physical member ports and add them to the corresponding logical stacking ports

```
<Huawei>system-view
[Huawei] sysname S5700-LI_1F_2
[S5700-LI_1F_2] interface stack-port 0/1
[S5700-LI_1F_2-stack-port0/1] port interface gigabitethernet 0/0/27 enable
Warning: Enabling stack function may cause configuration loss on the interface. Continue? [Y/N]:y
Info: This operation may take a few seconds. Please wait.
[S5700-LI_1F_2-stack-port0/1] quit
[S5700-LI_1F_2] interface stack-port 0/2
```

[S5700-LI_1F_2-stack-port0/2] port interface gigabitethernet 0/0/28 enable

Warning: Enabling stack function may cause configuration loss on the interface. Continue? [Y/N]:y

Info: This operation may take a few seconds. Please wait.

[S5700-LI_1F_2-stack-port0/2] quit

Configure the service ports GigabitEthernet0/0/27 and GigabitEthernet0/0/28 of the S5700-LI_1F_3 as physical member ports and add them to the corresponding logical stacking ports.

<Huawei>system-view

[Huawei] sysname S5700-LI_1F_3

[S5700-LI_1F_3] interface stack-port 0/1

[S5700-LI_1F_3-stack-port0/1] port interface gigabitethernet 0/0/27 enable

Warning: Enabling stack function may cause configuration loss on the interface. Continue? [Y/N]:y

Info: This operation may take a few seconds. Please wait.

[S5700-LI_1F_3-stack-port0/1] quit

[S5700-LI_1F_3] interface stack-port 0/2

[S5700-LI_1F_3-stack-port0/2] port interface gigabitethernet 0/0/28 enable

Warning: Enabling stack function may cause configuration loss on the interface. Continue? [Y/N]:y

Info: This operation may take a few seconds. Please wait.

[S5700-LI_1F_3-stack-port0/2] quit

(2) Configure the stack ID and stack priority.

[S5700-LI_1F_1] stack slot 0 priority 200#Configure the stacking priority of the S5700-LI_1F_1 to 200

Warning: Please do not frequently modify Priority because it will make the stack split. Continue? [Y/N]:y

[S5700-LI_1F_2] stack slot 0 renumber 1#Configure the stack ID for S5700-LI_1F_2 to 1

Warning: All the configurations related to the slot ID will be lost after the slot ID is modified.

Please do not frequently modify slot ID because it will make the stack split. Continue? [Y/N]:y

Info: Stack configuration has been changed, and the device needs to restart to make the configuration effective.

[S5700-LI_1F_3] stack slot 0 renumber 2#Configure the stack ID of the S5700-LI_1F_3 to 2

Warning: All the configurations related to the slot ID will be lost after the slot ID is modified.

Please do not frequently modify slot ID because it will make the stack split. Continue? [Y/N]:y

Info: Stack configuration has been changed, and the device needs to restart to make the configuration effective.

(3) SwitchA, SwitchB, and SwitchC are powered down, connected with SFP+ cables, and then powered on. This is shown in Figure 7.7.

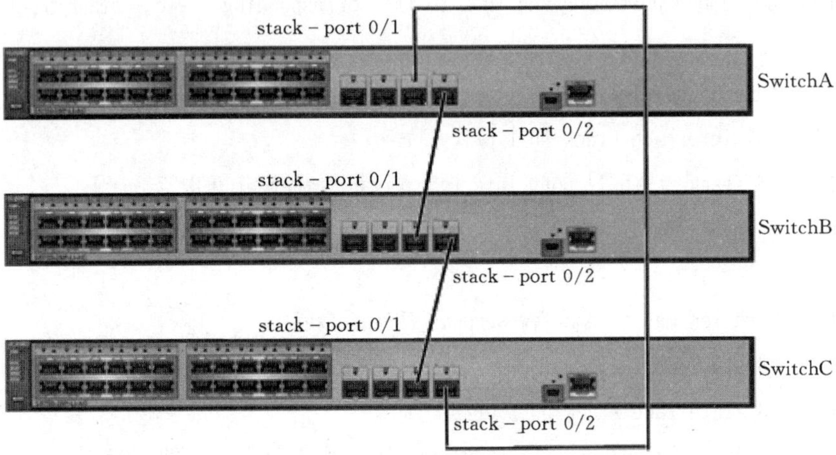

Figure 7.7 Stacking rendering

(4) Configure cross-device Eth-Trunk.

Create an eth-trunk on the stacking system and set the upstream physical port to the eth-trunk member interface

```
[S5700-LI_1F] interface eth-trunk 1
[S5700-LI_1F-Eth-Trunk1] trunkport gigabitethernet 0/0/26
[S5700-LI_1F-Eth-Trunk1] trunkport gigabitethernet 1/0/26
[S5700-LI_1F -Eth-Trunk1] trunkport gigabitethernet 2/0/26
[S5700-LI_1F -Eth-Trunk1]port trunk allow-pass vlan 10 20 30 40 255
[S5700-LI_1F -Eth-Trunk1] quit
```

(5) Configure proxy mode multi-master detection.

On the stacked system, configure the multi-master detection function of the proxy mode of the cross-device eth-trunk

```
[S5700-LI_1F] interface eth-trunk 1
[S5700-LI_1F -Eth-Trunk10] mad detect mode relay
[S5700-LI_1F -Eth-Trunk10] return
```

(6) Configure the port VLAN and join to the edge port.

```
[S5700-LI_1F] interface gigabitethernet 0/0/1
[S5700-LI_1F _S7706-GigabitEthernet0/0/1] port link-type access
[S5700-LI_1F -GigabitEthernet0/0/1] port default vlan 10
[S5700-LI_1F -GigabitEthernet0/0/1] stp edged-port enable
[S5700-LI_1F -GigabitEthernet0/0/1] quit
```

(7) Configure the management IP address.

```
[S5700-LI_1F] interface vlanif 255
[S5700-LI_1F -Vlanif255] ip address 10.10.255.2 24
[S5700-LI_1F -Vlanif255]description to_Manager
[S5700-LI_1F -Vlanif255] quit
```

7.4 Ideological and political integration

China has made great progress in the field of information technology, and the gap with developed countries is narrowing day by day, and computer network engineering has formed a separate discipline, developing in different fields in the direction of high precision. In the process of network engineering project construction, we must constantly pursue perfection and extreme, and integrate the meticulous and excellent craftsman spirit into every link. In the process of learning, college students run through the skill training process with the "four ones" meticulous requirements of "Write every line of code, splice every fibre, install every piece of equipment, do every project well", cultivating the ultimate craftsman's spiritual awareness and professional behavior habits in the spiral process of practical skills and theoretical knowledge, paying attention to the cultivation of core professional qualities, and making first-class network engineering projects.

7.5 Thinking and training

1. What is stacking and what are the advantages of stacking?
2. What can be achieved with switch stacking?
3. What is an edge port? What are the advantages of setting up edge ports?
4. What are the purposes of policy routing?
5. LAN users can use NAT translation to convert a private IP address to a public IP address to access the Internet. Write out the configuration steps.

项目 8

项目 8　某地风力发电厂网络项目

8.1　项目导入

随着现代科学技术的发展与进步，依托计算机网络技术成功地应用到了各个领域当中。在风力发电企业的生产管理中运用计算机网络技术能够高效地完成工作，而且能够实时监测生产，并且收集现场进行生产的数据，及时地了解发电厂的运行情况，提高了工作效率。

8.2　相关知识点

8.2.1　电厂网络相关知识介绍

电厂业务分为综合数据网和调度数据网两大部分。本案例介绍的是调度数据网中的业务配置模式，也是所有电厂中使用最多的模式。电厂网络业务如图 8.1 所示。

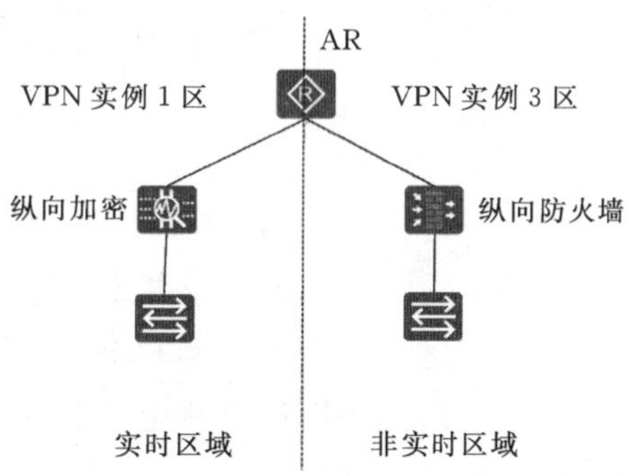

图 8.1　电厂网络业务

调度数据网分为两大区域，实时区域和非实时区域，实时区域包含了出口路由器，纵向加密装置和实时区域交换机，非实时区域包含出口路由器纵向防火墙和实时区域交换机。

实时区域的业务为：远动装置，计量装置等。

非实时区域业务为：风功率装置，遥测装置等。

本案例主要介绍 AR 设备的相关配置。

8.2.2　多区域 OSPF

8.2.2.1　多区域 OSPF 优点

在 OSPF 单区域中，每台路由器都要收集其他所有路由器的链路状态信息，如果网路规模不断扩大，链路状态信息也会随之不断增多，这将使得单台路由器上链路状态数据库非常庞大，导致路由器负担加重，也不便于维护管理。

在一个大型 OSPF 网络中，SPF 算法的反复计算，庞大的路由表和拓扑表的维护以及 LSA 的泛洪等都会占用路由器的资源，因而会降低路由器的运行效率。OSPF 协议可以利用区域的概念来减小这些不利的影响。因为在一个区域内的路由器将不需要了解它们所在区域外的拓扑细节。OSPF 多区域的拓扑结构有如下的优势：①降低 SPF 计算频率；②减小路由表；③降低了通告 LSA 的开销；④将不稳定限制在特定的区域。

8.2.2.2　OSPF 路由器类型

当一个 AS 划分成几个 OSPF 区域时，根据一个路由器在相应的区域之内的作用，可以将 OSPF 路由器作如下分类，如图 8.2 所示。

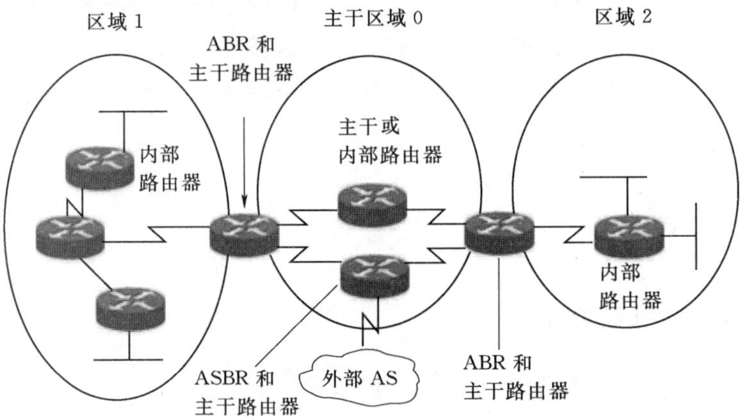

图 8.2　OSPF 路由器类型

（1）内部路由器：OSPF 路由器上所有直连的链路都处于同一个区域。

（2）主干路由器：具有连接区域 0 接口的路由器。

（3）区域边界路由器 (ABR)：路由器与多个区域相连。

（4）自治系统边界路由器 (ASBR)：与 AS 外部的路由器相连并互相交换路由信息。

8.2.2.3　LSA 类型

一台路由器中所有有效的 LSA 通告都被存放在它的链路状态数据库中，正确的 LSA 通告可以描述一个 OSPF 区域的网络拓扑结构。常见的 LSA 有 6 类，相应的描述见表 8.1。

8.2.2.4　区域类型

一个区域所设置的特性控制着它所能接收到的链路状态信息的类型。区分不同 OSPF 区域类型的关键在于它们对外部路由的处理方式。OSPF 区域类型如下：

表 8.1　　LSA 类型及相应的描述

类型代码	名称及路由代码	描　述
1	路由器 LSA（O）	所有的 OSPF 路由器都会产生这种数据包，用于描述路由器上连接到某一个区域的链路或是某一接口的状态信息。该 LSA 只会在某一个特定的区域内扩散，而不会扩散至其他的区域
2	网络 LSA（O）	由 DR 产生，只会在包含 DR 所处的广播网络的区域中扩散，不会扩散至其他的 OSPF 区域
3	网络汇总 LSA（O IA）	由 ABR 产生，描述 ABR 和某个本地区域的内部路由器之间的链路信息。这些条目通过主干区域被扩散到其他的 ABR
4	ASBR 汇总 LSA（O IA）	由 ABR 产生，描述到 ASBR 的可达性，由主干区域发送到其他 ABR
5	外部 LSA（O E1 或 E2）	由 ASBR 产生，含有关于自治系统外的链路信息
6	NSSA 外部 LSA（O N1 或 N2）	由 ASBR 产生的关于 NSSA 的信息，可以在 NSSA 区域内扩散，ABR 可以将类型 7 的 LSA 转换为类型 5 的 LSA

（1）标准区域：可以接收链路更新信息和路由汇总。

（2）主干区域：连接各个区域的中心实体，所有其他的区域都要连接到这个区域上交换路由信息。

（3）末节区域（Stub Area）：不接受外部自治系统的路由信息。

8.2.3　BGP

8.2.3.1　BGP 概述

BGP（Border Gateway Protocol）是一种自治系统间的动态路由发现协议，它的基本功能是在自治系统间自动交换无环路的路由信息，通过交换带有自治系统号（AS）序列属性的路径可达信息，来构造自治区域的拓扑图，从而消除路由环路并实施用户配置的路由策略。与 OSPF 和 RIP 等在自治区域内部运行的协议对应，BGP 是一类 EGP（Exterior Gateway Protocol）协议，而 OSPF 和 RIP 等为 IGP（Interior Gateway Protocol）协议。BGP 协议经常用于 ISP 之间。与 OSPF、RIP 等的内部路由协议不同，其着眼点不在于发现和计算路由，而在于控制路由的传播和选择最好的路由。路由更新时，BGP 只发送增量路由，大大减少了 BGP 传播路由所占用的带宽，适用于在 Internet 上传播大量的路由信息。

8.2.3.2　自治系统

自治系统指由同一个技术管理机构管理、使用统一选路策略的一些路由器的集合。每个自治系统都有唯一的自治系统编号，这个编号是由因特网授权的管理机构分配的。引入自治系统的基本思想：就是通过不同的编号来区分不同的自治系统。这样，当网络管理员不期望自己的通信数据通过某个自治系统时，这种编号方式就十分有用了。或许，该网络管理员的网络完全可以访问这个自治系统，但由于它可能是由竞争对手在管理，或是缺乏足够的安全机制，因此，可能要回避它。通过采用路由协议和自治系统编号，路由器就可以确定彼此间的路径和路由信息的交换方法。

自治系统的编号范围是 1～65535，其中 1～65411 是注册的因特网编号，65412～65535 是专用网络编号。

8.2.3.3 BGP 的工作机制

BGP 系统作为应用层协议运行在一个特定的路由器上。系统初启时通过发送整个 BGP 路由表交换路由信息，之后为了更新路由表只交换更新消息（Update Message）。系统在运行过程中，是通过接收和发送 keep-alive 消息来检测相互之间的连接是否正常。

发送 BGP 消息的路由器称为 BGP 发言人（Speaker），它不断的接收或产生新路由信息，并将该路由信息宣告给其他 BGP 发言人。当 BGP 发言人收到来自其他自治系统的新路由广告时，如果该路由比当前已知路由好或者当前还没有可接受路由，它就把这个路由广告给自治系统内所有其他的 BGP 发言人。

8.2.3.4 BGP 路由属性

BGP 路由属性是一套参数，它对特定的路由进行了进一步的描述，使得 BGP 能够对路由进行过滤和选择。在配置路由策略时我们将广泛地使用路由属性，但是不是所有路由属性都要被用上。常见的 BGP 属性如下：

（1）Origin 起点属性：定义路径信息的来源，标记一条路由是怎样成为 BGP 路由的，如 IGP、EGP、Incomplete 等。

（2）As-Path AS 路径属性：是路由经过的 AS 的序列，即列出在到达所通告的网络之前所经过的 AS 的清单。BGP 发言者将自己的 AS 前置到接收到的 AS 路径的头部，它可以防止路由循环，并用于路由的过滤和选择。

（3）Next hop 下一跳属性：包含到达更新消息所列网络的下一跳边界路由器的 IP 地址。BGP 的下一跳与 IGP 有所不同，它可以是通告此路由的对等体的地址，如 EBGP，这同 IGP 是相同的。而在其他情况下，BGP 使用第三方的下一跳，如 IBGP 对从 EBGP 对等体获得的下一跳不加改变地在自治系统内传递；在多路访问媒体上，BGP 以路由的实际来源为下一跳，即使它不是 BGP 对等体。

（4）MED（Multi-Exit-Discriminators）属性：当某个 AS 有多个入口时，可以用 MED 属性来帮助其外部的邻居路由器选择一个较好的入口路径。一条路由的 MED 值越小，其优先级越高。

（5）Local-Preference 本地优先属性：本地优先属性用于在自治系统内优选到达某一个目的地的路由。反映了 BGP 发言人对每个外部路由的偏好程度。本地优先属性值越大，路由的优选程度就越高。

（6）Community 团体属性：团体属性标识了一组具有相同特征的路由信息，与它所在的 IP 子网和自治系统无关。公认的团体属性值有：NO-EXPORT、NO-ADVERTISE、LOCAL-AS 和 INTERNET。

8.2.3.5　BGP 路由选择过程

本地 BGP 路由选择的过程如下：

（1）如果此路由的下一跳不可达，忽略此路由。

（2）选择本地优先级较大的路由。

（3）选择本地路由器始发的路由（本地优先级相同）。

（4）选择 AS 路径较短的路由。

（5）依次选择起点类型为 IGP、EGP、INCOMPLETE 类型的路由。

（6）选择 MED 较小的路由。

（7）选择 Router ID 较小的路由。

8.2.4　MPLS VPN

MPLS VPN 是指采用 MPLS（多协议标记转换）技术在骨干的宽带 IP 网络上构建企业 IP 专网，实现跨地域、安全、高速、可靠的数据、语音、图像多业务通信，并结合差别服务、流量工程等相关技术，将公众网可靠的性能、良好的扩展性、丰富的功能与专用网的安全、灵活、高效结合在一起。

8.2.4.1　MPLS VPN 提出的意义

传统的 IP 数据转发是基于逐跳式的，每个转发数据的路由器都要根据 IP 包头的目的地址查找路由表来获得下一跳的出口，这是个繁琐又效率低下的工作，主要的原因有以下两个：

（1）有些路由的查询必须对路由表进行多次查找，这就是所谓的递归搜索。

（2）由于路由匹配遵循最长匹配原则，所以迫使几乎所有的路由器的交换引擎必须用软件来实现，用软件实现的交换引擎和 ATM 交换机上用硬件来实现的交换引擎在效率上无法相抗衡。

当今的互联网应用需求日益增多，对带宽、对时延的要求也越来越高。如何提高转发效率，各个路由器生产厂家做了大量的改进工作。但这些修补并不能完全解决目前互联网所面临的问题。IP 和 ATM 曾经是两个互相对立的技术，各个 IP 设备制造商和 ATM 设备制造商都曾努力想“吃掉”对方，IP 想一统天下，ATM 也想一家独秀。但是最终是这两种技术的融合，那就是 MPLS（Multi- Protocol Label Switching）技术做到了这一点。因为 MPLS 技术结合了 IP 技术信令简单和 ATM 交换引擎高效的优点。MPLS 的标签转发，通过事先分配好的标签，为报文建立了一条标签转发通道（LSP），在通道经过的每一台设备中，只需要一次查找，进行快速的标签交换即可。

8.2.4.2　基于 MPLS 的 VPN

VPN 的历史 VPN 服务是很早就提出的概念，不过以前电信提供商提供 VPN 是在传输网上提供覆盖型的 VPN 服务。电信运营商给用户出租线路，用户上层使用何种的路由协议、路由怎么走等，这些电信运营商不管。这种租用线路来搭建 VPN 的好处是安全，但是价格昂贵，线路资源浪费严重。后来随着 IP 网络的全面铺开，电信服务提供商在竞争的压力下，不得不提供更加廉价的 VPN 服务，也就是三层 VPN 服务。这种服务提供给用户一个 IP 平台，用户通过 IP Over IP 的封装格式在公网上打隧道，同时也提供了加密等手段提供安全保障。这类 VPN 用户目前在网络上的数量还是相当巨大的。但是这类 VPN 服务因大量的加密工作、传统路由器根据 IP 包头的目的地址转发效率不高等原因不是非常令人满意。MPLS 技术的出现和 BGP 协议的改进，让大家看到了另一种实现 VPN 的曙光。MPLS 是 IP 与 ATM 技术更好地结合，其以短的、固定长度的标签代替 IP 头作为转发依据，提高转发速度，提供了增值业务，同时不损坏效率。

MPLS 结合了网络层的灵活连接和可扩展性，以及 ATM 的标签转发的可靠传输和 QoS 支持多种标准的路由协议，如 BGP、OSPF 等，支持多种标签生成协议，如 LDP、RSVP，支持多种网络层协议，包括 IPv4、IPv6、IPX 等，有效地解决 QoS 问题，具有标签转发的高性能。

部署了 MPLS 的网络最终都以提供具备某些特性的报文转发为其服务形式。对于各种基于 MPLS 的应用而言，具体的 LSR(Label Switching Router，标记转发路由器及具备标记转发特性的路由器) 在实施报文转发时所采用的核心转发技术事实上是一致的。具体到一个网络层报文而言，当它第一次进入 MPLS 网络时，首先处理它的那个 LSR 必须根据它的网络层头部信息决定对于它的转发动作，这个转发动作包括给它打上 MPLS 标记和从某个接口转发出去；对于网络内部 LSR 而言，只需要以输入报文中的标记信息为索引查找某个预先建立起来的转发信息表得出转发动作，进而根据这个转发动作进行转发；MPLS 网络中最后一个处理这个报文的 LSR 对这个报文的转发决定就可能包含去掉标记这样的动作。

8.3　项目实施

8.3.1　项目需求

（1）全网使用 OSPF 交互路由器信息。

（2）使用 BGP/MPLS VPN 构建专用的 VPN 网络。

8.3.2　设备清单

本项目主要用到路由器、接入层交换机等设备，具体的设备型号和数量见表 8.2。

表 8.2　　网络设备清单表

序号	设备型号	数量	功能	备注
1	AR2240	1	出口路由器	
2	S5700-52P-SI-AC	2	接入交换机	

8.3.3　项目拓扑

本项目的网络拓扑结构如图 8.3 所示。

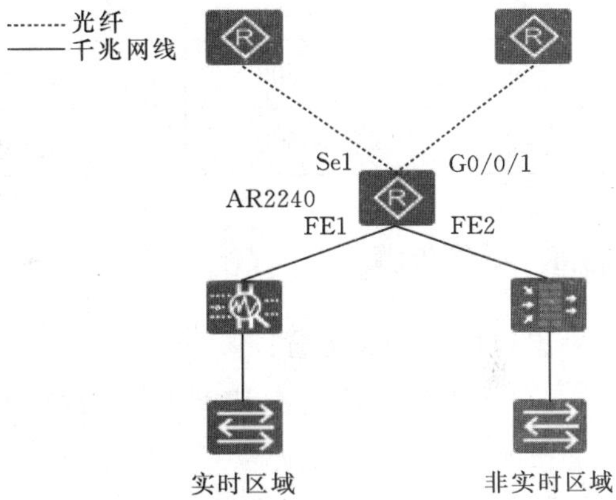

图 8.3　某风力发电厂网络拓扑结构图

8.3.4　配置思路

项目采用如下的思路配置：

（1）出口路由器使用 OSPF 交互相关 BGP 所需的地址信息。

（2）组建 VPN1 为实时区域 VPN。

（3）组建 VPN3 为非实时区域 VPN。

（4）电厂侧的网关放置在 AR 设备上。

（5）两条出口优选走 E1 端口，以太网口为备份端口。

8.3.5　相关信息规划

出口路由器信息规划见表 8.3，OSPF 信息规划见表 8.4，VPN 信息规划见表 8.5。

表 8.3　　出口路由器信息规划表

序号	接口	IP 地址	子网掩码	备注
1	E1	10.15.52.144	255.255.255.252	
2	G0/0/1	10.15.54.144	255.255.255.252	
3	VPN 接口 1	10.66.106.126	255.255.255.192	
4	VPN 接口 3	10.67.106.126	255.255.255.192	
5	本端	10.15.53.141	255.255.255.255	
6	对端 1	10.15.52.1	255.255.255.255	E1
7	对端 2	10.15.52.3	255.255.255.255	G1

表 8.4　　OSPF 信息规划表

序号	OSPF ID	area	主备	备注
1	1	52	主	
2	1	54	备	

表 8.5　　VPN 信息规划表

序号	AS	RD	RT	备注
1	64600	64600:10	64600:10	VPN1
2	64600	64600:30	64600:30	VPN3

8.3.6　操作步骤

出口设备相关配置：

（1）外网出口配置 IP 地址，方法如下：

```
<Huawei> system-view
[Huawei] sysname YCFDC_AR2240
[YCFDC_AR2240] interface gigabitethernet 0/0/1
[YCFDC_AR2240-GigabitEthernet0/0/1]ip address 10.15.54.144  255.255.255.252
[YCFDC_AR2240]controller E1 1/0/0
[YCFDC_AR2240-E1 1/0/0] using e1
[YCFDC_AR2240]interface Serial1/0/0:0
[YCFDC_AR2240- Serial1/0/0:0]link-protocol ppp
[YCFDC_AR2240- Serial1/0/0:0]ip address 10.15.52.144  255.255.255.252
```

（2）配置 router ID，方法如下：

```
[YCFDC_AR2240]interface LoopBack0
[YCFDC_AR2240- LoopBack0] ip address 10.15.53.141 255.255.255.255
[YCFDC_AR2240- LoopBack0] ospc cost 1
[YCFDC_AR2240] router id10.15.53.141
```

（3）配置 MPSL LDP，方法如下：

```
[YCFDC_AR2240]mpls lsr-id  10.15.53.141
[YCFDC_AR2240]mpls
[YCFDC_AR2240]mpls ldp
[YCFDC_AR2240] interface gigabitethernet 0/0/1
```

```
[YCFDC_AR2240-GigabitEthernet0/0/1]mpls
[YCFDC_AR2240-GigabitEthernet0/0/1]mpls ldp
[YCFDC_AR2240]interface Serial1/0/0:0
[YCFDC_AR2240- Serial1/0/0:0]mpls
[YCFDC_AR2240- Serial1/0/0:0]mpls ldp
```

（4）配置 OSPF，方法如下：

```
[YCFDC_AR2240]ospf 1
[YCFDC_AR2240-ospf-1] import-route direct
[YCFDC_AR2240-ospf-1] import-route static
[YCFDC_AR2240-ospf-1] area 52
[YCFDC_AR2240-1-area-0.0.0.52]network 10.15.52.112 0.0.0.3
[YCFDC_AR2240-ospf-1] area54
[YCFDC_AR2240-1-area-0.0.0.54] network 10.15.54.112 0.0.0.3
```

（5）配置 BGP，方法如下：

```
[YCFDC_AR2240]ip vpn-instance vpn1
[YCFDC_AR2240-vpn-instance-vpn1]ipv4-family
[Huawei-vpn-instance-vpn1-af-ipv4]route-distinguisher 64600:10
[Huawei-vpn-instance-vpn1-af-ipv4]vpn-target 64600:10 both
[YCFDC_AR2240]ip vpn-instance vpn3
[YCFDC_AR2240-vpn-instance-vpn1]ipv4-family
[YCFDC_AR2240-vpn-instance-vpn1-af-ipv4]route-distinguisher 64600:30
[YCFDC_AR2240-vpn-instance-vpn1-af-ipv4]vpn-target 64600:30 both

[YCFDC_AR2240]bgp  64600
[YCFDC_AR2240-bgp] peer 10.15.52.1 as-number 64600
[YCFDC_AR2240-bgp] peer 10.15.52.1 connect-interface LoopBack0
[YCFDC_AR2240-bgp] peer 10.15.52.3 as-number 64600
[YCFDC_AR2240-bgp] peer 10.15.52.3 connect-interface LoopBack0

[YCFDC_AR2240-bgp]ipv4-family unicast
[YCFDC_AR2240-bgp-af-ipv4]import-route static
[YCFDC_AR2240-bgp]ipv4-family vpnv4
[YCFDC_AR2240-bgp-af-vpnv4]policy vpn-target
[YCFDC_AR2240-bgp-af-vpnv4] peer 10.15.52.1 enable
```

```
[YCFDC_AR2240-bgp-af-vpnv4]peer 10.15.52.3 enable

[YCFDC_AR2240-bgp] ipv4-family vpn-instance vpn1
[YCFDC_AR2240-bgp-vpn1]import-route direct
[YCFDC_AR2240-bgp-vpn1]import-route static

[YCFDC_AR2240-bgp] ipv4-family vpn-instance vpn3
[YCFDC_AR2240-bgp-vpn3]import-route direct
[YCFDC_AR2240-bgp-vpn3]import-route static
```

（6）配置 VPN 业务网关，方法如下：

```
[YCFDC_AR2240] vlan batch 101 301
[YCFDC_AR2240]interface Vlanif101
[YCFDC_AR2240-Vlanif101]ip binding vpn-instance vpn1
[YCFDC_AR2240-Vlanif101]ip address 10.66.106.126 255.255.255.192

[YCFDC_AR2240]interface Vlanif301
[YCFDC_AR2240-Vlanif101]ip binding vpn-instance vpn3
[YCFDC_AR2240-Vlanif101]ip address 10.67.106.126 255.255.255.192
```

（7）配置接口 VLAN，方法如下：

```
[YCFDC_AR2240]interface Ethernet0/0/1
[YCFDC_AR2240-Ethernet0/0/1]port link-type access
[YCFDC_AR2240-Ethernet0/0/1]port default vlan 101
[YCFDC_AR2240]interface Ethernet0/0/2
[YCFDC_AR2240-Ethernet0/0/2]port link-type access
[YCFDC_AR2240-Ethernet0/0/2]port default vlan 301
```

8.4　思政融入

习近平总书记指出“核心技术是国之重器”。在运用网络为社会发展服务的同时，也必须通过自主创新、技术研发掌握更多的核心技术。在重压下的华为，启用了海思、麒麟芯片，研发了鸿蒙操作系统，在各方面完善其创新生态体系，已走出自主创新之路。当代大学生要树立正确的爱国观念，要有为国家发展作贡献的热情，只有把自己的理想与国家的命运紧密联系在一起，不断努力学习和积累，才能实现厚积而后薄发，在竞争中居于优势地位，企业发展如此，自己现在的学习和未来发展也需如此。

8.5 思考与训练

1. 什么是实时区域和非实时区域？它们有什么区别？
2. 多区域 OSPF 的优点是什么？
3. BGP 路由协议的基本功能是什么？
4. 什么是自治系统？
5. 请简述 BGP 的工作机制，写出基本的 BGP 配置步骤。

Project 8

Project 8 Network project for wind prower plants

8.1 Project import

With the development and progress of modern science and technology, computer network technology has been successfully applied to various fields. In the production management of wind power enterprises, the use of computer network technology can efficiently complete the work, and can monitor the production in real time, collect the production data on site, timely understand the operation of the power plant, and improve the work efficiency.

8.2 Relevant knowledge

8.2.1 Introduction to relevant knowledge of power plant network

The power plant business is divided into two parts: integrated data network and dispatching data network; This case introduces the service configuration mode in the dispatching data network, which is also the most used mode in all power plants.

The relevant topology is shown in Figure 8.1.

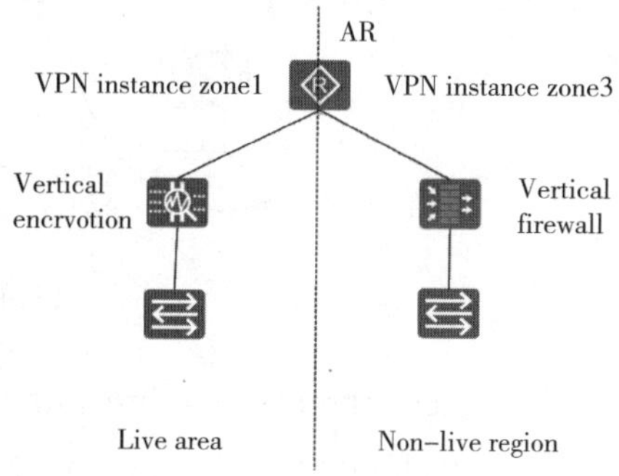

Figure 8.1 grid network business

The dispatching data network is divided into two areas, real-time area and non real-time area. The real-time area includes export router, longitudinal encryption device and real-time area switch, and the non real-time area includes export router, longitudinal firewall and real-time area switch.

The business of real-time area is: telecontrol device, metering device, etc.

Non real-time regional services include: wind power device, telemetry device, etc.

This case mainly introduces the relevant configuration of AR equipment.

8.2.2 Multi area OSPF

8.2.2.1 Advantages of multi area OSPF

In the OSPF single area, each router needs to collect the link state information of all other routers. If the network scale continues to expand, the link state information will continue to increase, which will make the link state database on a single router very large, which will increase the burden on the router and make it difficult to maintain and manage.

In a large OSPF network, the repeated calculation of SPF algorithm, the maintenance of huge routing table and topology table and the flooding of LSA will occupy the resources of the router, which will reduce the operation efficiency of the router. OSPF protocol can use the concept of region to reduce these adverse effects. Because routers in an area will not need to know the topology details outside their area. OSPF multi region topology has the following advantages: ① Reduce SPF calculation frequency; ② Reduce routing table; ③ Reduce the overhead of advertising LSA; ④ Limit instability to specific areas.

8.2.2.2 OSPF router type

When an AS is divided into several OSPF areas, OSPF routers can be classified as follows according to the role of a router in the corresponding area, as shown in Figure 8.2.

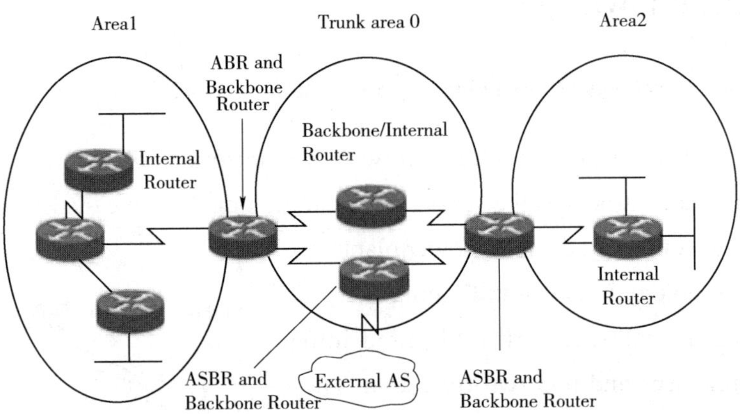

Figure 8.2 OSPF router type

(1) Internal router: all direct links on OSPF router are in the same area.

(2) Backbone router: router with connection area 0 interface.

(3) Area boundary router (ABR): the router is connected with multiple areas.

(4) Autonomous system boundary router (ASBR): connect with the router outside the AS and exchange routing information with each other.

8.2.2.3 LSA type

All valid LSA advertisements in a router are stored in its link state database. The correct LSA advertisement can describe the network topology of an OSPF area. There are 6 types of common LSAS, and the corresponding description is shown in Table 8.1.

Table 8.1 LSA type and corresponding description

Type code	Name and routing code	Describe
1	Router LSA（O）	All OSPF routers will generate this kind of data packet, which is used to describe the status information of the link or interface connected to a certain area on the router. The LSA will only spread in a specific area, not to other areas
2	Network LSA（O）	Generated by DR, it will only spread in the area of the broadcast network containing DR, and will not spread to other OSPF areas
3	Network summary LSA（O IA）	It is generated by ABR and describes the link information between ABR and the internal router in a local area. These entries are diffused to other ABRs through the trunk area
4	ASBR summary LSA（O IA）	Generated by ABR, describing the accessibility to ASBR, and sent by the trunk area to other ABRs
5	External LSA (O E1 or E2)	It is generated by ASBR and contains link information outside the autonomous system
6	NSSA external LSA（O N1 or N2）	The information about NSSA generated by ASBR can spread in the NSSA area, and ABR can convert type 7 LSA to type 5 LSA

8.2.2.4 Area type

The characteristics set by an area control the type of link state information it can receive. The key to distinguishing different OSPF area types lies in their processing methods of external routing. OSPF area types are as follows:

（1）Standard area: it can receive link update information and route summary.

（2）Backbone area: the central entity connecting each area. All other areas should be connected to this area to exchange routing information.

（3）Stub area: does not accept routing information from external autonomous systems.

8.2.3 BGP

8.2.3.1 BGP overview

BGP (Border Gateway Protocol) is a dynamic routing discovery protocol between autonomous systems. Its basic function is to automatically exchange loop free routing information between autonomous systems, and construct the topology of the autonomous region by exchanging path reachability information with autonomous system number (AS) sequence attribute, so as to eliminate the routing loop and implement the user configured routing strategy. Corresponding to OSPF, RIP and other protocols running within the autonomous region, BGP is a kind of EGP (Exterior Gateway Protocol) protocol, while OSPF, RIP and other protocols are IGP (Interior Gateway Protocol). BGP protocol is often used between ISPs. Different from OSPF, RIP and other internal routing protocols, its focus is not to discover and calculate routes, but to control the propagation of routes and select the best routes. When the route is updated, BGP only sends incremental routes, which greatly reduces the bandwidth occupied by BGP propagation routes and is suitable for transmitting a large amount of routing information on the Internet.

8.2.3.2 Autonomous system

Autonomous system refers to a collection of routers managed by the same technical management organization and using unified routing strategy. Each autonomous system has a unique autonomous system number, which is assigned by the management organization authorized by the Internet. The basic idea of introducing autonomous system is to distinguish different autonomous systems by different numbers. In this way, this numbering method is very useful when network administrators do not expect their communication data to pass through an autonomous system. Perhaps, the network administrator's network can fully access the autonomous system, but it may be managed by competitors or lack of sufficient security mechanism, so it may be avoided. By using routing protocol and autonomous system number, routers can determine the path between each other and the exchange method of routing information.

The number range of autonomous systems is 1 to 65535, where 1 to 65411 are registered Internet numbers and 65412 to 65535 are private network numbers.

8.2.3.3 Working mechanism of BGP

BGP system runs on a specific router as an application layer protocol. At initial startup, the system exchanges routing information by sending the whole BGP routing table, and then only exchange update messages in order to update the routing table. During operation, the system detects whether the connection between them is normal by receiving and sending keep-alive messages.

The router that sends BGP messages is called BGP speaker. It constantly receives or generates new routing information and announces the routing information to other BGP speaker. When the BGP speaker receives a new route advertisement from other autonomous systems, if the route is better than the currently known route or there is no acceptable route, it will advertise the route to all other BGP speaker in the autonomous system.

8.2.3.4 BGP routing attribute

BGP routing attribute is a set of parameters, which further describes the specific route, so that BGP can filter and select the route. We will widely use routing attributes when configuring routing policies, but not all routing attributes should be used. Common BGP attributes are:

(1) Origin origin attribute: defines the source of path information and marks how a route becomes a BGP route. Such as IGP, EGP, incomplete, etc.

(2) As-Path attribute: it is the sequence of as passed by the route, that is, it lists the list of as passed before reaching the advertised network. BGP speakers put their own as in front of the head of the received as path, which can prevent routing cycle and be used for routing filtering and selection.

(3) Next hop attribute: contains the IP address of the next hop boundary router reaching the network listed in the update message. The next hop of BGP is different from IGP. It can be the address of the peer that advertises this route, such as EBGP, which is the same as IGP. In other cases, BGP uses the next hop of the third party, for example, iBGP transmits the next hop obtained from EBGP peer unchanged in the autonomous system; On multi access media, BGP takes the actual source of the route as the next hop, even if it is not a BGP peer.

(4) MED (Multi-Exit-Discriminators) attribute: when an AS has multiple entries, the Med attribute can be used to help its external neighbor router choose a better entry path. The smaller the MED value of a route, the higher its priority.

(5) Local-Preference attribute: the Local-Preference attribute is used to select the route to a destination in the autonomous system. It reflects the BGP speaker preference for each external route. The greater the local priority attribute value, the higher the optimization degree of routing.

(6) Community attribute: the community attribute identifies a group of routing information with the same characteristics, which is independent of its IP subnet and autonomous system. Recognized community attribute values include: NO-EXPORT, NO-ADVERTISE, LOCAL-AS and INTERNET.

8.2.3.5 BGP routing process

The process of local BGP routing is as follows:

(1) If the next hop of this route is unreachable, ignore this route.

(2) Select the route with higher local priority.

(3) Select the route originating from the local router (the local priority is the same).

(4) Select the route with shorter as path.

(5) Select routes with IGP, EGP and INCOMPLETE starting point types in turn.

(6) Select the route with smaller MED.

(7) Select a route with a smaller Router ID.

8.2.4 MPLS VPN

MPLS VPN refers to using MPLS (multi protocol label conversion) technology to build an enterprise IP private network on the backbone broadband IP network to realize cross regional, safe, high-speed and reliable multi service communication of data, voice and image. Combined with differential service, traffic engineering and other related technologies, MPLS VPN combines the reliable performance, good scalability and rich functions of the public network with the security, flexibility and efficiency of the private network.

8.2.4.1 The significance of MPLS VPN

The traditional IP data forwarding is based on hop by hop. Each router forwarding data must look up the routing table according to the destination address of the IP packet header to obtain the exit of the next hop. This is a cumbersome and inefficient work. The main reasons are two:

(1) Some route queries must search the route table many times, which is the so-called recursive search.

(2) Because the route matching follows the longest matching principle, almost all router switching engines must be realized by software. The switching engine realized by software and the switching engine realized by hardware on ATM switch cannot compete in efficiency.

Nowadays, the demand for Internet applications is increasing, and the requirements for bandwidth and delay are also higher and higher. How to improve the forwarding efficiency, various router manufacturers have done a lot of improvements. However, these fixes can not completely solve the problems faced by the Internet. IP and ATM used to be two opposing technologies. Various IP equipment manufactures and ATM equipment manufactures have tried to annex each other, IP want to dominate the world, and ATM also wants to be a one-man show. But in the end, it is the integration of these two technologies, that is, MPLS (Multi-Protocol Label Switching) technology, because MPLS technology combines the advantages of simple signaling of IP technology and efficient ATM switching engine. In MPLS label forwarding, a labelled forwarding channel (LSP) is established for the message by means of a pre-assigned label. Each device passing through the channel only needs one search for fast label exchange.

8.2.4.2 VPN based on MPLS

The history of VPN service is a concept put forward for a long time, but in the past, VPN provided by telecom providers was an overlay VPN service provided on the transmission network. Telecom operators rent lines to users, which routing protocol the upper layer of users uses, how to route, etc. These telecom operators don't care. This kind of leased line to build VPN has the advantage of security, but it is expensive and a serious waste of line resources. Later, with the full spread of IP network, telecom service providers had to provide cheaper VPN services, that is, Layer-3 VPN services, under the pressure of competition. By providing users with an IP platform, users can tunnel on the public network through the encapsulation format of IP over IP. At the same time, encryption and other means are also provided to provide security. The number of such VPN users on the current network is still quite huge. However, this kind of VPN service is not very satisfactory because of a lot of encryption work and the low forwarding efficiency of traditional routers according to the destination address of IP packet header. The advent of MPLS technology and improvements in the BGP protocol have shown another way to implement VPN. MPLS is a better combination of IP and ATM technology. It uses short and fixed length tags instead of IP headers as the forwarding basis to improve the forwarding speed and provide value-added services without damaging the efficiency.

MPLS combines the flexible connection and scalability of the network layer and the reliable transmission and QoS of ATM label forwarding. It supports a variety of standard routing protocols, such as BGP and OSPF. It supports a variety of label generation protocols, such as LDP and RSVP. It supports a variety of network layer protocols, including IPv4, IPv6 and IPX. It effectively solves the QoS problem and has the high performance of label forwarding.

The network deployed with MPLS finally provides message forwarding with some characteristics as its service form. For various MPLS based applications, the core forwarding technology adopted by the specific LSR (Label Switching Router) in the implementation of message forwarding is actually the same. For a network layer message, when it enters the MPLS network for the first time, the LSR that first processes it must determine its forwarding action according to its network layer header information. This forwarding action includes marking it with MPLS mark and forwarding it from an interface; For the internal LSR of the network, it only needs to look up a pre established forwarding information table with the tag information in the input message as the index to obtain the forwarding action, and then forward according to the forwarding action; The forwarding decision of the last LSR to process the message in MPLS network may include the action of removing the mark.

8.3 Project implementation

8.3.1 Project requirements

(1) The entire network uses OSPF to interact with router information.

(2) Build a dedicated VPN network with BGP/MPLS VPN.

8.3.2 Device inventory

This project mainly uses routers, access layer switches and other equipment,the specific equipment model and quantity are shown in Table 8.2.

Table 8.2 List of network devices

Serial No.	Device model	Quantity	Function	Remark
1	AR2240	1	Egress router	
2	S5700–52P–SI–AC	2	Access switch	

8.3.3 Project topology

The network topology of this project is shown in Figure 8.3.

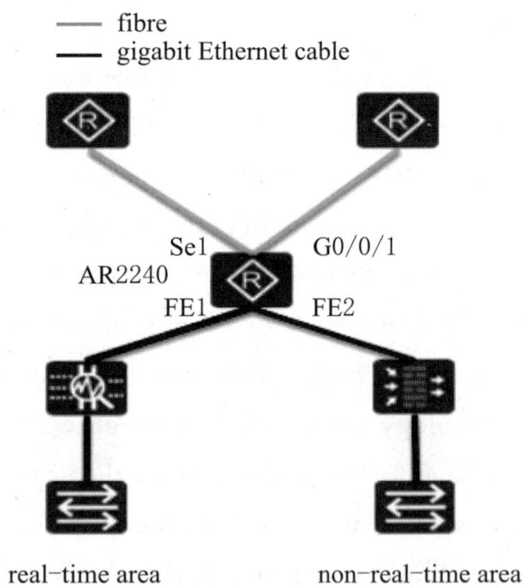

Figure 8.3 Topology diagram of a wind farm network

8.3.4 Configuration ideas

The project adopts the following idea configuration:

(1) The exit router uses OSPF to interact with the address information required for the relevant BGP.

(2) VPN1 is formed as a real-time area VPN.

(3) VPN3 is formed as a non real-time area VPN.

(4) The gateway on the power plant side is placed on the AR device.

(5) The two exit ports are preferably E1 ports, and the Ethernet port is the backup port.

8.3.5 Planning of relevant information

The information planning of the exit router is shown in Table 8.3, Table 8.4 shows the OSPF information plan, and Table 8.5 shows the VPN information plan.

Table 8.3 Exit router information planning table

Serial No.	Interface	IP address	Mask	Remark
1	E1	10.15.52.144	255.255.255.252	
2	G0/0/1	10.15.54.144	255.255.255.252	
3	VPN interface1	10.66.106.126	255.255.255.192	
4	VPN interface3	10.67.106.126	255.255.255.192	
5	Local	10.15.53.141	255.255.255.255	
6	Peer1	10.15.52.1	255.255.255.255	E1
7	Peer2	10.15.52.3	255.255.255.255	G1

Table 8.4 OSPF information planning table

Serial No.	OSPF ID	Area	Primary and Reserve	Remark
1	1	52	Primary	
2	1	54	Reserve	

Table 8.5 VPN information planning table

Serial No.	AS	RD	RT	Remark
1	64600	64600:10	64600:10	VPN1
2	64600	64600:30	64600:30	VPN3

8.3.6 Procedure

Exit device–related configuration:

(1) Configure the IP address for the exit of the Internet.

```
<Huawei> system-view
[Huawei] sysname YCFDC_AR2240
[YCFDC_AR2240] interface gigabitethernet 0/0/1
[YCFDC_AR2240-GigabitEthernet0/0/1]ip address 10.15.54.144  255.255.255.252
[YCFDC_AR2240]controller E1 1/0/0
[YCFDC_AR2240-E1 1/0/0] using e1
[YCFDC_AR2240]interface Serial1/0/0:0
[YCFDC_AR2240- Serial1/0/0:0]link-protocol ppp
```

```
[YCFDC_AR2240- Serial1/0/0:0]ip address 10.15.52.144  255.255.255.252
```

（2）Configure router ID.

```
[YCFDC_AR2240]interface LoopBack0
[YCFDC_AR2240- LoopBack0] ip address 10.15.53.141 255.255.255.255
[YCFDC_AR2240- LoopBack0] ospc cost 1
[YCFDC_AR2240] router id10.15.53.141
```

（3）Configure MPSL LDP.

```
[YCFDC_AR2240]mpls lsr-id  10.15.53.141
[YCFDC_AR2240]mpls
[YCFDC_AR2240]mpls ldp
[YCFDC_AR2240] interface gigabitethernet 0/0/1
[YCFDC_AR2240-GigabitEthernet0/0/1]mpls
[YCFDC_AR2240-GigabitEthernet0/0/1]mpls ldp
[YCFDC_AR2240]interface Serial1/0/0:0
[YCFDC_AR2240- Serial1/0/0:0]mpls
[YCFDC_AR2240- Serial1/0/0:0]mpls ldp
```

（4）Configure OSPF.

```
[YCFDC_AR2240]ospf 1
[YCFDC_AR2240-ospf-1] import-route direct
[YCFDC_AR2240-ospf-1] import-route static
[YCFDC_AR2240-ospf-1] area 52
[YCFDC_AR2240-1-area-0.0.0.52]network 10.15.52.112 0.0.0.3
[YCFDC_AR2240-ospf-1] area54
[YCFDC_AR2240-1-area-0.0.0.54] network 10.15.54.112 0.0.0.3
```

（5）Configure BGP .

```
[YCFDC_AR2240]ip vpn-instance vpn1
[YCFDC_AR2240-vpn-instance-vpn1]ipv4-family
[Huawei-vpn-instance-vpn1-af-ipv4]route-distinguisher 64600:10
[Huawei-vpn-instance-vpn1-af-ipv4]vpn-target 64600:10 both
[YCFDC_AR2240]ip vpn-instance vpn3
[YCFDC_AR2240-vpn-instance-vpn1]ipv4-family
[YCFDC_AR2240-vpn-instance-vpn1-af-ipv4]route-distinguisher 64600:30
[YCFDC_AR2240-vpn-instance-vpn1-af-ipv4]vpn-target 64600:30 both
```

```
[YCFDC_AR2240]bgp  64600
[YCFDC_AR2240-bgp] peer 10.15.52.1 as-number 64600
[YCFDC_AR2240-bgp] peer 10.15.52.1 connect-interface LoopBack0
[YCFDC_AR2240-bgp] peer 10.15.52.3 as-number 64600
[YCFDC_AR2240-bgp] peer 10.15.52.3 connect-interface LoopBack0

[YCFDC_AR2240-bgp]ipv4-family unicast
[YCFDC_AR2240-bgp-af-ipv4]import-route static
[YCFDC_AR2240-bgp]ipv4-family vpnv4
[YCFDC_AR2240-bgp-af-vpnv4]policy vpn-target
[YCFDC_AR2240-bgp-af-vpnv4] peer 10.15.52.1 enable
[YCFDC_AR2240-bgp-af-vpnv4]peer 10.15.52.3 enable

[YCFDC_AR2240-bgp] ipv4-family vpn-instance vpn1
[YCFDC_AR2240-bgp-vpn1]import-route direct
[YCFDC_AR2240-bgp-vpn1]import-route static

[YCFDC_AR2240-bgp] ipv4-family vpn-instance vpn3
[YCFDC_AR2240-bgp-vpn3]import-route direct
[YCFDC_AR2240-bgp-vpn3]import-route static
```

（6）Configure the VPN service gateway.

```
[YCFDC_AR2240] vlan batch 101 301
[YCFDC_AR2240]interface Vlanif101
[YCFDC_AR2240-Vlanif101]ip binding vpn-instance vpn1
[YCFDC_AR2240-Vlanif101]ip address 10.66.106.126 255.255.255.192

[YCFDC_AR2240]interface Vlanif301
[YCFDC_AR2240-Vlanif101]ip binding vpn-instance vpn3
[YCFDC_AR2240-Vlanif101]ip address 10.67.106.126 255.255.255.192
```

（7）Configure the interface VLAN.

```
[YCFDC_AR2240]interface Ethernet0/0/1
[YCFDC_AR2240-Ethernet0/0/1]port link-type access
[YCFDC_AR2240-Ethernet0/0/1]port default vlan 101
[YCFDC_AR2240]interface Ethernet0/0/2
```

```
[YCFDC_AR2240-Ethernet0/0/2]port link-type access
[YCFDC_AR2240-Ethernet0/0/2]port default vlan 301
```

8.4 Ideological and political integration

"Core technology is the most important weapon of the country". While using the network to serve social development, we must also master more core technologies through independent innovation and technology research and development. Under the heavy pressure, Huawei has enabled HiSilicon and Kirin chips, developed the Hongmeng operating system, improved its innovation ecosystem in all aspects, and has embarked on the road of independent innovation. Contemporary college students should establish a correct patriotic concept and have the enthusiasm to contribute to the development of the country, only by closely linking their ideals with the fate of the country, Study and accumulate hard in order to achieve that accumulation of thick and thin, and to have an advantage in the competition, and occupy an advantageous position in the competition.

8.5 Thinking and training

1. What are real time and non real-time areas? What are the differences between them?
2. What are the advantages of multi-region OSPF?
3. What are the basic functions of BGP routing protocols?
4. What is an autonomous system?
5. Please briefly describe the working mechanism of BGP and write out the basic BGP configuration steps.

项目 9　安装防火墙加固内网安全项目

9.1　项目导入

当前，随着网络技术的不断发展，计算机网络为人们的生活带来了极大的方便，企业信息化、数字化转型升级进一步提速。然而，利用计算机网络进行的各种违法犯罪活动也在迅速增长，计算机犯罪、黑客、有害程序和后门等问题严重威胁着网络安全。作为内部网络与外部公共网络之间的第一道屏障，防火墙是最先受到人们重视的网络安全产品之一。

防火墙是网络基础设施中用于保护网络安全的设备，是网络安全的第一道防线。防火墙可以由硬件组成，也可以由软件组成，也可以集成在路由器中，借助路由器对网络的路由功能对流经的数据包进行分析和监控。防火墙一般部署在企业内部网络和外部网络之间互联的中间设备上，在内网和外网之间、专用网与公共网之间建立保护屏障的作用。防火墙的作用是检查通过防火墙的数据包并根据预设的安全策略决定数据包的流向，可保护内部用户不被互联网非法用户侵入，也可阻止内部重要信息向外流出，有双向监督功能，如图 9.1 所示。

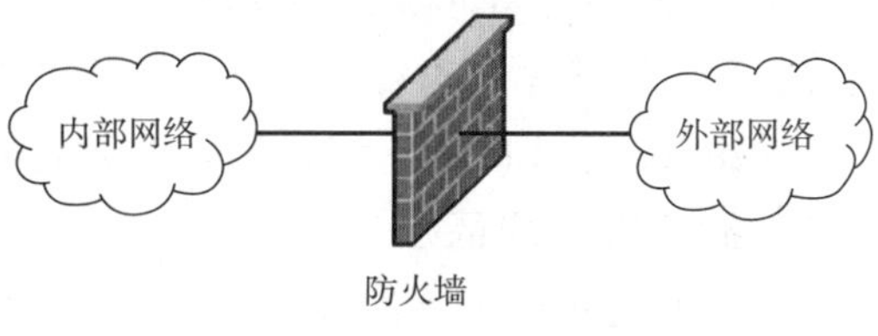

图 9.1　防火墙部署位置

9.2　相关知识点

9.2.1　防火墙的概念

防火墙（Firewall）也称防护墙，是一种位于内部网络与外部网络之间的网络安全系统。它是一项信息安全的防护系统，依照特定的规则，允许或是限制传输的数据通过。

防火墙起源于建筑领域，原本是指房屋之间修建的一道墙，用以防止火灾发生时火势蔓延，阻止火势从一个区域蔓延到另一个区域。在网络中，防火墙是指一种将内部网和外部网（如互联网）分开的方法，它实际上是一种隔离技术。防火墙是在两个网络通信时执行的一种访问控制尺度，它能允许用户“同意”的人和数据进入他的网络，同时将用户“不同意”的人和数据拒之门外，最大限度地阻止网络中的黑客来访问自己的网络。换句话说，如果不通过防火墙，公司内部的人就无法访问 Internet，Internet 上的人也无法和公司内部的人进行通信。

9.2.2　防火墙的分类

9.2.2.1　按软、硬件形式区分

按软、硬件形式的不同，防火墙可分为软件防火墙、硬件防火墙、芯片级防火墙。

（1）软件防火墙：安装在操作系统上的软件，软件防火墙通过软件去实现隔离内部网与外部网之间的一种保护屏障。由于它安装在系统主机上，因此需要依赖计算机硬件资源来工作，需要消耗系统的内存和 CPU 资源。软件防火墙像其他软件产品一样需要在每台计算机上安装并做好配置才可以使用。

（2）硬件防火墙：本质上是把软件防火墙嵌入在硬件中，硬件防火墙的硬件和软件都需要单独设计，使用专用网络芯片来处理数据包，同时，采用专门的操作系统平台，从而避免通用操作系统的安全漏洞导致内网安全受到威胁。

（3）芯片级防火墙：芯片级防火墙以专用的硬件作为基础，不需要专门的操作系统的支持。专用的芯片因为自身特点，具有更快的工作速度、更高的处理能力以及更高的效率。专有的 ASIC 芯片促使它们比其他种类的防火墙速度更快、处理能力更强、性能更高。

9.2.2.2　按过滤技术区分

按过滤技术区分，防火墙分为包过滤技术、应用代理技术、状态、检测技术和完全内容检测技术等几大类型。

（1）包过滤技术。包过滤技术是一种简单、有效的安全控制技术，它工作在网络层，通过在网络间相互连接的设备上加载允许、禁止来自某些特定的源地址、目的地址、TCP 端口号等规则，对通过设备的数据包进行检查，限制数据包进出内部网络。

包过滤的最大优点是对用户透明，传输性能高。但由于安全控制层次在网络层、传输层，安全控制的力度也只限于源地址、目的地址和端口号，因而只能进行较为初步的安全控制，对于恶意的拥塞攻击、内存覆盖攻击或病毒等高层次的攻击手段，则无能为力。

（2）应用代理技术。应用代理防火墙工作在 OSI 的第七层，它通过检查所有应用层的信息包，并将检查的内容信息放入决策过程，从而提高网络的安全性。

应用网关防火墙是通过打破客户机 / 服务器模式实现的。每个客户机 / 服务器通信需要两个连接：一个是从客户端到防火墙，另一个是从防火墙到服务器。另外，每个代

理需要一个不同的应用进程，或一个后台运行的服务程序，对每个新的应用必须添加针对此应用的服务程序，否则不能使用该服务。所以，应用网关防火墙具有可伸缩性差的缺点。

（3）状态检测技术。状态检测防火墙工作在 OSI 的第二至四层，采用状态检测包过滤的技术，由传统包过滤功能扩展而来。状态检测防火墙在网络层有一个检查引擎截获数据包并抽取出与应用层状态有关的信息，并以此为依据决定对该连接是接受还是拒绝。这种技术提供了高度安全的解决方案，同时具有较好的适应性和扩展性。状态检测防火墙一般也包括一些代理级的服务，它们提供附加的对特定应用程序数据内容的支持。

状态检测防火墙基本保持了简单包过滤防火墙的优点，性能比较好，同时对应用是透明的，在此基础上，对于安全性有了大幅提升。这种防火墙摒弃了简单包过滤防火墙仅仅考察进出网络的数据包、不关心数据包状态的缺点，在防火墙的核心部分建立状态连接表，维护了连接，将进出网络的数据当成一个个的事件来处理。主要特点是由于缺乏对应用层协议的深度检测功能，无法彻底识别数据包中大量的垃圾邮件、广告以及木马程序等。

（4）完全内容检测技术。完全内容检测技术防火墙综合状态检测与应用代理技术，并在此基础上进一步基于多层检测架构，把防病毒、内容过滤、应用识别等功能整合到防火墙里，其中还包括 IPS 功能，多单元融为一体，在网络界面对应用层扫描，把防病毒、内容过滤与防火墙结合起来，这体现了网络与信息安全的新思路（因此也被称为“下一代防火墙技术”）。它在网络边界实施 OSI 第七层的内容扫描，实现了实时在网络边缘部署病毒防护、内容过滤等应用层服务措施。完全内容检测技术防火墙可以检查整个数据包内容，根据需要建立连接状态表，并且拥有网络层保护强、应用层控制细等优点，但由于功能集成度高，对产品硬件的要求比较高。

9.2.3　防火墙的安全区域

安全区域（Security Zone）简称为区域（Zone）。安全区域是一个或多个接口的集合，防火墙通过安全区域来划分网络、标识报文流动的“路线”。默认情况下，报文在同一个安全区域内流动时不受控制，在不同的安全区域之间流动时受到控制。

华为防火墙上默认提供了 4 个安全区域，分别是 Trust、DMZ、Untrust 和 Local，每个安全区域都有一个唯一的安全级别，用 1～100 的数字表示，数字越大，则代表该区域内的网络越可信。通过把防火墙的不同接口划分到不同的安全区域中，就可以在防火墙上划分出不同的安全区域网络。

9.2.3.1　Trust 区域

Trust 区域的默认安全级别是 85，该区域内网络的受信任程度高，通常用来定义内部用户所在的网络。

9.2.3.2　DMZ 区域

DMZ 区域的默认安全级别是 50，该区域内网络的受信任程度中等，通常用于定义内网服务器所在区域。因为这种设备虽然部署在内网，但是经常需要被外网访问，存在较大安全隐患，同时一般又不允许其主动访问外网，所以将其部署在一个优先级比 Trust 低、但是比 Untrust 高的安全区域中。DMZ（Demilitarized Zone）起源于军方，是介于严格的军事管制区和松散的公共区域之间的一种有着部分管制的区域。防火墙设备引用了这一术语，指的是介于内部网络和外部网络之间的安全区域。

9.2.3.3　Untrust 区域

Untrust 区域的默认安全级别是 5，该区域代表的是不受信任的网络，通常用来定义 Internet 等不安全的网络。

9.2.3.4　Local 区域

Local 区域的默认安全级别是 100，代表防火墙本身。Local 区域中不能添加任何接口，但防火墙上所有接口本身都隐含属于 Local 区域。

图 9.2 给出了防火墙部署在企业内部的真实环境组网图。从图中可以看出，企业内部网络中的用户、服务器以及位于外部的 Internet，都被划分到不同的安全区域中了，防火墙对各个安全区域之间流动的报文进行安全检查。

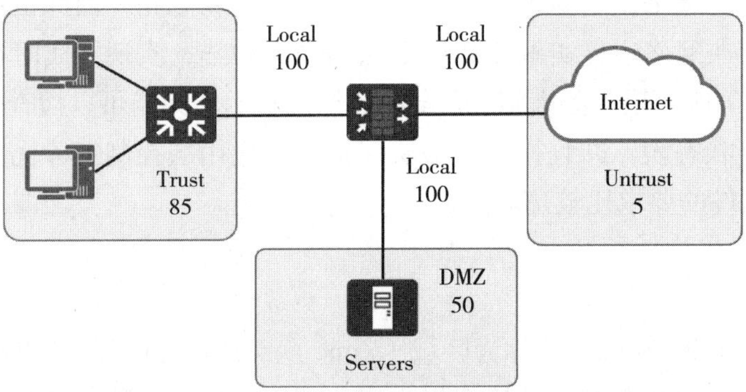

图 9.2　防火墙组网拓扑图

在没有防火墙的情况下，路由器会在内部网络和外部网络之间直接传递流量且没有过滤机制，而防火墙技术的功能主要在于及时发现并处理计算机网络运行时可能存在的安全风险、数据传输等问题，其中处理措施包括隔离与保护，同时可对计算机网络安全当中的各项操作实施记录与检测，以确保计算机网络运行的安全性，保障用户资料与信息的完整性，为用户提供更好、更安全的计算机网络使用体验。

9.3　项目实施

9.3.1　项目需求

（1）部署防火墙，实现内网用户能访问互联网。

（2）外网用户能访问服务器区的 Web 服务。

（3）内网用户能访问服务器区的 ICMP、Web 服务。

9.3.2　设备清单

本项目主要用到防火墙、接入层交换机等设备，具体的设备型号和数量见表 9.1。

表 9.1　网络设备清单表

序号	设备型号	数量	功能	备注
1	S5700	1	接入交换机	
2	USG6000V	1	防火墙	
3	AR2220	1	路由器	运营商设备

9.3.3　项目拓扑

本项目的网络拓扑结构如图 9.3 所示。

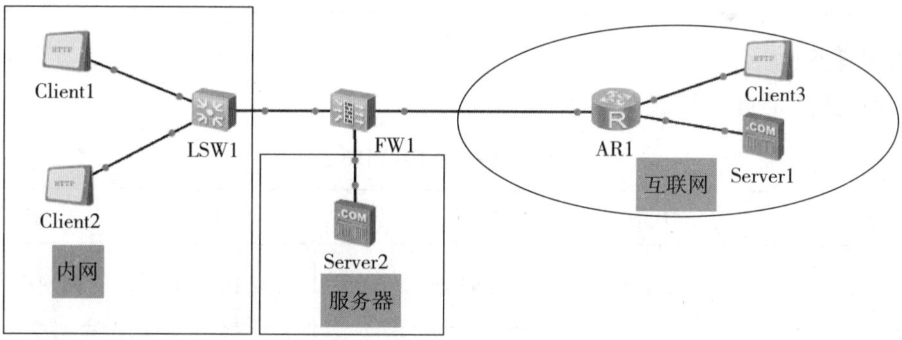

图 9.3　网络拓扑图

9.3.4　配置思路

项目采用如下的思路配置：

（1）初始化防火墙，允许管理员通过 Web 管理防火墙。

（2）配置防火墙各接口地址及所属的安全区域。

（3）配置安全策略，满足数据流向要求。

（4）配置 NAT，满足内网访问外网需求。

（5）配置服务器映射，使得外网用户能正常访问服务器区。

9.3.5　相关信息规划

9.3.5.1　防火墙

防火墙的地址规划见表 9.2。

表 9.2　防火墙地址规划表

序号	接口	IP 地址	子网掩码	备注
1	G1/0/1	209.165.222.1	255.255.255.248	连接外网
2	G1/0/2	192.168.1.1	255.255.255.0	连接内网
3	G1/0/3	192.168.2.1	255.255.255.0	连接服务器

9.3.5.2　路由器

运营商的设备 AR1 地址规划见表 9.3。

表 9.3　路由器地址规划表

序号	接口	IP 地址	子网掩码	备注
1	G0/0/0	209.165.222.2	255.255.255.248	连接防火墙
2	G0/0/1	8.8.8.1	255.255.255.0	连接互联网服务器 Server1

9.3.5.3　其他设备

其他设备的地址规划见表 9.4。

表 9.4　其他设备地址规划表

序号	设备名称	IP 地址	子网掩码	网关	备注
1	内网 PC Client1	192.168.1.2	255.255.255.0	192.168.1.1	内网用户 PC
2	内网 PC Client2	192.168.1.3	255.255.255.0	192.168.1.1	内网用户 PC
3	服务器区 Sever2	192.168.3.2	255.255.255.0	192.168.3.1	服务器区服务器
4	互联网服务器 Server1	8.8.8.8	255.255.255.0	8.8.8.1	互联网服务器
5	互联网用户 Client3	9.9.9.2	255.255.255.0	9.9.9.1	互联网用户

9.3.6　操作步骤

9.3.6.1　初始化防火墙，允许管理员通过 Web 管理防火墙

启动防火墙，通过配置线登录防火墙的 CLI 界面后，进行防火墙的初始化。

（1）输入默认用户名 admin，密码 Admin@123，登录防火墙。

（2）出现提示是否需要修改密码的提示，选择 y。

（3）输入旧密码 Admin@123，然后输入新密码，再重复输入一遍新密码，即可登录防火墙，新密码要求满足密码复杂性要求，即最低 8 位字符，包含大写字母、小写字母、数字、特殊字符中的 3 种，如图 9.4 所示。

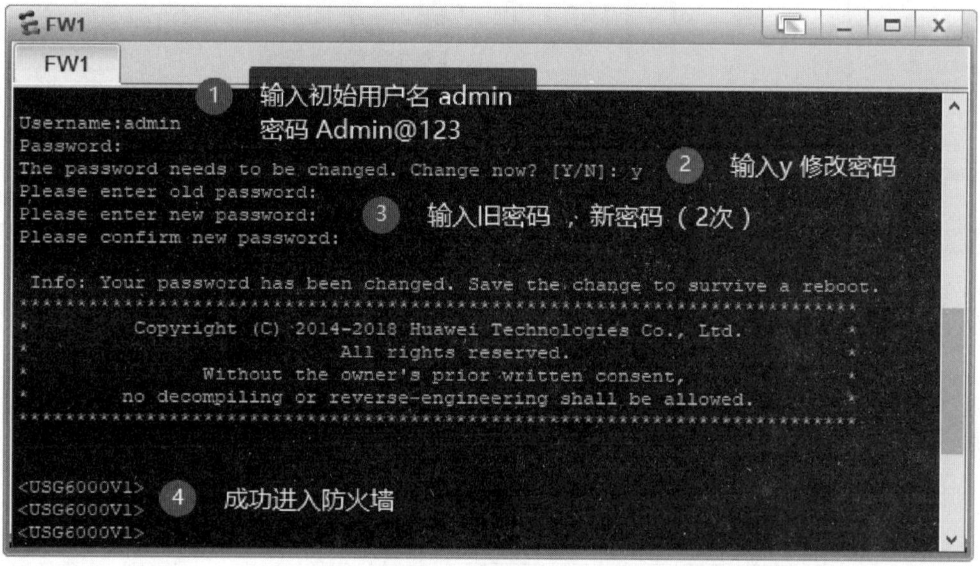

图 9.4　通过 CLI 登录防火墙后修改密码

（4）进入防火墙的管理口，修改管理口的 IP 地址，允许使用 Web 方式管理防火墙。命令如下：

```
<USG6000V1>system-view
[USG6000V1]interface GigabitEthernet 0/0/0
[USG6000V1-GigabitEthernet0/0/0]ip address 192.168.100.100 24  #修改防火墙的管理地址为192.168.100.100，具体的IP地址需要根据现场的管理PC确定
[USG6000V1-GigabitEthernet0/0/0]service-manage https permit  #允许管理员使用Web方式管理防火墙
```

9.3.6.2　连接防火墙的 G0/0/0 口，打开浏览器，通过 Web 方式管理防火墙

（1）在现场部署安装防火墙时，需要使用双绞线把管理 PC 和防火墙的 G0/0/0 口连接。为了满足在实验环境中学习的需要，我们给防火墙 eNSP 添加一朵云，然后修改云的参数配置，把云连接到防火墙 G0/0/0 口，使之能连接防火墙。如图 9.5 和图 9.6 所示。

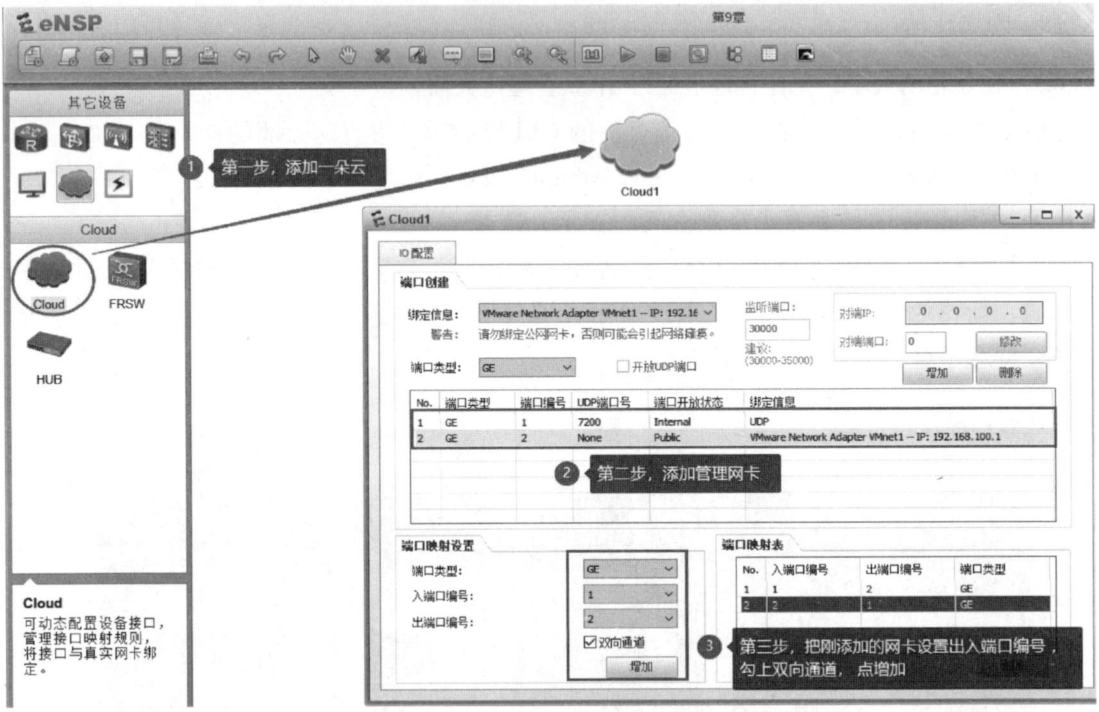

图 9.5　在 eNSP 中添加管理防火墙的网卡

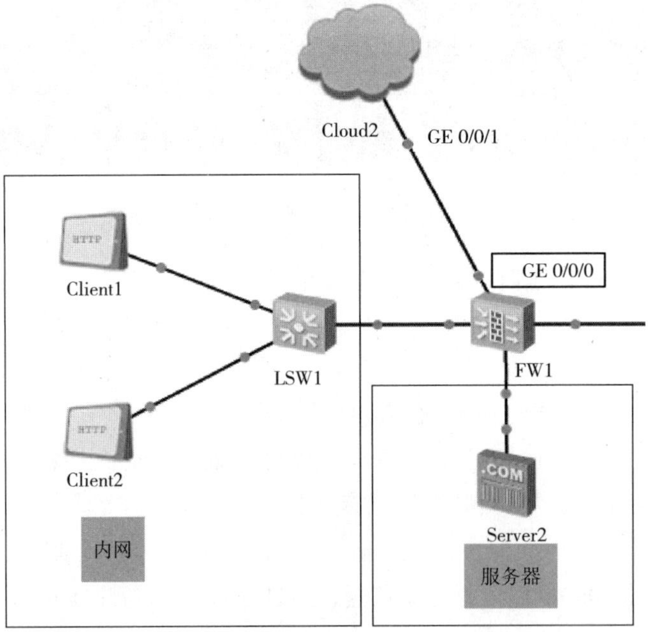

图 9.6　把云连接到防火墙的管理口

（2）在浏览器中输入 https://192.168.100.100:8443，使用 Web 方式登录防火墙，如图 9.7 所示。

图 9.7　通过网页登录防火墙

9.3.6.3　配置防火墙各接口地址及所属的安全区域

（1）配置防火墙外网口 IP 地址和安全区域。步骤：输入用户名和密码登录防火墙后，点击网络→接口→ GE1/0/1 →选择外网口 GE1/0/1 归属的安全区域（untrust）→输入外网口 IP 地址→输入默认网关→确定，如图 9.8 所示。

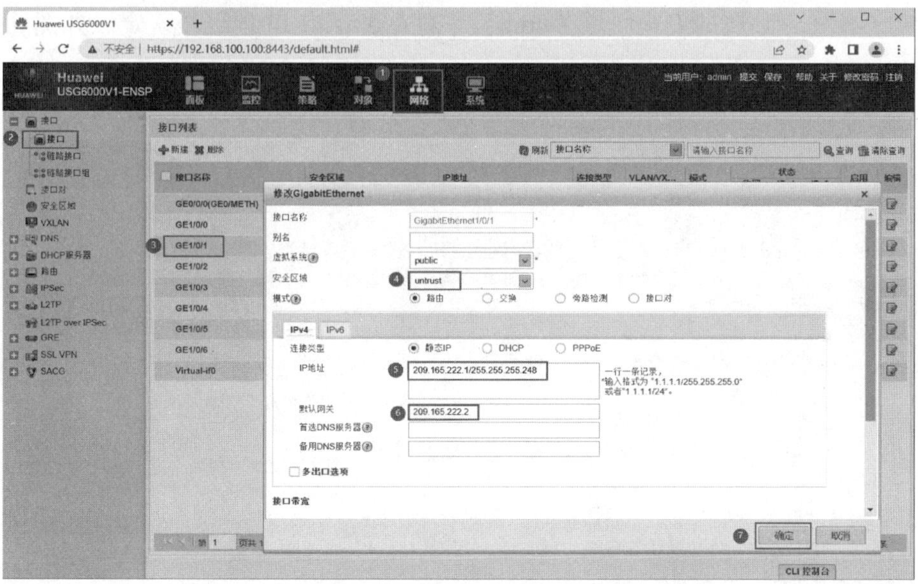

图 9.8　设置防火墙外网口 IP 地址及安全区域

（2）配置防火墙内网口 IP 地址和安全区域。步骤：输入用户名和密码登录防火墙后，点击网络→接口→ GE1/0/2 →选择内网口 GE1/0/2 归属的安全区域（trust）→输入内网口 IP 地址→确定，如图 9.9 所示。

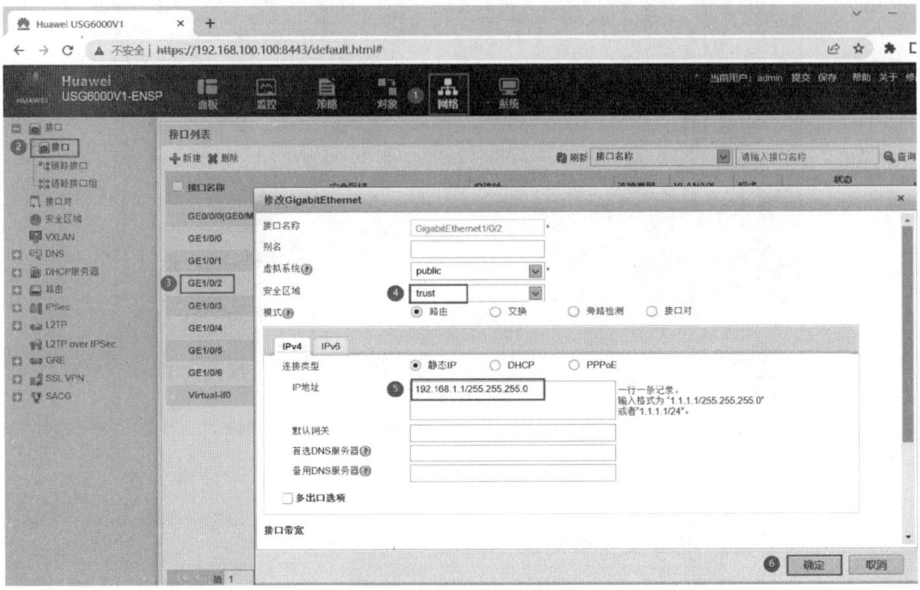

图 9.9　设置防火墙内网口 IP 地址及安全区域

（3）配置防火墙 DMZ 口 IP 地址和安全区域。步骤：点击网络→接口→ GE1/0/3 →选择 DMZ 口 GE1/0/3 归属的安全区域（dmz）→输入 dmz 口 IP 地址→确定，如图 9.10 所示。

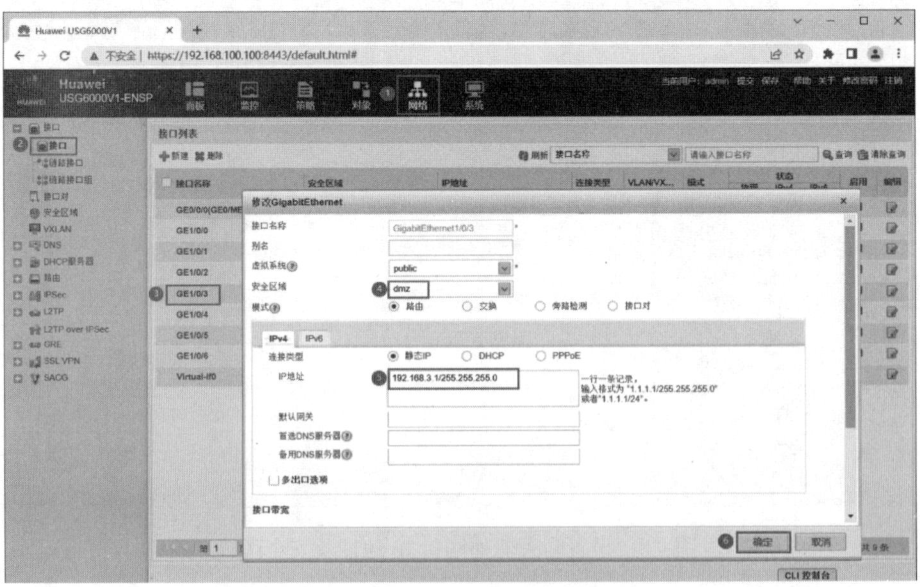

图 9.10　设置防火墙 DMZ 口 IP 地址及安全区域

9.3.6.4　配置安全策略，满足数据流向要求

（1）添加允许内网访问外网的安全策略。步骤：点击策略→安全策略→新建安全策略→策略名称输入：允许内网访问外网→源安全区域选择：trust →目的安全区域选择：untrust →动作选择：允许→确定，如图 9.11 所示。

图 9.11　添加允许内网访问外网的安全策略

（2）添加允许内网访问服务器区的安全策略。步骤：点击策略→安全策略→新建安全策略→策略名称输入：允许内网访问服务器区→源安全区域选择：trust→目的安全区域选择：dmz →服务选择：icmp、http →动作选择：允许→确定，如图 9.12 所示。

（3）添加允许外网访问服务器区的安全策略。步骤：点击策略→安全策略→新建安全策略→策略名称输入：允许外网访问服务器区→源安全区域选择：untrust →目的安全区域选择：dmz →服务选择：http →动作选择：允许→确定，如图 9.13 所示。

9.3.6.5　配置 NAT，满足内网访问外网需求

由于内网使用的都是私有 IP 地址，要访问外网，必须经过网络地址转换，把私有地址转换为公有地址才能访问互联网。

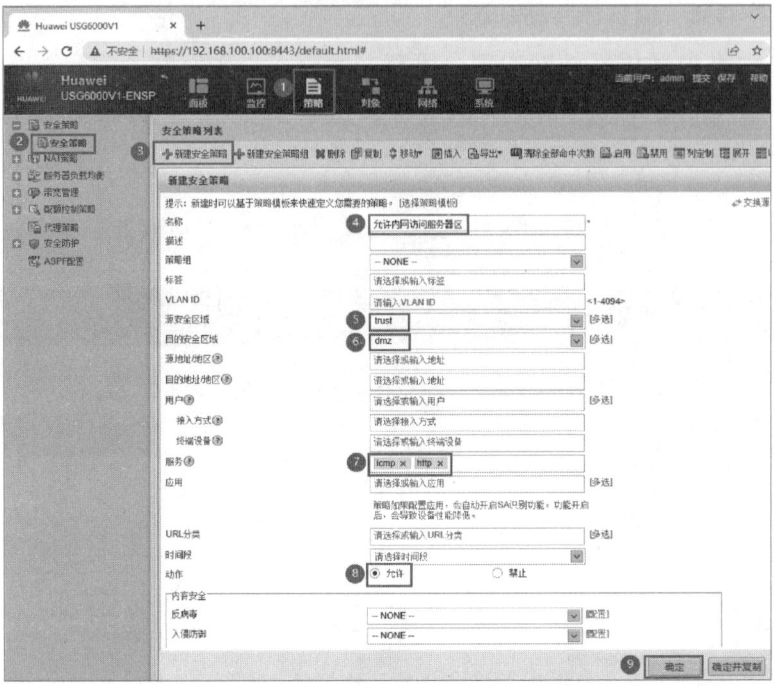

图 9.12　添加允许内网访问服务器区的安全策略

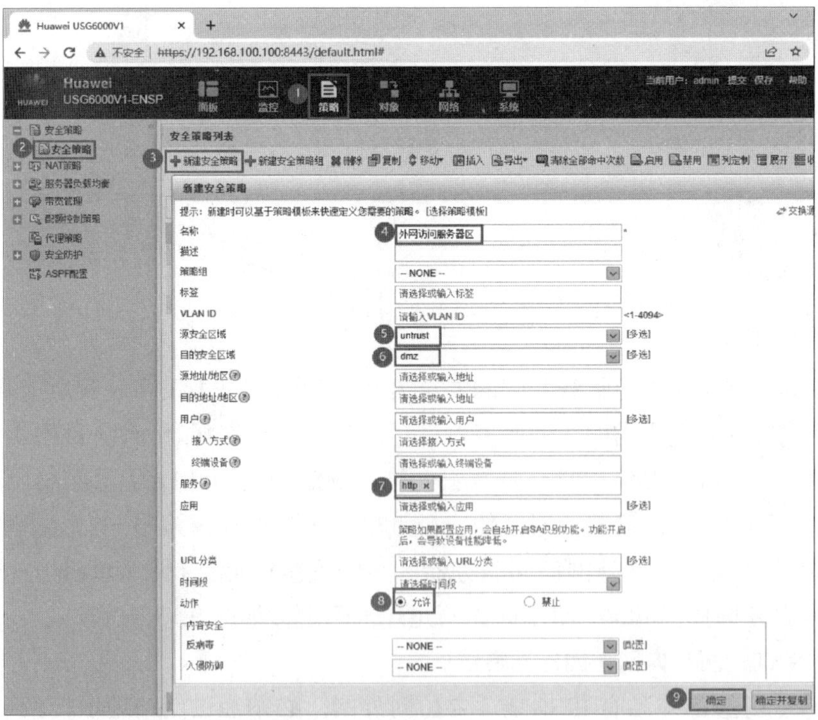

图 9.13　添加允许外网访问服务器区的安全策略

步骤：点击策略→ NAT 策略→新建→名称输入：允许内网私有地址转换→源安全区域选择：trust →目的类型选择：目的安全区域，untrust →转换后的数据包选择：出接口地址→确定，如图 9.14 所示。

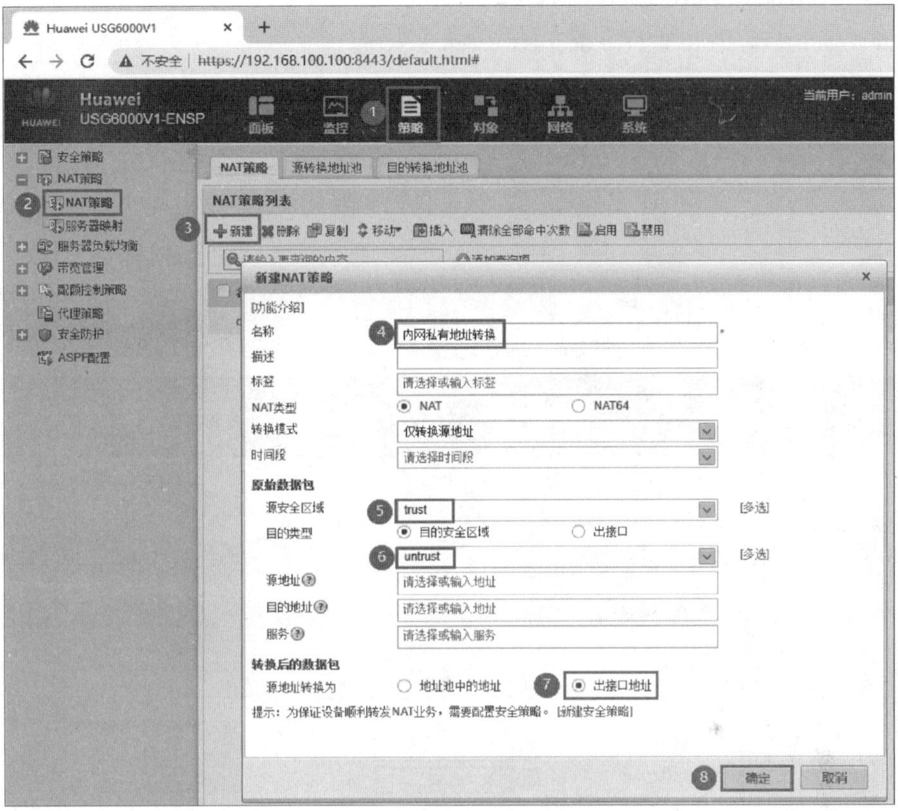

图 9.14　配置 NAT 满足内网访问外网需求

9.3.6.6　配置服务器映射，满足外网用户访问服务器区 Web 服务需求

由于服务器区的服务器使用的都是私有 IP 地址，外网用户要访问服务器区资源，必须经过服务器地址映射，把私有地址映射为公有地址后，才能被访问。

步骤：点击策略→ NAT 策略→服务器映射→新建→名称输入：允许外网用户访问 Web 服务→公网地址输入：209.165.222.10 →私网地址输入：192.168.3.2 →指定协议，公网端口 80，私网端口 80 →确定，如图 9.15 所示。

9.3.6.7　给其他设备配置 IP 地址

（1）互联网路由器 IP 地址配置（在虚拟仿真实验中为了测试方便需要配置，在实践工程项目中，互联网区域不需要进行配置）。

```
<Huawei>system-view
Enter system view, return user view with Ctrl+Z
```

```
[Huawei]interface GigabitEthernet 0/0/0
[Huawei-GigabitEthernet0/0/0]ip address 209.165.222.2 24
[Huawei-GigabitEthernet0/0/0]quit
[Huawei]interface GigabitEthernet 0/0/1
[Huawei-GigabitEthernet0/0/1]ip address 8.8.8.1 24
[Huawei-GigabitEthernet0/0/1]quit
[Huawei]interface GigabitEthernet 0/0/2
[Huawei-GigabitEthernet0/0/2]ip address 9.9.9.2 24
[Huawei-GigabitEthernet0/0/2]quit
```

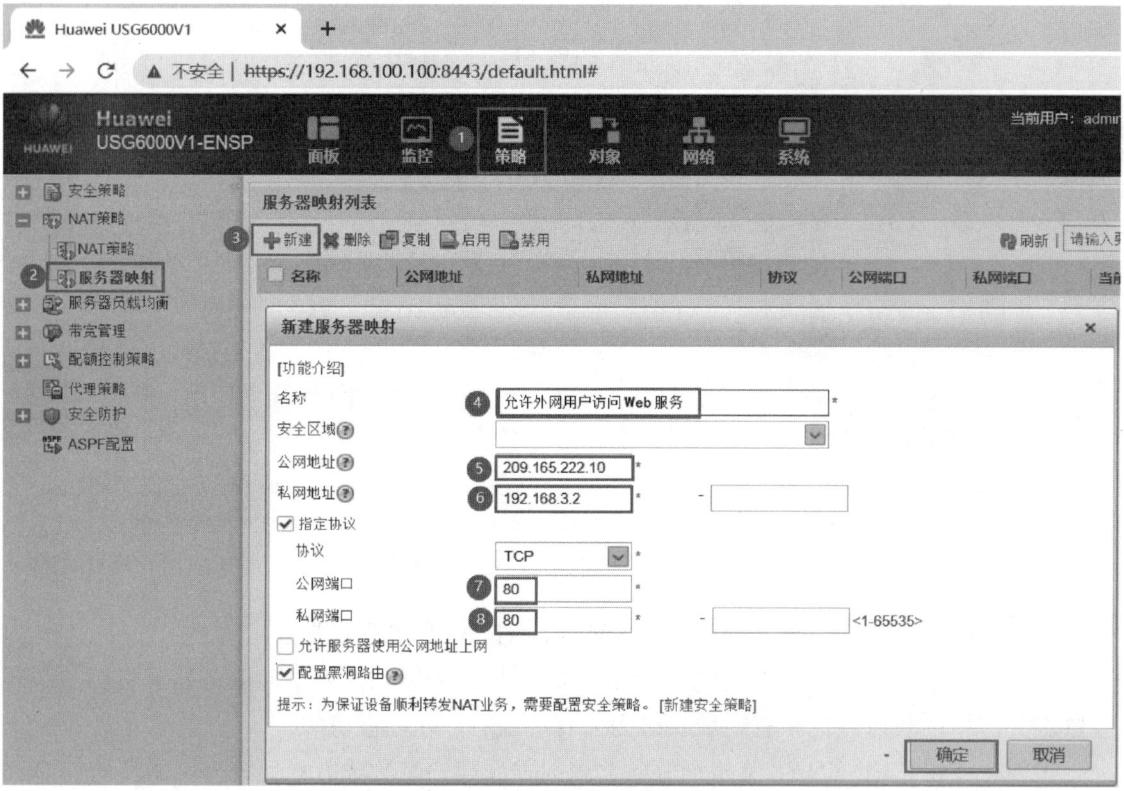

图 9.15　配置服务器映射满足外网用户访问服务器区 Web 服务需求

（2）终端设备配置 IP 地址。步骤：双击终端设备→输入本机（IP）地址、子网掩码、网关等参数→保存，如图 9.16 所示。

（3）启动服务器区 Web 服务。步骤：双击服务器设备→点击服务器信息→点击 HttpServer →选择文件根目录（建议在文件根目录中创建一个 aa.txt 文本文件，便于测试）→启动→确定，如图 9.17 所示。

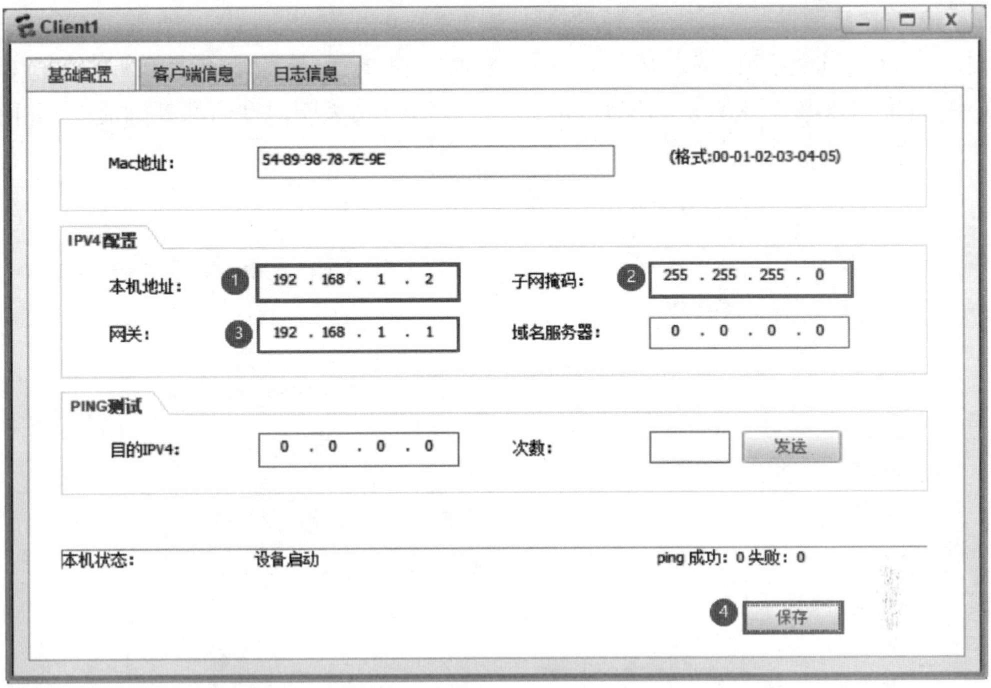

图 9.16　终端设备配置 IP 地址

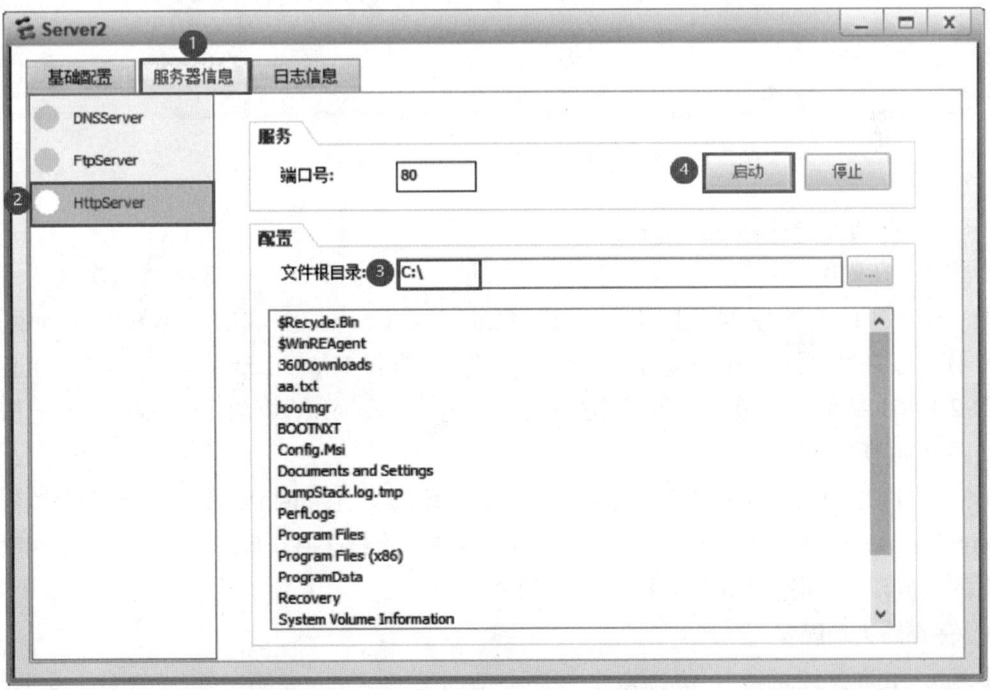

图 9.17　启动服务器区 Web 服务

9.3.6.8　测试网络连通性

（1）内网访问外网测试。步骤：在内网 PC 中，输入外网服务器的 IP 地址（目的 IPV4）→选择测试次数 5→发送→出现 ping 成功：5 的显示，表明内网访问外网成功，如图 9.18 所示。

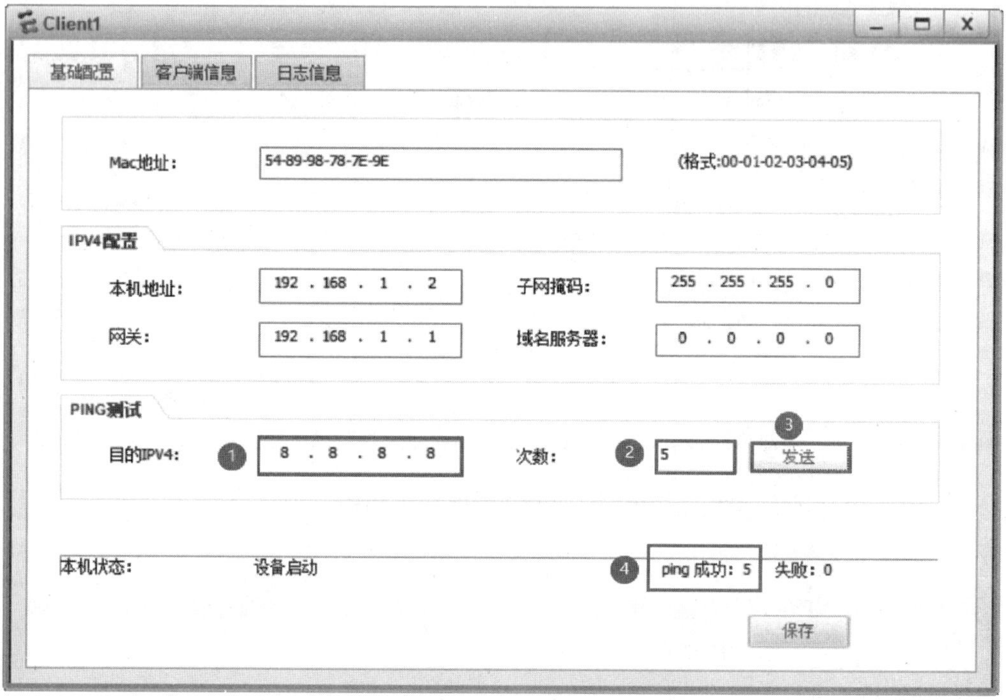

图 9.18　内网访问外网测试

（2）内网访问服务器区服务器测试。

步骤 1：在内网 PC 中，输入服务器的 IP 地址（目的 IPV4）→选择测试次数 5→发送→出现 ping 成功：5 的显示，表明内网访问服务器区 ICMP 服务成功，如图 9.19 所示。

步骤 2：在内网 PC 中，选择客户端信息→选择 HttpClient→输入访问的地址：http://192.168.3.2/aa.txt→获取→出现访问成功的显示，表明内网访问服务器区 Web 服务成功，如图 9.20 所示。

（3）外网用户访问服务器区 Web 服务测试。步骤：在外网 PC 中，选择客户端信息→选择 HttpClient→输入访问的地址：http://209.165.222.10/aa.txt→获取→出现访问成功的显示，表明内网访问服务器区 Web 服务成功，如图 9.21 所示。

9.3.6.9　保存防火墙配置

步骤：在防火墙的右上角，点击保存→确定，保存防火墙的配置信息，如图 9.22 所示。

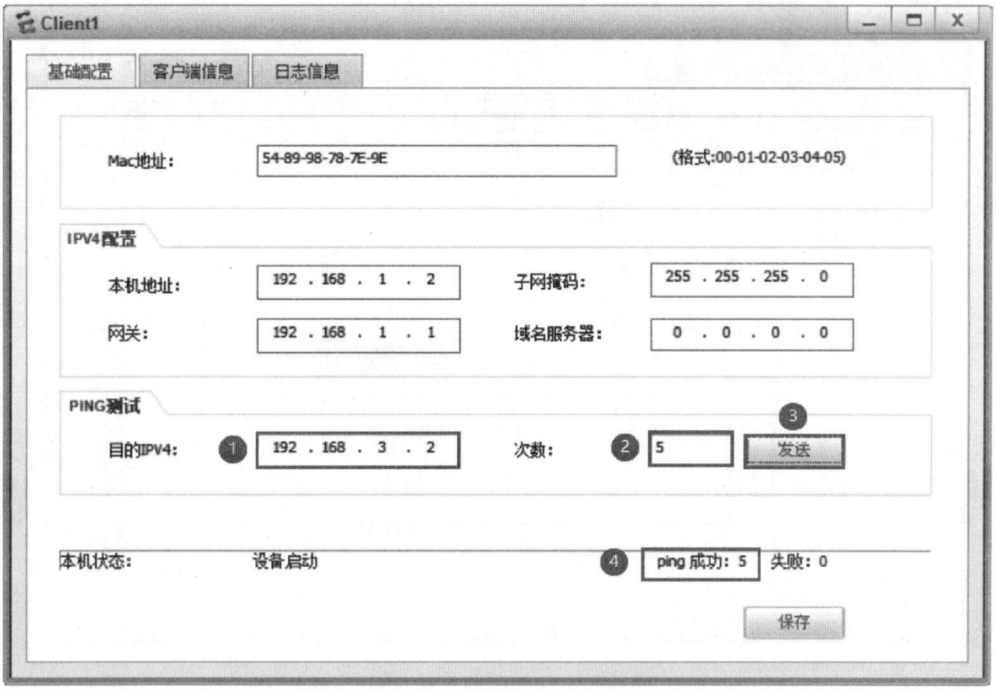

图 9.19　内网访问服务器区服务器 ICMP 服务测试

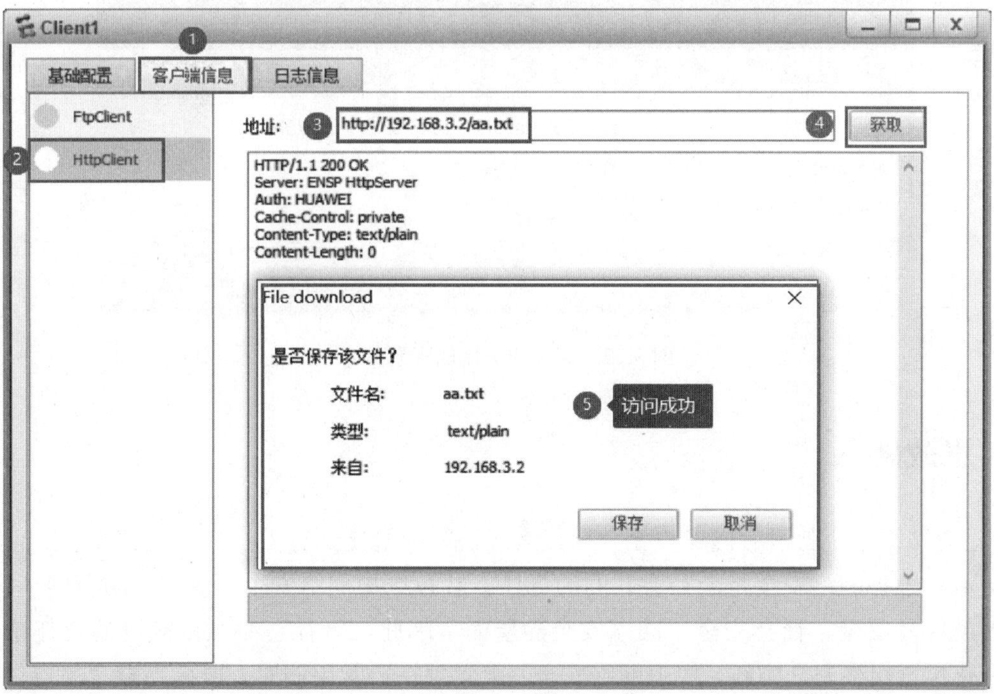

图 9.20　内网访问服务器区服务器 Web 服务测试

图 9.21　外网用户访问服务器区 Web 服务测试

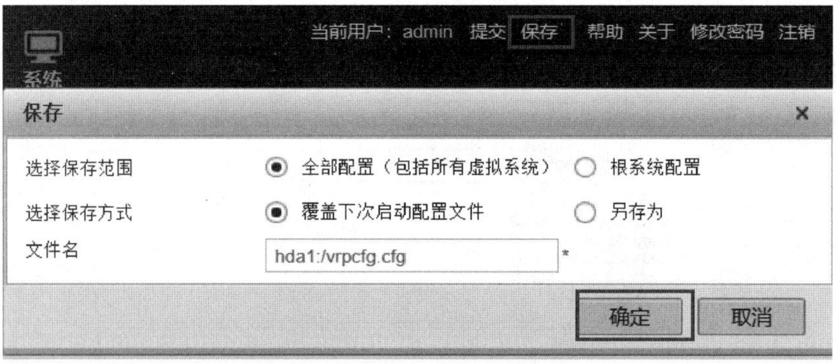

图 9.22　保存防火墙的配置信息

9.4　思政融入

“没有网络安全就没有国家安全，就没有经济社会稳定运行，广大人民群众利益也难以得到保障。”维护网络安全是全社会共同的责任。当前，维护网络安全不仅仅是防范网络病毒，它已涉及经济安全、社会安全、国家安全的层面，因此，全社会必须以高度的责任感和历史使命感构筑网络安全防线，捍卫网络安全。大学生在日常用网时，要提高自身的责任感，安全使用网络，时刻注意规范自身的网络行为。

9.5　思考与训练

1. 软件防火墙和硬件防火墙有什么区别?
2. 防火墙的过滤技术有哪些?
3. 防火墙的安全区域分为哪几个？各有什么作用?
4. 防火墙的初始用户名和密码是什么？
5. 请简述配置防火墙实现通过静态公网 IP 访问互联网的基本步骤。

Project 9 Installation of firewall to strengthen intranet security

9.1 Project import

At present, with the continuous development of network technology, computer networks have brought great convenience to people's lives, and the transformation and upgrading of enterprise informatization and digitalization have been further accelerated. However, various illegal and criminal activities using computer networks are also growing rapidly, and computer crimes, hackers, harmful programs and backdoors pose a serious threat to network security. As the first barrier between the internal network and the external public network, firewall is one of the first network security products that people attach importance to.

Firewall is a device used to protect network security in network infrastructure, and it is the first line of defense for network security. The firewall can be composed of hardware or software, or it can be integrated in the router to analyze and monitor the data packets flowing through the network with the router's routing function. The firewall is generally deployed on the intermediate equipment connecting the internal network and external network of the enterprise to establish a protective barrier between the internal network and the external network, and between the private network and the public network. The function of the firewall is to check the data packets passing through the firewall and decide the flow direction of the data packets according to the preset security policy, which can protect internal users from being invaded by illegal Internet users, and also prevent important internal information from flowing out. It has a two-way monitoring function. As shown in Figure 9.1.

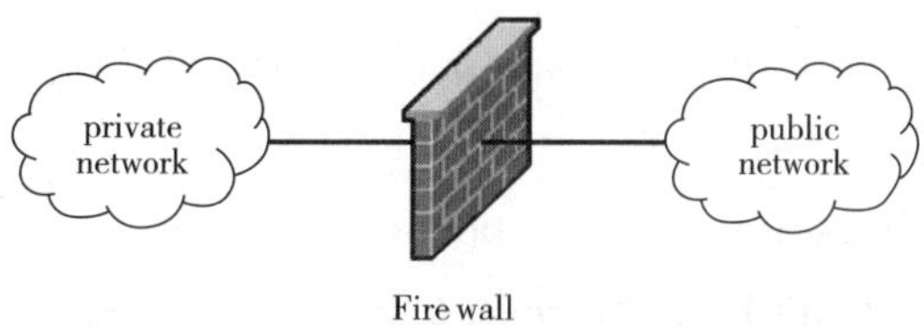

Figure 9.1 Firewall Deployment Location

9.2 Relevant knowledge points

9.2.1 The concept of firewall

Firewall, also known as protective wall, is a network security system between internal network and external network. It is an information security protection system that allows or restricts the transmission of data according to specific rules.

The firewall originated from the construction field. It originally refers to a wall built between houses to prevent the spread of fire from one area to another.In the network, firewall refers to a method to separate the internal network from the external network (such as the Internet). It is actually an isolation technology.The firewall is an access control measure implemented when two networks communicate. It can allow people and data you "agree" to enter your network, and at the same time keep people and data you "disagree" out of the door, so as to prevent hackers in the network from accessing your network to the maximum extent.In other words, if you do not pass through the firewall, people inside the company cannot access the Internet, and people on the Internet cannot communicate with people inside the company.

9.2.2 Classification of firewalls

9.2.2.1 Divide according to the different forms of software and hardware

According to the different forms of software and hardware, firewalls can be divided into software firewalls, hardware firewalls and chip level firewalls.

(1) Software firewall: software installed on the operating system. The software firewall is a protective barrier to isolate the internal network from the external network through software. Since it is installed on the system host, it needs to rely on the computer hardware resources to work, and it needs to consume the memory and CPU resources of the system. Like other software products, software firewalls can only be used after being installed and configured on each computer.

(2) Hardware firewall: in essence, embeds the software firewall in the hardware. The hardware and software of the hardware firewall need to be designed separately, using a dedicated network chip to process data packets, and at the same time, using a dedicated operating system platform, so as to avoid threats to the security of the internal network caused by the security vulnerabilities of the general operating system.

(3) Chip level firewall: Chip level firewall is based on dedicated hardware, and does not require the support of a dedicated operating system. Due to its own characteristics, special chips have faster working speed, higher processing capacity, and more efficient performance. Proprietary

ASIC chips enable them to be faster, more powerful, and have higher performance than other types of firewalls.

9.2.2.2 Divide according to the filtering technology

According to the filtering technology, firewalls are divided into packet filtering, application level gateways and proxy servers.

(1) Packet filtering technology. Packet filtering technology is a simple and effective security control technology. It works on the network layer. It checks the data packets passing through the devices by loading rules such as allow and prohibit from certain specific source addresses, destination addresses, and TCP port numbers on the devices connected to each other in the network, and restricts the data packets from entering and leaving the internal network。

The biggest advantages of packet filtering are transparency to users and high transmission performance. However, since the security control layer is at the network layer and transport layer, the strength of security control is also limited to the source address, destination address and port number, so only preliminary security control can be carried out, and there is nothing to do for high-level attacks such as malicious congestion attacks, memory coverage attacks or viruses.

(2) Application agent technology. Application proxy firewall works in the Layer-7 of OSI. It improves network security by checking all application layer packets and putting the checked content information into the decision-making process.

The application gateway firewall is implemented by breaking the client/server model. Each client/server communication requires two connections: one is from the client to the firewall, and the other is from the firewall to the server. In addition, each agent needs a different application process or a background running service program. For each new application, a service program for this application must be added, or the service cannot be used. Therefore, the application gateway firewall has the disadvantage of poor scalability.

(3) State detection technology. The state detection firewall works in the Layer-2 to Layer-7 of OSI. It uses the state detection packet filtering technology, which is an extension of the traditional packet filtering function. The state detection firewall has a check engine at the network layer to intercept data packets and extract information related to the application layer state, and decide whether to accept or reject the connection based on this. This technology provides a highly secure solution with good adaptability and scalability. The state detection firewall generally includes some proxy level services, which provide additional support for specific application data content.

The state detection firewall basically maintains the advantages of the simple packet filtering firewall, with good performance and transparency to applications. On this basis, the security has been greatly improved. This kind of firewall discards the disadvantage that the simple packet filtering firewall only inspects the data packets entering and exiting the network, and does not care about the status of the data packets. It establishes a status connection table in the core part of the firewall, maintains the connection, and treats the data entering and exiting the network as a single event. The main feature is that due to the lack of deep detection function of application layer protocol, it is unable to completely identify a large number of spam, advertisements and Trojan programs in the data package.

(4) Full content detection technology. Full content detection technology firewall integrates state detection and application proxy technology, and further integrates anti-virus, content filtering, application identification and other functions into the firewall based on multi-layer detection architecture, including IPS function, multi unit integration, application layer scanning in the network world, and anti-virus, content filtering and firewall integration, which reflects the new idea of network and information security (Therefore, it is also called the "next generation firewall technology"). It implements OSI layer-7 content scanning at the network boundary, and implements real-time deployment of virus protection, content filtering and other application layer service measures at the network edge. The full content detection technology firewall can check the entire packet content, establish the connection status table as required, and has the advantages of strong network layer protection and fine application layer control. However, due to the high functional integration, it has high requirements on the product hardware.

9.2.3 Security zone of firewall

Security zone, referred to as zone for short. A security zone is a collection of one or more interfaces. The firewall divides the network and identifies the "route" of message flow through the security zone. By default, messages are not controlled when they flow in the same security area, but are controlled when they flow between different security areas.

Huawei firewall provides four security zones by default, namely Trust, DMZ, Untrust and Local. Each security zone has a unique security level, which is represented by a number of 1 to 100. The larger the number, the more trusted the network in the zone. By dividing different interfaces of the firewall into different security zones, you can divide different security zone networks on the firewall.

9.2.3.1 Trust region

The default security level of the Trust region is 85. The network in this region is highly trusted. It is usually used to define the network where internal users are located.

9.2.3.2 DMZ region

The default security level of the DMZ region is 50. The network in this region is moderately trusted. It is usually used to define the region where the intranet server is located. Although this kind of equipment is deployed in the internal network, it often needs to be accessed by the external network, which has great security risks. At the same time, it is generally not allowed to actively access the external network, so it is deployed in a security area with a lower priority than Trust but higher priority than Untrust. DMZ (Demilitarized Zone), originated from the military, is a partially controlled area between a strict military control zone and a loose public area. The firewall device refers to the security zone between the internal network and the external network.

9.2.3.3 Untrust zone

The default security level of the Untrust zone is 5. This zone represents an untrusted network. It is usually used to define insecure networks such as the Internet.

9.2.3.4 Local Area

The default security level of the local area is 100, representing the firewall itself. You can not add any interfaces to the local zone, but all interfaces on the firewall themselves implicitly belong to the local zone.

Figure 9.2 shows the networking diagram of the real environment in which the firewall is deployed inside the enterprise. From the figure, we can see that users, servers and the Internet in the enterprise's internal network are divided into different security zones. The firewall performs security checks on the packets flowing between the security zones.

In the absence of a firewall, the router will directly transfer traffic between the internal network and the external network without a filtering mechanism. The function of the firewall technology is to timely discover and handle the security risks, data transmission and other problems that may exist during the operation of the computer network. The processing measures include isolation and protection, and can record and detect various operations in the computer network security, To ensure the security of computer network operation, ensure the integrity of user data and information, and provide users with a better and more secure computer network experience.

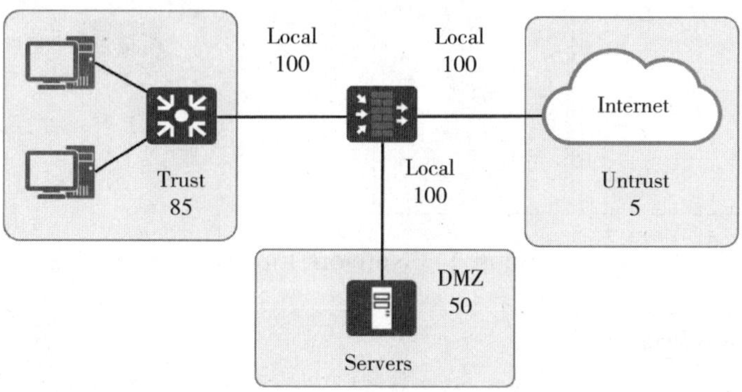

Figure 9.2 Firewall networking topology

9.3 Project implementation

9.3.1 Project requirements

(1) Deploy firewall to enable intranet users to access the Internet.

(2) Internet users can access the Web services in the server area.

(3) Intranet users can access ICMP and Web services in the server area.

9.3.2 Equipment list

This project mainly uses firewall, access layer switch and other equipment, and the specific equipment model and quantity are shown in Table 9.1.

Table 9.1 List of network equipment

Serial No.	Equipment model	Quantity	Function	Remarks
1	S5700	1	Access switch	
2	USG6000V	1	Firewall	
3	AR2220	1	Router	Operator equipment

9.3.3 Access switch project topology

The network topology of the project is shown in Figure 9.3.

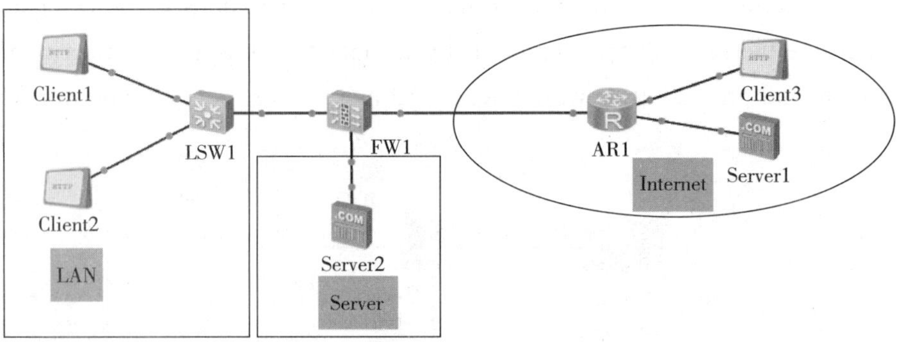

Figure 9.3 Network topology

9.3.4 Configuration ideas

The project adopts the following ideas:

(1) Initialize the firewall and allow the administrator to manage the firewall through the Web.

(2) Configure the interface addresses and security zones of the firewall.

(3) Configure security policies to meet data flow requirements.

(4) Configure NAT to meet the requirements of intranet access to the Internet.

(5) Configure the server mapping so that Internet users can access the server area normally.

9.3.5 Relevant information planning

9.3.5.1 firewall

The address planning of the firewall is shown in Table 9.2.

Table 9.2 Firewall address planning table

Serial No.	Interface	IP address	Mask	Remarks
1	G1/0/1	209.165.222.1	255.255.255.248	Connect to the Internet
2	G1/0/2	192.168.1.1	255.255.255.0	Connect to intranet
3	G1/0/3	192.168.2.1	255.255.255.0	Connect server

9.3.5.2 Router

The AR1 address plan of the operator's equipment is shown in Table 9.3.

Table 9.3 Router address planning table

Serial No.	Interface	IP address	Mask	Remarks
1	G0/0/0	209.165.222.2	255.255.255.248	Connection firewall
2	G0/0/1	8.8.8.1	255.255.255.0	Connect to the Internet server Server1

9.3.5.3 Other equipment

The address planning of other equipment is shown in Table 9.4.

Table 9.4 Address Planning of Other Equipment

Serial No.	Interface	IP address	Mask	Gateway	Remarks
1	Intranet PC Client1	192.168.1.2	255.255.255.0	192.168.1.1	Intranet user PC
2	Intranet PC Client2	192.168.1.3	255.255.255.0	192.168.1.1	Intranet user PC
3	Sever2 in server area	192.168.3.2	255.255.255.0	192.168.3.1	Server area server
4	Sever1 in server area	8.8.8.8	255.255.255.0	8.8.8.1	Internet server
5	Internet user Client3	9.9.9.2	255.255.255.0	9.9.9.1	Internet user

9.3.6 Operating steps

9.3.6.1 Initialize the firewall and allow the administrator to manage the firewall through the Web

Start the firewall, log in to the CLI interface of the firewall through the configuration line, and then initialize the firewall.

(1) Enter the default username admin and password Admin@123 , log in to the firewall.

(2) When prompted whether to change the password, select y.

(3) Enter old password Admin@123 , then enter a new password, and then enter the new password again to log in to the firewall. The new password must meet the password complexity requirements, that is, at least 8 characters, including 3 types of capital letters, lowercase letters, numbers, and special characters. As shown in Figure 9.4.

(4) Enter the management port of the firewall, modify the IP address of the management port, and allow the firewall to be managed in the Web mode. The command is as follows:

Figure 9.4 Modifying the password after logging in to the firewall through CLI

```
<USG6000V1>system-view
[USG6000V1]interface GigabitEthernet 0/0/0
[USG6000V1-GigabitEthernet0/0/0]ip address 192.168.100.100 24  # The management address of the firewall is modified to 192.168.100.100, and the specific IP address needs to be determined according to the on-site management PC.
[USG6000V1-GigabitEthernet0/0/0]service-manage https permit  # Allow administrators to manage firewalls in Web mode
```

9.3.6.2 Connect to the G0/0/0 port of the firewall, open the browser, and manage the firewall through Web

(1) When deploying and installing firewalls on site, twisted pair cables are required to

connect the management PC to the G0/0/0 port of the firewall.

In order to learn in the experimental environment, we add a cloud to the firewall eNSP, and then modify the cloud parameter configuration to connect the cloud to the firewall G0/0/0 port, so that it can connect to the firewall. As shown in Figures 9.5 and 9.6.

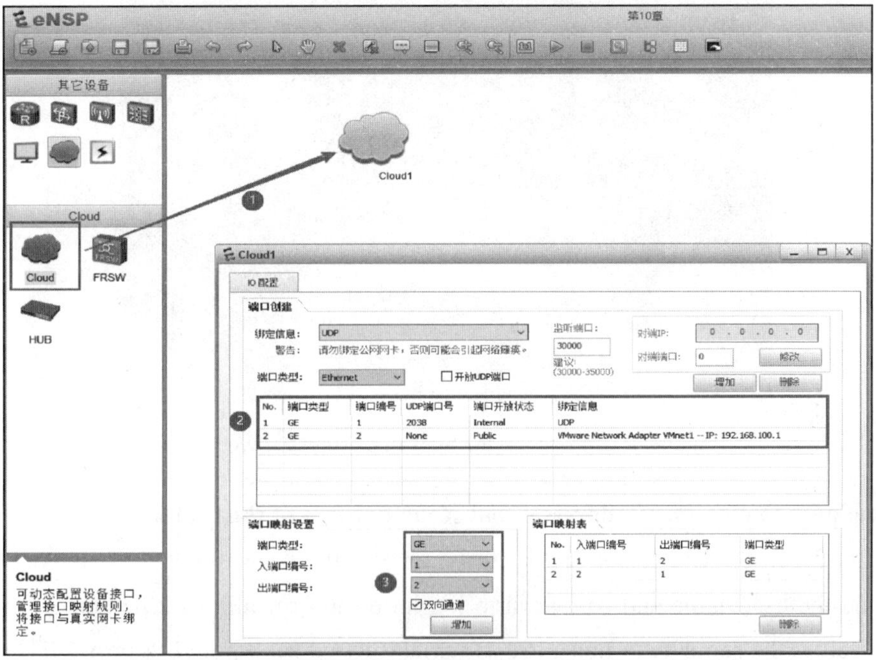

Figure 9.5 Add the network card of the management firewall in eNSP

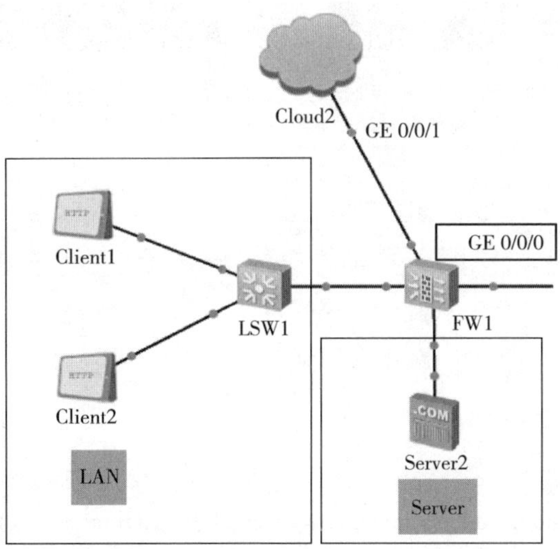

Figure 9.6 Management port for connecting cloud to firewall

(2) Enter in browser https://192.168.100.100:8443, log in to the firewall in Web mode. As shown in Figure 9.7.

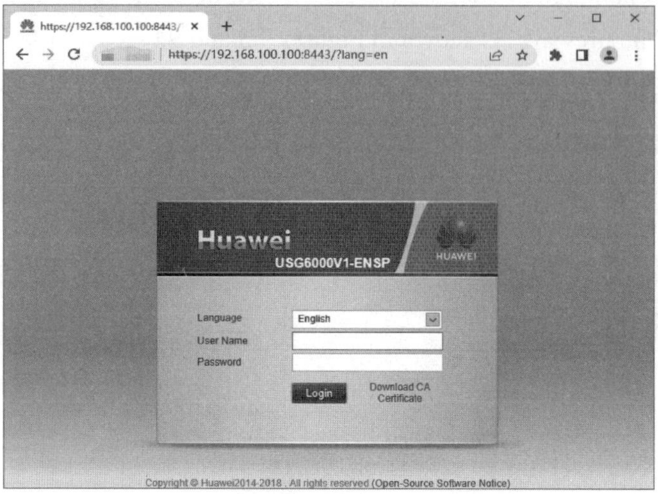

Figure 9.7 Log in to the firewall through the Webpage

9.3.6.3 Configure the interface addresses and security zones of the firewall

(1) Configure the IP address and security zone of the firewall external network port. Step: After entering the user name and password to log in to the firewall, click network →Interface → GE1/0 1 →Select the security zone (untrust) which the external network port GE1/0/1 belongs to →Enter the IP address of the external network port →Enter default gateway →OK, as shown in Figure 9.8.

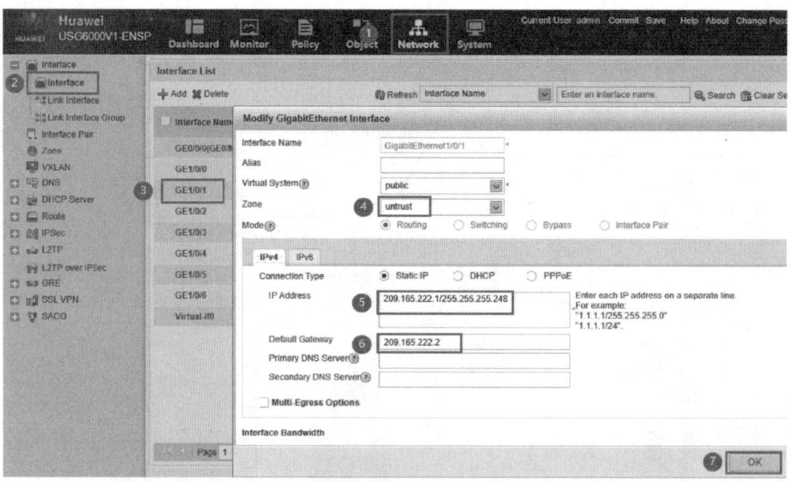

Figure 9.8 Setting the IP address and security area of the external network port of the firewall

(2) Configure the IP address and security zone of the network port in the firewall. Step:

After entering the user name and password to log in to the firewall, click network →Interface →GE1/0/2 →Select the security zone (trust) which the internal network port GE1/0/2 belongs to →Enter the IP address of the internal network port →OK, as shown in Figure 9.9.

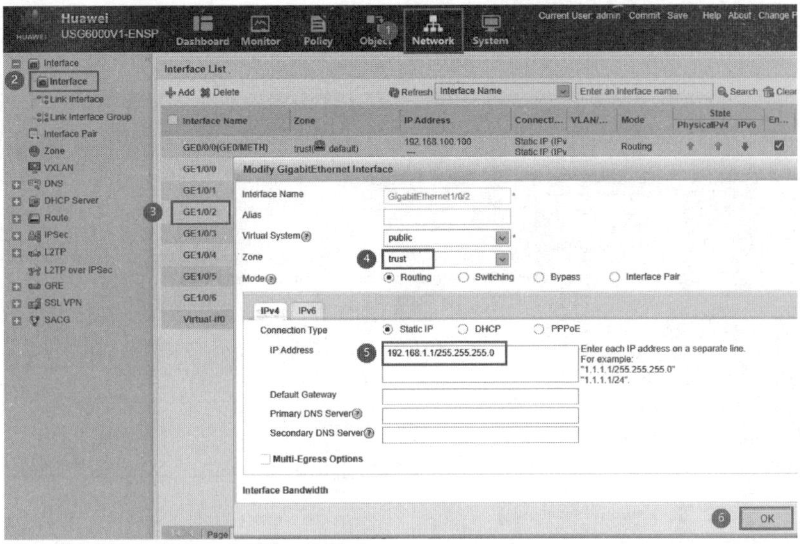

Figure 9.9 Setting the IP address and security area of the network port in the firewall

(3) Configure the IP address and security zone of the firewall DMZ port. Step: Click network →Interface →GE1/0/3 →Select the security zone (dmz) which the DMZ port GE1/0/3 belongs to →Enter the IP address of the dmz port →OK, as shown in Figure 9.10.

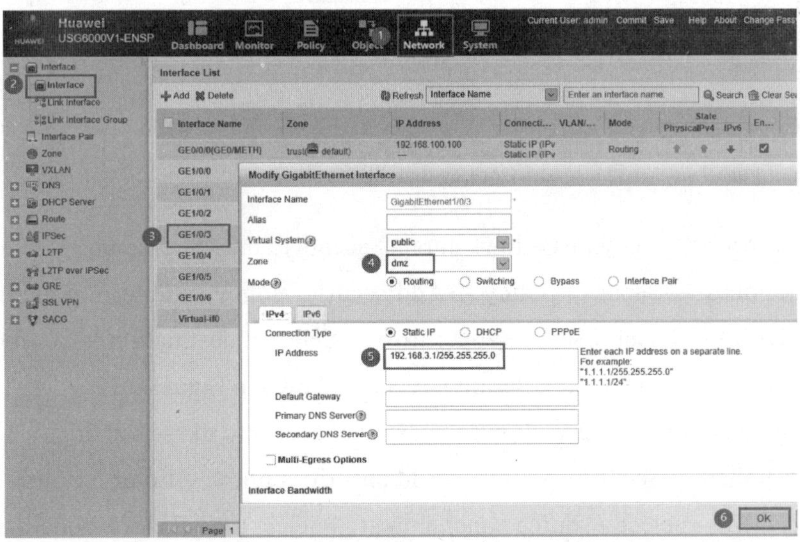

Figure 9.10 Setting the IP address and security area of the firewall DMZ port

9.3.6.4 Configure security policies to meet data flow requirements

(1) Add a security policy that allows intranet access to the internet. Step: Click policy →Security policy →Add security policy →Enter a policy name: Allow LAN access to WAN→Select a source zone: trust →Select a destination zone: untrust →Select an action: Permit→OK, as shown in Figure 9.11.

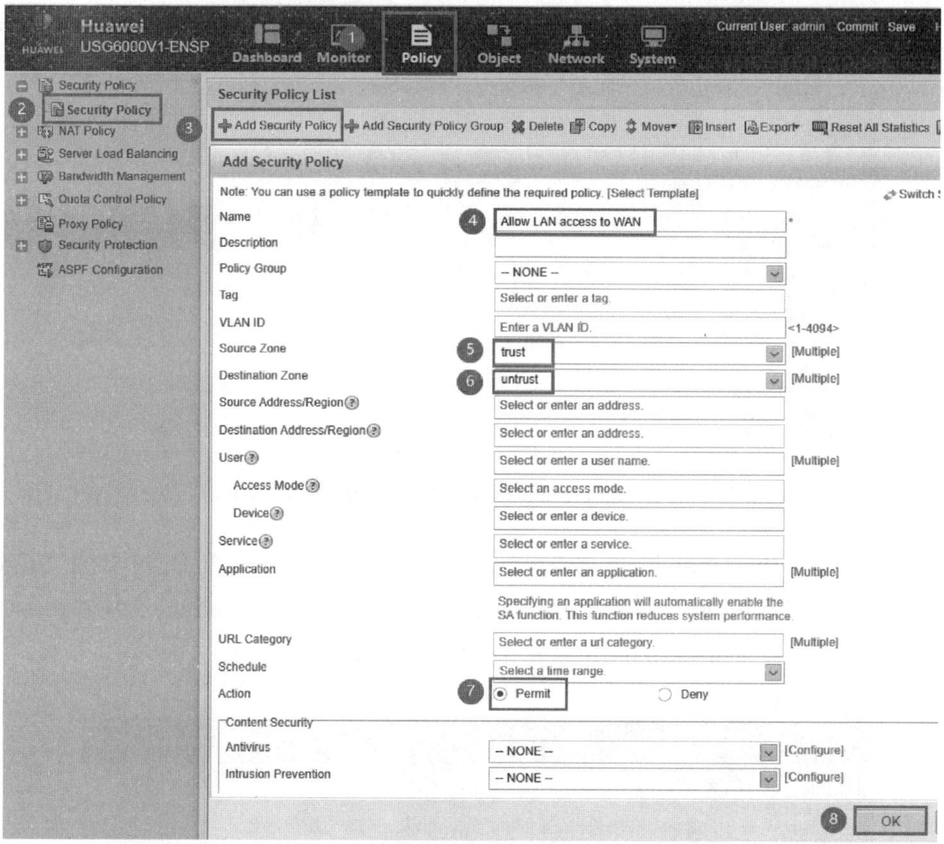

Figure 9.11 Add a security policy that allows intranet access to the extranet

(2) Add a security policy that allows intranet access to the server zone.

Step: Click policy →Security policy →Add security policy →Enter a policy name: Allow LAN access to DMZ →Select a source zone: trust →Select a destination zone: dmz →Select a service: icmp, http →Select an action: Permit →OK, as shown in Figure 9.12.

(3) Add a security policy that allows the extranet access to the server area.

Step: Click policy →Security policy →Add security policy →Enter a policy name: Allow WAN access to DMZ →Select a source zone: untrust →Select a destination zone: dmz →Select a service: http →Select an action: Permit →OK, as shown in Figure 9.13.

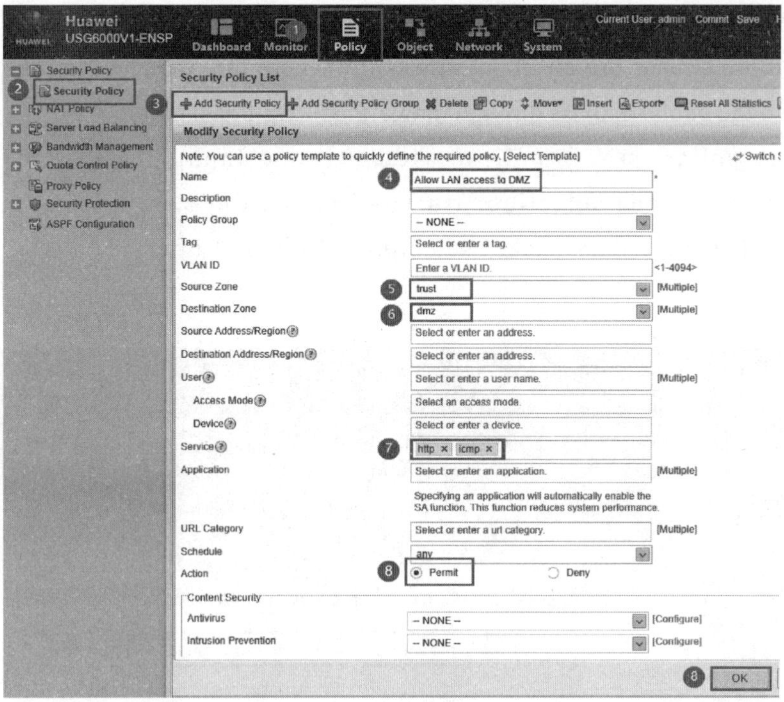

Figure 9.12 Add a security policy that allows intranet access to the server area

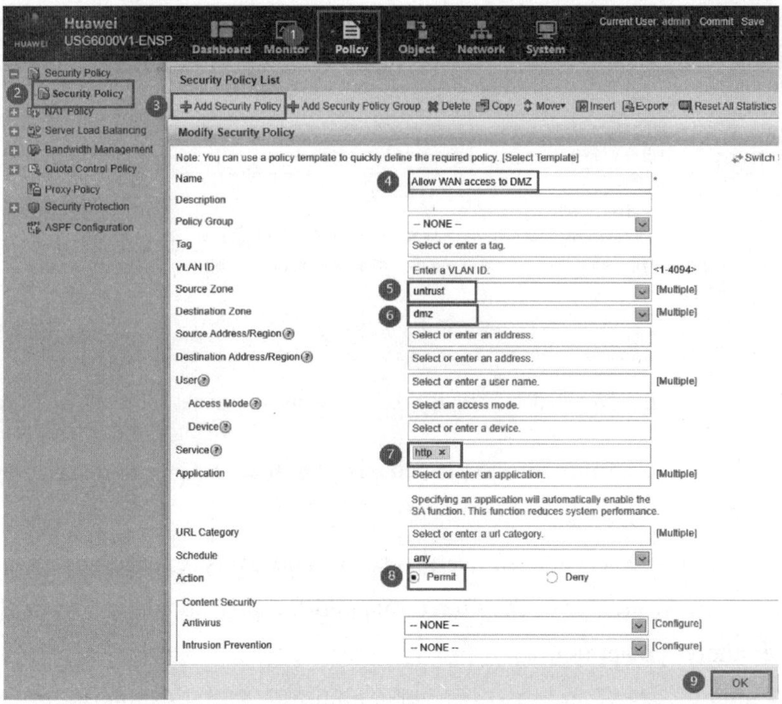

Figure 9.13 Add a security policy that allows the extranet access to the server area

9.3.6.5 Configure NAT to meet the requirements of intranet access to the extranet

Since the intranet uses private IP addresses, to access the extranet, you must convert the private address to the public address through network address translation.

Step: Click policy →NAT policy→Add →Name Input: Intranet private address translation →Source zone selection: trust →Destination type selection: Destination zone, untrust →Translate data packet selection: Outbound interface →OK, as shown in Figure 9.14.

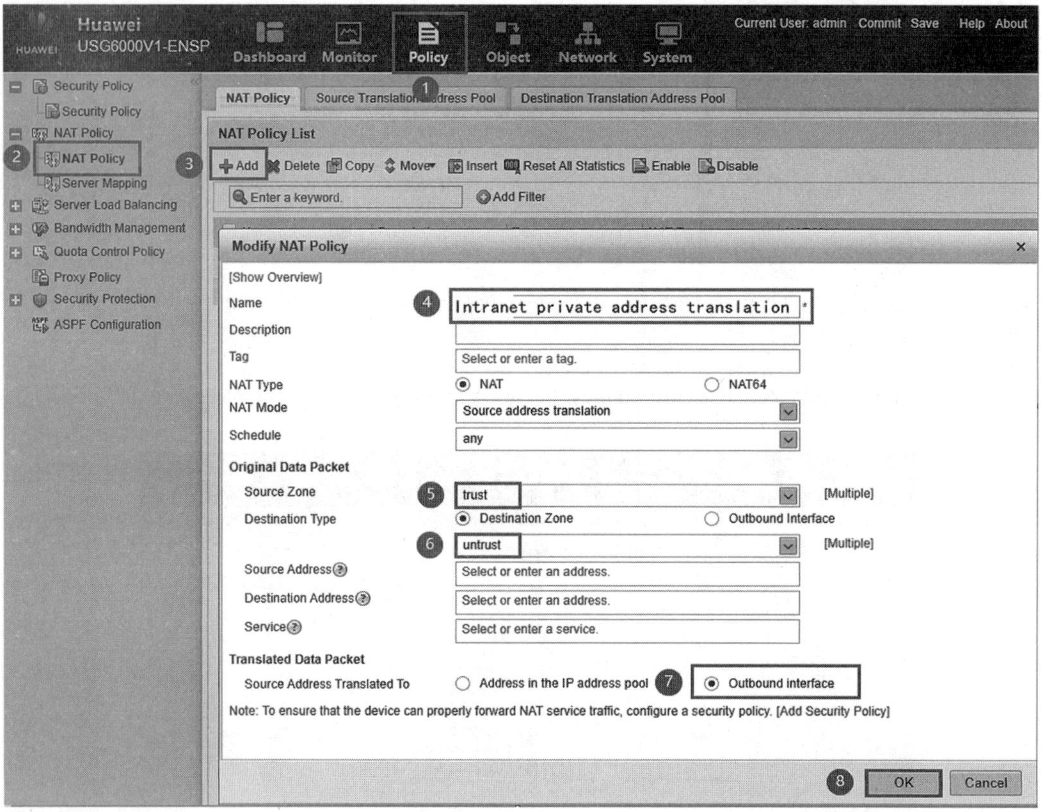

Figure 9.14 Configuring NAT to meet the requirements of intranet access to the extranet

9.3.6.6 Configure server mapping to meet the needs of extranet users to access the Web services in the server area

Since the servers in the server area use private IP addresses, extranet users can only access the resources in the server area after they have mapped the private address to the public address through the server address mapping.

Step: Click policy →NAT policy →Server mapping→Add →Name input: Allow external users to access Web services →Public IP address input: 209.165.222.10 →Private IP address input:

192.168.3.2 →Specify protocol, public port 80, private port 80 →OK, as shown in Figure 9.15.

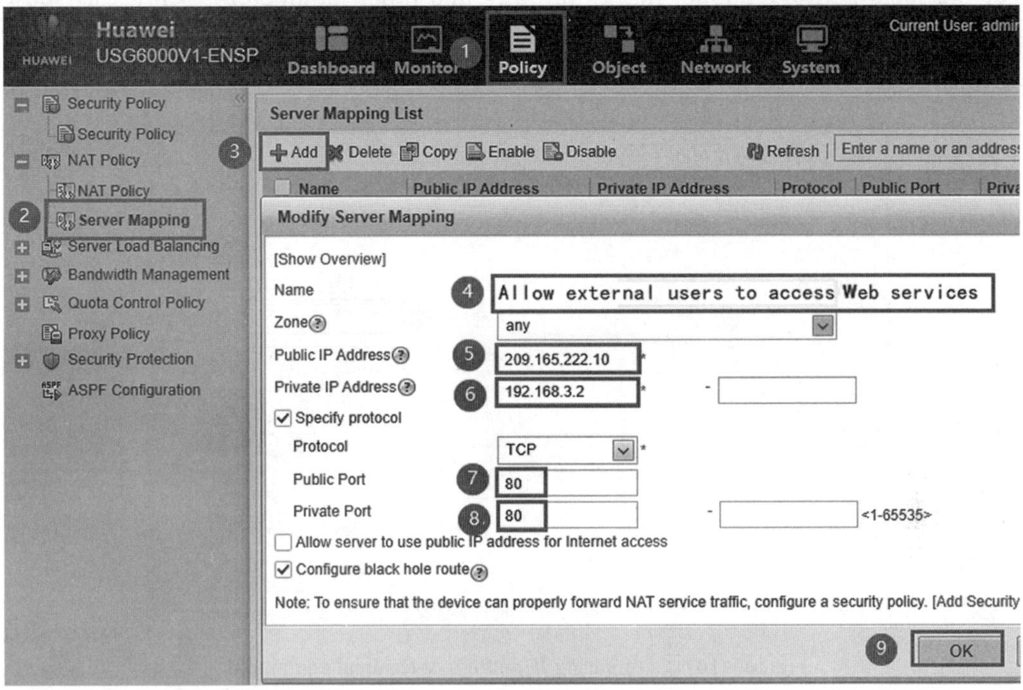

Figure 9.15 Configure the server mapping to meet the needs of extranet users to access the Web services in the server area

9.3.6.7 Configure IP address for other devices

(1) IP address configuration of Internet router (configuration is required for testing convenience in virtual simulation experiment, but it is not required for Internet area in practical engineering project).

```
<Huawei>system-view
Enter system view, return user view with Ctrl+Z
[Huawei]interface GigabitEthernet 0/0/0
[Huawei-GigabitEthernet0/0/0]ip address 209.165.222.2 24
[Huawei-GigabitEthernet0/0/0]quit
[Huawei]interface GigabitEthernet 0/0/1
[Huawei-GigabitEthernet0/0/1]ip address 8.8.8.1 24
[Huawei-GigabitEthernet0/0/1]quit
[Huawei]interface GigabitEthernet 0/0/2
[Huawei-GigabitEthernet0/0/2]ip address 9.9.9.2 24
[Huawei-GigabitEthernet0/0/2]quit
```

(2) Configure IP address of terminal equipment. Step: Double click the terminal device →Enter the local address, subnet mask, gateway and other parameters →Save, as shown in Figure 9.16.

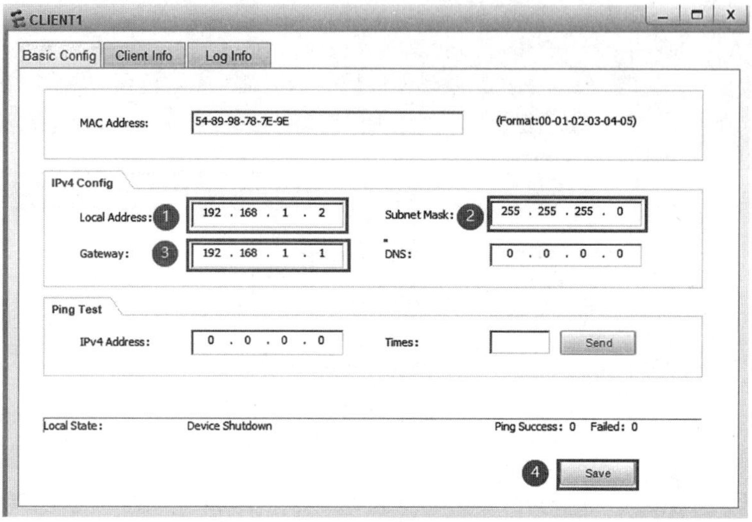

Figure 9.16 Configuring IP address of terminal equipment

(3) Start the Web service in the server area. Step: Double click the server device →Click the server information →Click httpserver →select the file root directory (it is recommended to create an aa.txt text file in the file root directory for testing) →Start →Confirm, as shown in Figure 9.17.

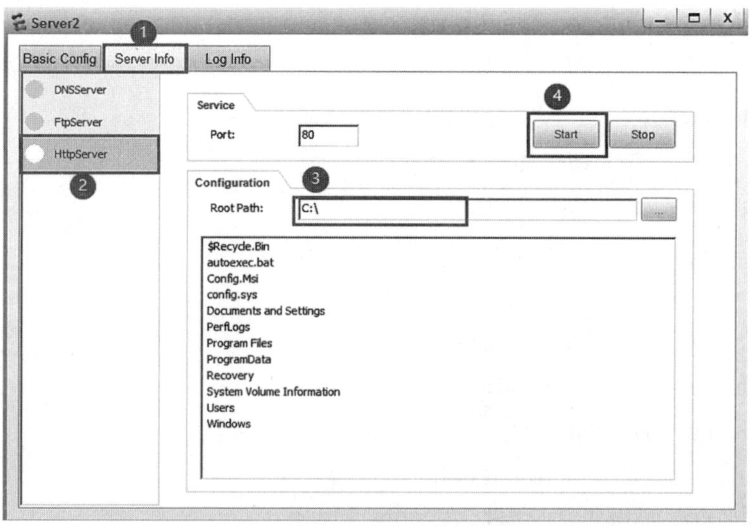

Figure 9.17 Start the Web service in the server area

9.3.6.8 Test network connectivity

(1) Test of intranet access to the extranet. Step: In the intranet PC, enter the IPV4 address of the external network server →Select 5 test times →Send →A successful ping is displayed, indicating that the internal network accesses the external network successfully. As shown in Figure 9.18.

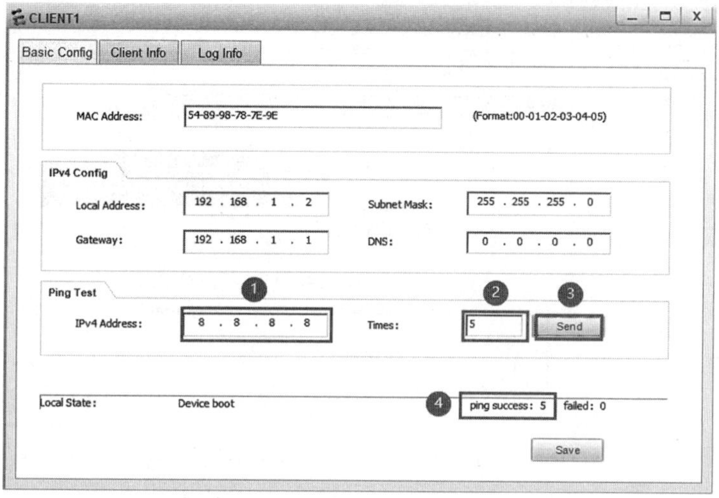

Figure 9.18 Test of intranet access to extranet

(2) Testing the server in the server area for intranet access.

Step 1: In the intranet PC, enter the IPV4 address of the server →Select 5 test times →Send →A successful ping display appears, indicating that the intranet accesses the ICMP service in the server area successfully. As shown in Figure 9.19.

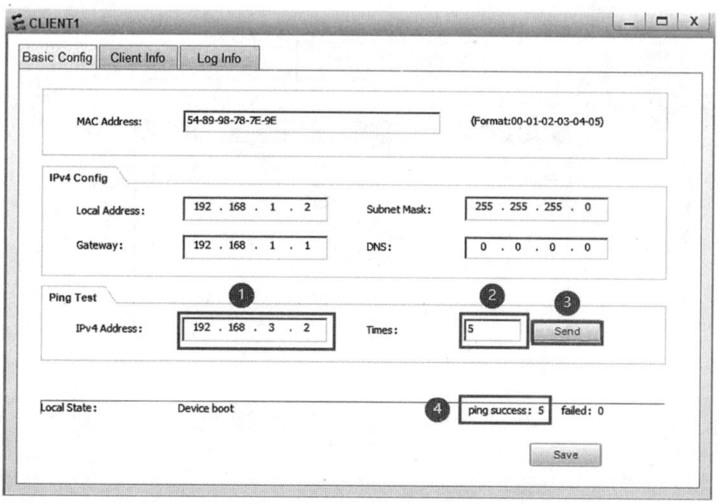

Figure 9.19 ICMP service test of server in intranet access server area

Step 2: In the intranet PC, select the client information →Select httpclient →Enter the access address: http://192.168.3.2/aa.txt →Go →The display of successful access indicates that the intranet has successfully accessed the Web service in the server area. As shown in Figure 9.20.

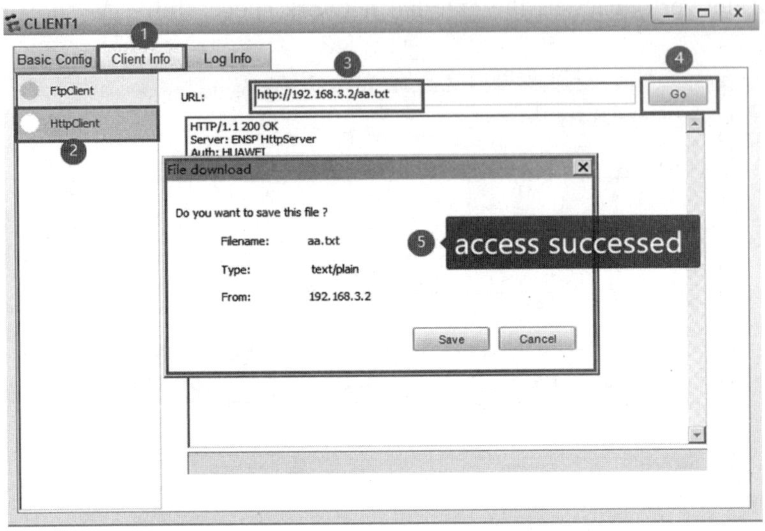

Figure 9.20 Web service test of server area server for intranet access

(3) Testing of extranet users' access to Web services in the server area. Step: In the extranet PC, select the client information →Select httpclient →Enter the access address: http://209.165.222.10/aa.txt →Go →The display of successful access indicates that the intranet has successfully accessed the Web service in the server area. As shown in Figure 9.21.

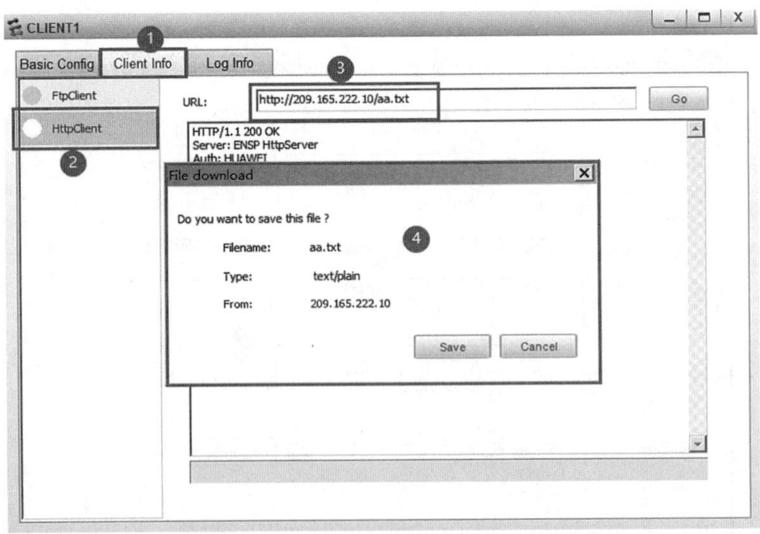

Figure 9.21 Test of extranet users' access to Web services in server area

9.3.6.9 Save firewall configuration

Step: In the upper right corner of the firewall, click save →OK to save the firewall configuration information. As shown in Figure 9.22.

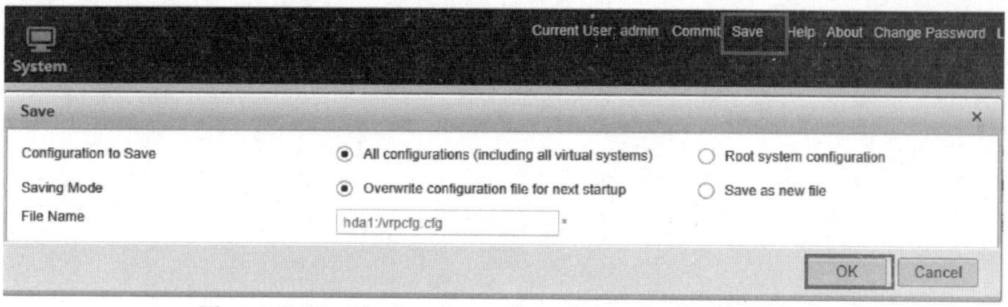

Figure 9.22 Saving firewall configuration information

9.4 Ideological and political integration

"Without cybersecurity, there will be no national security, there will be no stable economic and social operation, and it will be difficult to protect the interests of the broad masses of the people." Maintaining network security is the common responsibility of the whole society. At present, maintaining network security is not only about preventing network viruses, it has involved economic security, social security, and national security. Therefore, society as a whole must build a cybersecurity defense line and safeguard cybersecurity with a high sense of responsibility and historical mission. When college students use the Internet daily, they should improve their sense of responsibility, use the network safely, and always pay attention to standardizing their own network behavior.

9.5 Thinking and training

1. What is the difference between a software firewall and a hardware firewall?
2. What are the filtering technologies for firewalls?
3. What are the security zones of the firewall and what are the functions of each?
4. What is the initial username and password for firewall?
5. Please briefly describe the basic steps for configuring huawei firewall to access the Internet through a static public IP address.

项目 10

项目 10　部署虚拟专用网实现远程办公项目

10.1　项目导入

随着企业规模逐渐扩大，远程用户、远程办公人员、分支机构、合作伙伴也在不断增多，关键业务的需求也不断增加。出差在外的员工需要远程连入企业内网进行移动办公，位于不同城市的总部和分支机构需要进行远距离互联，企业与商业伙伴的网络之间也需要安全连接，对远程网络接入的需求越来越迫切。

在早期的解决方案中，员工通过电话拨号方式接入内部网络，不仅通信费用高，而且不安全。如果通过直接铺设网络线路或者租用运营商专线的方式，成本高，实现困难，令许多中小企业望而却步。

随着信息技术的发展，出现了一种通过公共网络 (如 Internet) 来建立自己专用网络的技术，这种技术就是虚拟专用网 (VPN)，虚拟专用网通过在公共的网络上建立逻辑的专用网络，建立可信的安全连接，保障数据的安全。虚拟专用网解决方案，大幅度减少企业在网络基础设施上的投入，还可以使企业将精力集中到自己的业务上，而不是网络上，提高工作效率。

10.2　相关知识点

10.2.1　虚拟专用网概述

虚拟专用网(Virtual Private Network，VPN)技术是指利用 Internet 等公共网络的基础设施，通过隧道技术，为用户提供一条与专用网络具有相同通信功能的安全数据通道，实现不同网络之间或者用户与内部网络之间的互联互通。虚拟是相对物理上端到端的专有连接而言的，其仅在用户通信时才在公共网络（如互联网）的基础设施上临时创建逻辑通道进行连接；专用是逻辑上的专用通道，而非物理上的专用线路。虚拟专用网是依靠 ISP(互联网服务提供商)，利用虚拟专用网技术建立的一个临时的、安全的、仅供特殊用户使用的网络。

目前，虚拟专用网技术是实现安全传输的重要手段之一，利用它可以在远程用户、公司分支机构、商业合作伙伴与公司的内部网之间建立可信的安全连接，并保护数据的安全传输。同时，通过将数据流转移到低成本的 IP 网络上而大幅度地减少用户在远程网络连接上的费用。正因为它的安全、费用低、灵活性大等特点，VPN 已成为网络技术领域中的一个热点。

10.2.2　虚拟专用网的特点

10.2.2.1　虚拟

与传统的网络不同，虚拟专用网不使用专用的物理线路，它并不是实际意义上的存在，而是利用现有的公共网络，通过配置而形成的一个虚拟网络，是一种逻辑上的网络。其使用公有网络进行通信，一方面，可显著降低企业网络连接的成本；另一方面，大大提高了网络资源的利用率。

10.2.2.2　专用

VPN 不是任何连接在公共网络上的用户都可使用，只为特定的集团或企业所专有使用。从 VPN 使用者角度来看，使用虚拟专用网络与传统专网没有什么本质上的区别。虚拟专用网络作为私有网络，一方面，与底层承载网络之间保持着独立的资源，只允许授权用户使用；另一方面，VPN 提供足够的安全性，确保虚拟专用网络内部信息不被外部入侵。在远端用户、供应商、合作伙伴、驻外机构等与公司总部之间建立可靠的安全连接，保证数据传输的安全性。

10.2.2.3　方便

对 VPN 授权用户而言，使用 VPN 与使用传统网络一样，简单方便。尽管从技术上 VPN 实现比较复杂，但从用户角度而言，是透明的。

10.2.3　虚拟专用网的安全机制

10.2.3.1　隧道机制

隧道技术作为虚拟专用网络技术的核心技术之一，通过将局域网络上的信息数据资源包进行重新包装，形成数据包或者压缩包，使其在计算机网络中顺利传输，为信息数据的安全提供了多一层的保障，降低了数据在传输过程中丢失的风险，增加了数据信息在传输时的安全性和可靠性。

10.2.3.2　加密保护

通过运用虚拟专用网络技术中的加密技术，把网络传输中的明文变成密文，对计算机网络上的数据信息的安全进行保护，保障用户的信息安全，也为隧道技术的安全提供了保障。

10.2.3.3　用户身份认证

身份认证主要用于移动办公用户远程接入的情况。通过对用户的身份进行认证，确保接入内部网络的用户是合法用户，而非恶意用户。合法用户可以成功访问企业资源，非法用户的访问会被禁止。

10.2.4　虚拟专用网与传统数据专网的优势

10.2.4.1　建网快速方便

用户只需将各网络节点采用专线方式本地接入公用网络，并对网络进行相关配置即可。

10.2.4.2 降低建网投资

由于是利用公用网络为基础而建立的虚拟专网，因而可以避免建设传统专用网络所需的高额软硬件投资。

10.2.4.3 节约使用成本

用户采用虚拟专用网组网，不需要像传统专用网那样租用专线，也无需自行建设专用网络线路，大大节约链路租用费用及网络维护费用，减少企业的运营成本。

10.2.4.4 网络安全可靠

实现虚拟专用网主要采用国际标准的网络安全技术，通过在公用网络上建立逻辑隧道及网络层的加密，避免网络数据被修改和盗用，保证了用户数据的安全性及完整性。

10.2.4.5 简化用户对网络的维护及管理

大量的网络管理及维护工作由互联网服务提供商来完成，基础性的网络管理工作由 ISP 统一完成，简化了企业的网络管理维护工作。

10.2.5 虚拟专用网的类型

根据组网方式的不同，虚拟专用网可分为两大类：

10.2.5.1 远程访问 VPN（Access VPN）

适合出差员工 VPN 拨号接入的场景。员工可以在任何能够接入公网的地方，通过远程拨号接入企业内网，从而访问内网资源，如图 10.1 所示。

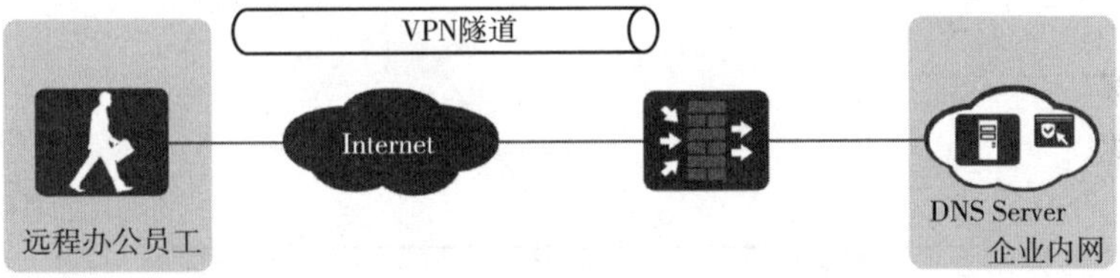

图 10.1 远程访问 VPN

10.2.5.2 局域网到局域网的 VPN（也称为网关到网关的 VPN）

它适用于公司两个异地机构的局域网互联，如图 10.2 所示。

图 10.2 局域网到局域网的 VPN

10.3　项目实施

10.3.1　项目需求

（1）部署虚拟专用网，实现企业总部和分部网络能互联互通。

（2）部署 SSL VPN，实现企业员工出差在外能访问企业内网。

10.3.2　设备清单

本项目主要用到防火墙、接入层交换机等设备，具体的设备型号和数量见表 10.1。

表 10.1　　网络设备清单表

序号	设备型号	数量	功能	备注
1	S5700	2	接入交换机	
2	USG6000V	2	防火墙	
3	AR2220	1	路由器	运营商设备

10.3.3　项目拓扑

本项目的网络拓扑结构如图 10.3 所示。

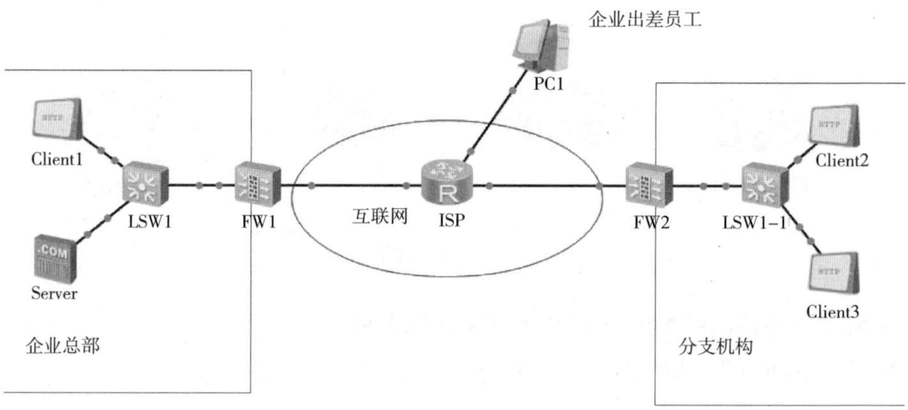

图 10.3　网络拓扑图

10.3.4　配置思路

项目采用如下的思路配置：

（1）初始化防火墙，允许管理员通过 Web 管理防火墙。

（2）配置防火墙各接口地址及所属的安全区域。

（3）配置安全策略，满足数据流向要求。

（4）配置 IPSec VPN，满足企业总部和分支机构的互访需求。

（5）配置 SSL VPN，使得出差用户能正常访问总部内网资源。

10.3.5　相关信息规划

10.3.5.1　总部防火墙

总部防火墙的地址规划见表 10.2。

表 10.2　总部防火墙 IP 地址规划表

序号	接口	IP 地址	掩码	备注
1	G1/0/1	1.1.1.1	255.255.255.252	连接企业总部外网
2	G1/0/2	192.168.1.1	255.255.255.0	连接企业总部内网

10.3.5.2　分支机构防火墙

分支机构防火墙的地址规划见表 10.3。

表 10.3　分支机构防火墙 IP 地址规划表

序号	接口	IP 地址	掩码	备注
1	G1/0/1	2.2.2.2	255.255.255.252	连接分支机构外网
2	G1/0/2	192.168.2.1	255.255.255.0	连接分支机构内网

10.3.5.3　路由器

运营商的设备 ISP 地址规划见表 10.4。

表 10.4　路 由 器 地 址 规 划 表

序号	接口	IP 地址	掩码	备注
1	G0/0/0	1.1.1.2	255.255.255.252	连接企业总部防火墙
2	G0/0/1	2.2.2.1	255.255.255.252	连接分支机构防火墙

10.3.5.4　其他设备

其他设备的地址规划见表 10.5。

表 10.5　其他设备地址规划表

序号	设备名称	IP 地址	掩码	网关	备注
1	企业总部 PC Client1	192.168.1.2	255.255.255.0	192.168.1.1	企业总部内网用户 PC
2	企业总部内网服务器 Server	192.168.1.3	255.255.255.0	192.168.1.1	企业总部内网服务器
3	分支机构 PC Client2	192.168.2.2	255.255.255.0	192.168.2.1	分支机构 PC

续表

序号	设备名称	IP 地址	掩码	网关	备注
4	分支机构 PC Client3	192.168.2.3	255.255.255.0	192.168.2.1	分支机构 PC
5	企业出差员工 PC	3.3.3.2	255.255.255.0	3.3.3.1	企业出差员工

10.3.6　操作步骤

10.3.6.1　初始化防火墙，允许管理员通过 Web 管理防火墙

启动防火墙，通过配置线登录防火墙的 CLI 界面后，进行防火墙的初始化。

（1）输入默认用户名 admin，密码 Admin@123，登录防火墙。

（2）出现是否需要修改密码的提示，选择 y。

（3）输入旧密码 Admin@123，然后输入新密码，再重复输入一遍新密码，即可登录防火墙。新密码要求满足密码复杂性要求，即最低 8 位字符，且同时包含大写字母、小写字母、数字、特殊字符中的 3 种要素，如图 10.4 所示。

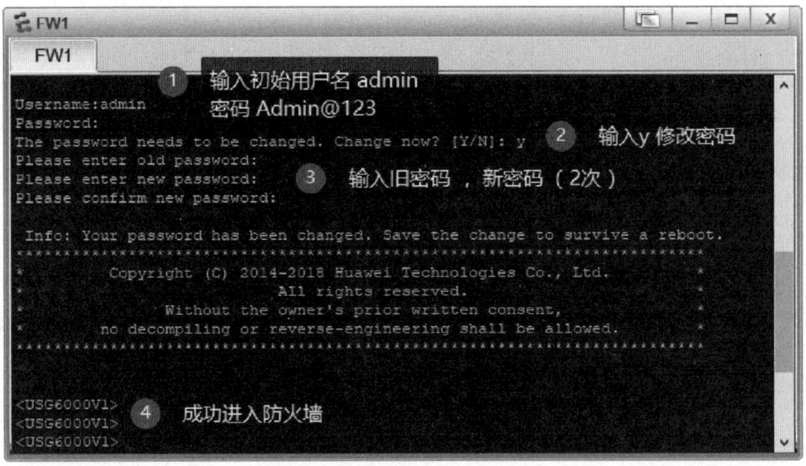

图 10.4　通过 CLI 登录防火墙后修改密码

（4）进入防火墙的管理口，修改管理口的 IP 地址，允许使用 Web 方式管理防火墙。命令如下：

```
<USG6000V1>system-view
[USG6000V1]interface GigabitEthernet 0/0/0
[USG6000V1-GigabitEthernet0/0/0]ip address 192.168.100.100 24  #修改防火墙的管理地址为192.168.100.100，具体的IP地址需要根据现场的管理PC确定。
[USG6000V1-GigabitEthernet0/0/0]service-manage https permit  #允许管理员使用Web方式管理防火墙
```

10.3.6.2　连接防火墙的 G0/0/0 口，打开浏览器，通过 Web 方式管理防火墙

（1）在现场部署安装防火墙时，需要使用双绞线把管理 PC 和防火墙的 G0/0/0 口连接。为了满足在实验环境中学习的需要，给防火墙 eNSP 添加一朵云，然后修改云的参数配置，把云连接到防火墙 G0/0/0 口，使之能连接防火墙，如图 10.5 和图 10.6 所示。

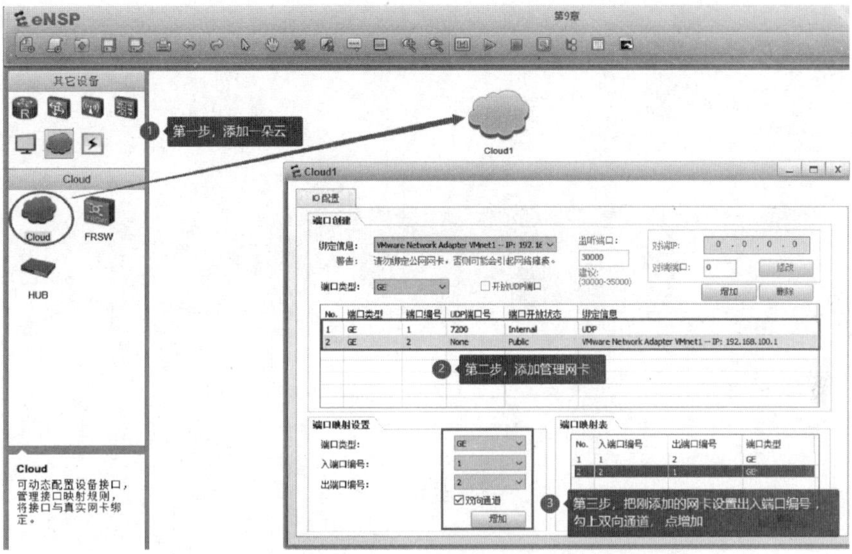

图 10.5　在 eNSP 中添加管理防火墙的网卡

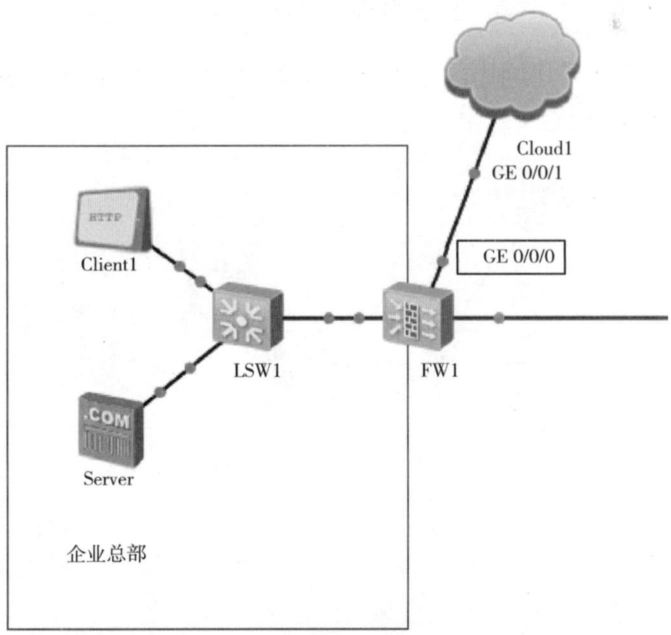

图 10.6　把云连接到防火墙的管理口

（2）在浏览器中输入 https://192.168.100.100:8443，使用 Web 方式登录防火墙，如图 10.7 所示。

图 10.7　通过网页登录防火墙

（3）分支机构防火墙采用同样的方法连接防火墙的管理口，便可登录，如图 10.8 所示。

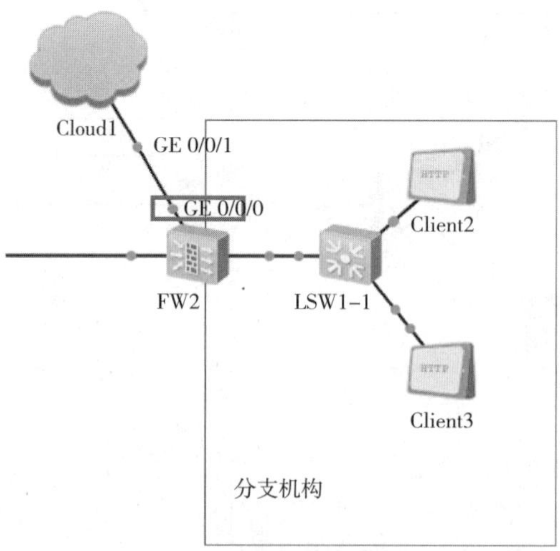

图 10.8　把云连接到分支机构防火墙的管理口

10.3.6.3　企业总部防火墙配置

1. 配置企业总部防火墙各接口地址及所属的安全区域

步骤一：配置防火墙外网口 IP 地址和安全区域。步骤：输入用户名和密码登录防火墙后，点击网络→接口→ GE1/0/1 →选择外网口 GE1/0/1 归属的安全区域（untrust）→输入外网口 IP 地址→输入默认网关→确定，如图 10.9 所示。

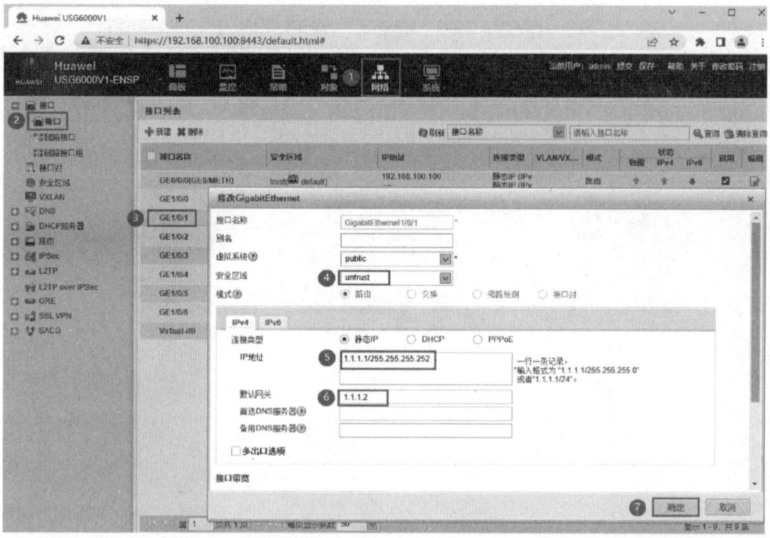

图 10.9　设置防火墙外网口 IP 地址及安全区域

步骤二：配置企业总部防火墙内网口 IP 地址和安全区域。步骤：点击网络→接口→ GE1/0/2 →选择内网口 GE1/0/2 归属的安全区域（trust）→输入内网口 IP 地址→确定，如图 10.10 所示。

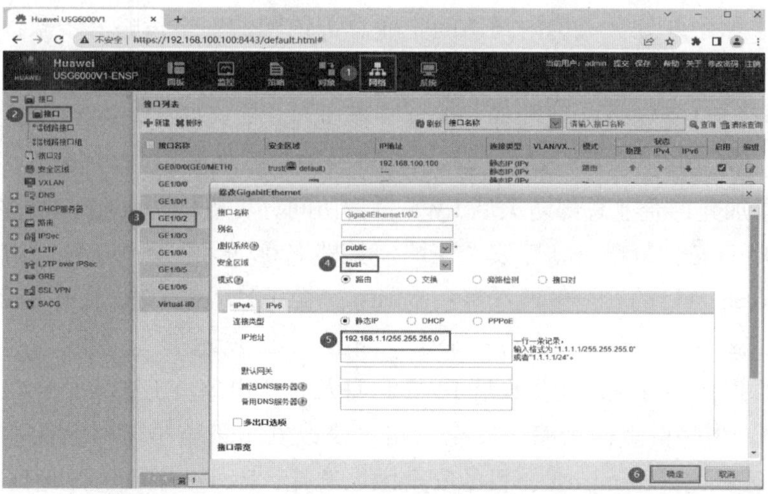

图 10.10　设置防火墙内网口 IP 地址及安全区域

2. 配置安全策略，满足数据流向要求

步骤一：添加允许内网访问外网的安全策略。步骤：点击策略→安全策略→新建安全策略→策略名称输入：允许内网访问外网→源安全区域选择：trust →目的安全区域选择：untrust →动作选择：允许→确定，如图 10.11 所示。

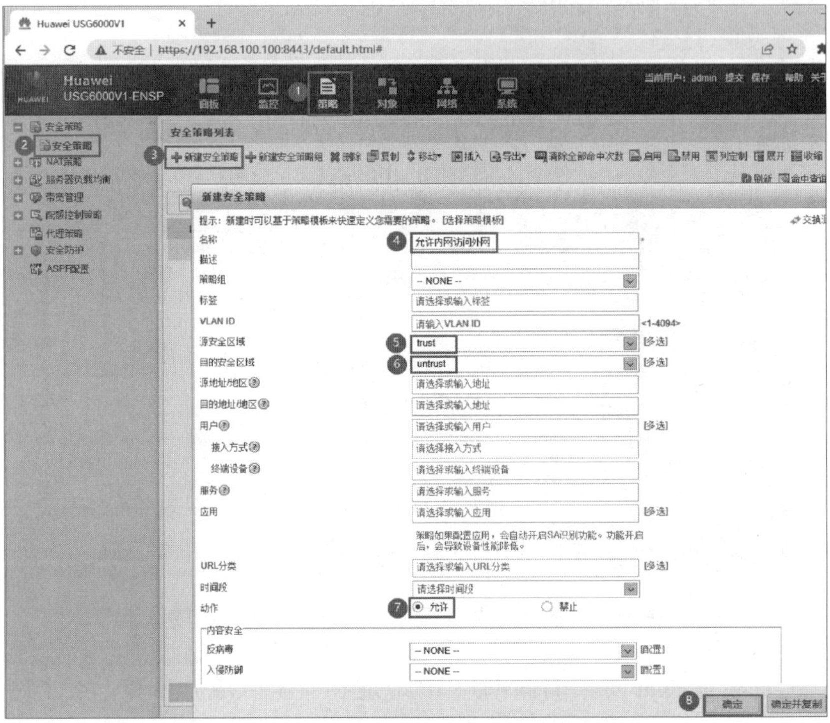

图 10.11　添加允许内网访问外网的安全策略

步骤二：添加允许分支机构内网访问企业总部内网的安全策略。步骤：点击策略→安全策略→新建安全策略→策略名称输入：允许分支机构内网访问企业总部内网→源安全区域选择：untrust →目的安全区域选择：trust →源地址 / 地区输入：192.168.2.0/24 →目的地址 / 地区输入：192.168.1.0/24 →动作选择：允许→确定，如图 10.12 所示。

步骤三：添加允许企业总部防火墙 FW1 自身访问分支机构防火墙 FW2 的公网 IP 地址。步骤：点击策略→安全策略→新建安全策略→策略名称输入：允许 FW1 访问 FW2 →源安全区域选择：local →目的安全区域选择：untrust →源地址 / 地区输入：1.1.1.1 →目的地址 / 地区输入：2.2.2.2 →动作选择：允许→确定，如图 10.13 所示。

步骤四：添加允许分支机构防火墙 FW2 自身及企业出差员工访问企业总部防火墙 FW1 的公网 IP 地址。步骤：点击策略→安全策略→修改安全策略→策略名称输入：允许访问 FW1 →源安全区域选择：untrust →目的安全区域选择：local →源地址 / 地区输入：any（因为除了分支机构防火墙需要访问总部防火墙的公网 IP 地址外，企业出差员工也需要访问总部防火墙的公网地址，而企业出差员工的 IP 地址是不固定的）→目的地址 / 地区输入：1.1.1.1 →动作选择：允许→确定，如图 10.14 所示。

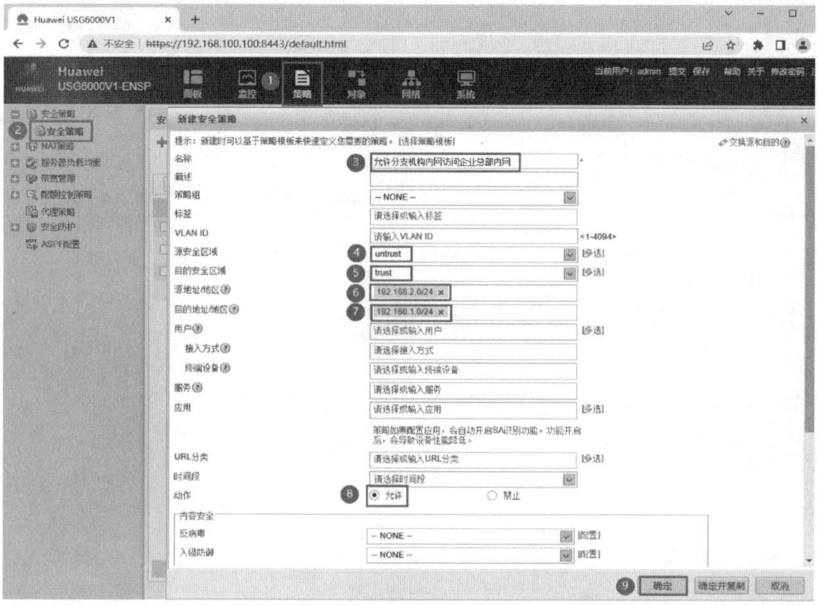

图 10.12　添加允许分支机构内网访问企业总部内网的安全策略

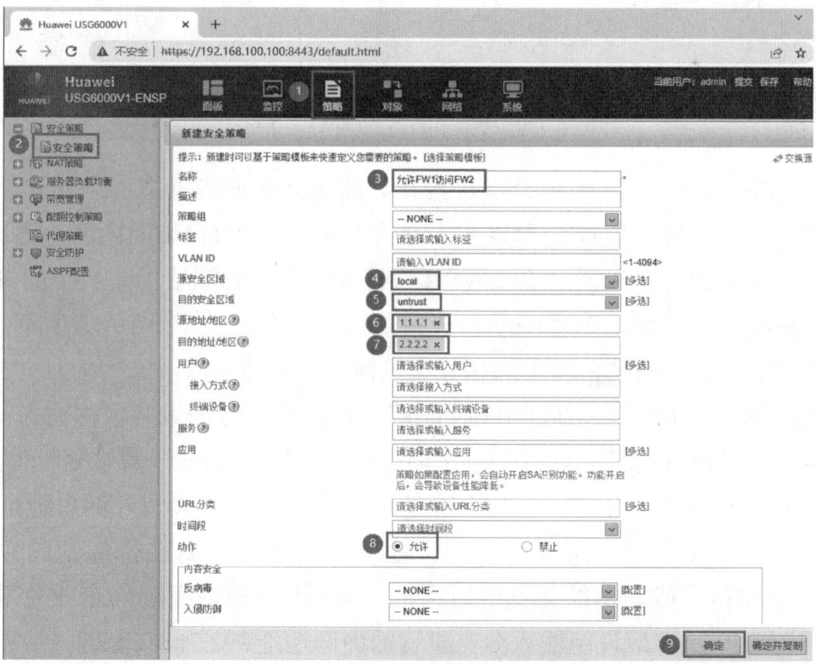

图 10.13　允许 FW1 访问 FW2

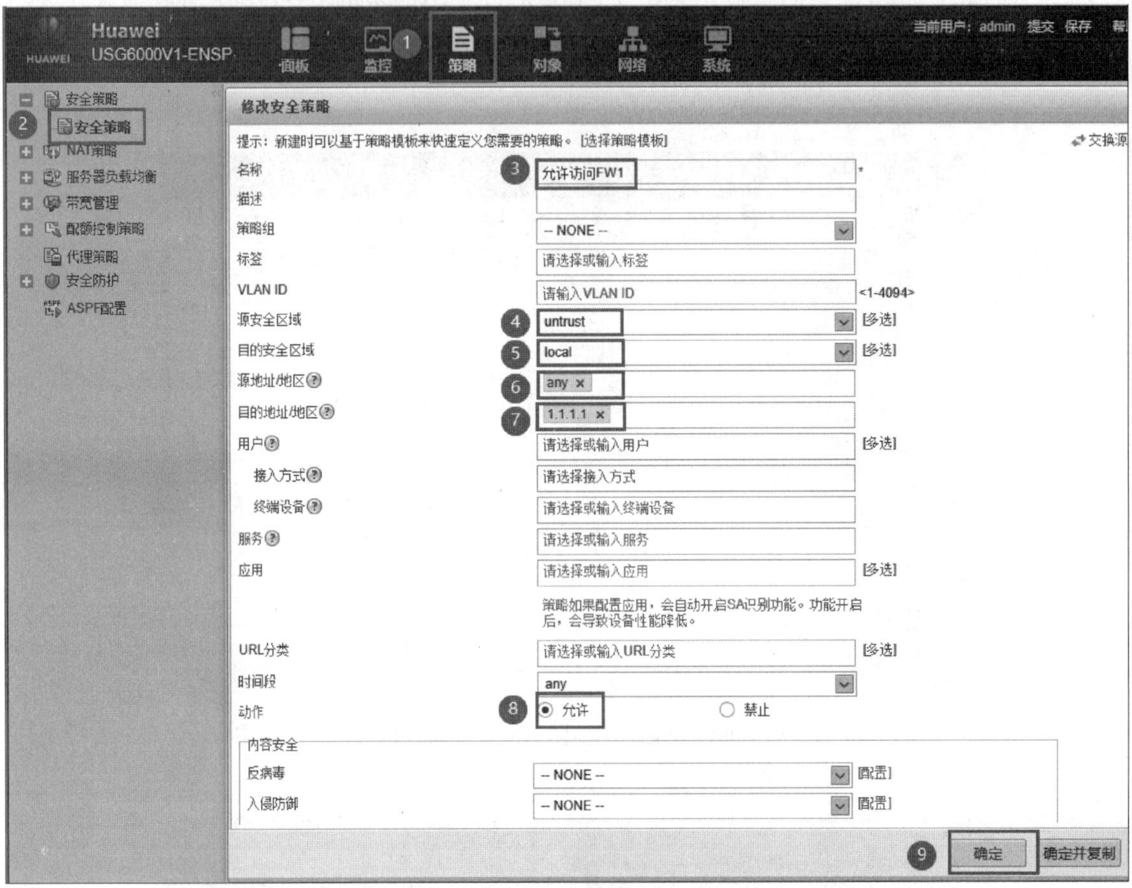

图 10.14　允许外网访问 FW1

步骤五：添加允许企业出差员工经过授权后能访问企业内网资源。步骤：点击策略→安全策略→新建安全策略→策略名称输入：允许 SSL VPN 用户访问内网→源地址 / 地区输入：10.1.1.0/24(企业出差员工通过 SSL VPN 认证授权后，分配给出差员工的 IP 地址)→目的地址 / 地区输入：192.168.1.0/24 →动作选择：允许→确定，如图 10.15 所示。

3. 配置 IPSec VPN，满足企业总部和分支机构互联互通的需求

步骤：点击网络→ IPSec →新建 IPSec 策略→选择点到点→策略名称输入：企业总部到分支机构 VPN →本端接口选择: GE1/0/1 →本端地址输入: 1.1.1.1 →对端地址输入: 2.2.2.2 →认证方式选择预共享密钥→本地 ID 选择 IP 地址，输入本端防火墙外网口地址 1.1.1.1 →对端 ID 选择 IP 地址，输入对端防火墙外网口地址 2.2.2.2 →然后在待加密的数据流新建需要加密的数据→在新建待加密的数据流对话框，源地址 / 地址组输入企业总部的内网地址 192.168.1.0/24，目的地址 / 地址组输入分支机构的内网地址 192.168.2.0/24 →最后确定、应用，如图 10.16 和图 10.17 所示。

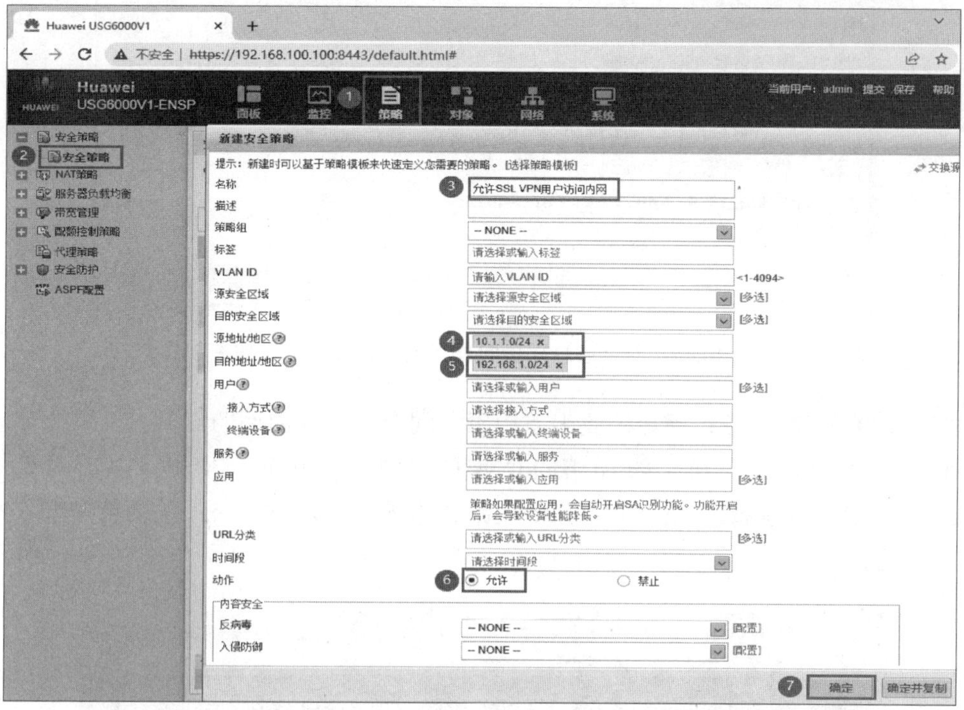

图 10.15　添加允许企业出差员工经过授权后能访问企业内网资源

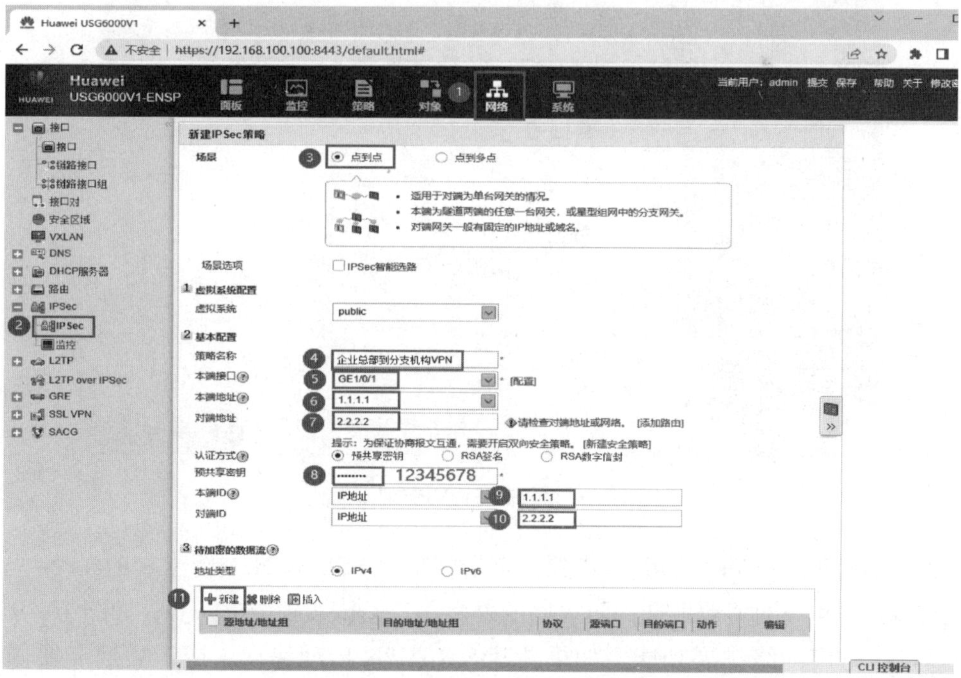

图 10.16　添加点到点的 IPSec VPN

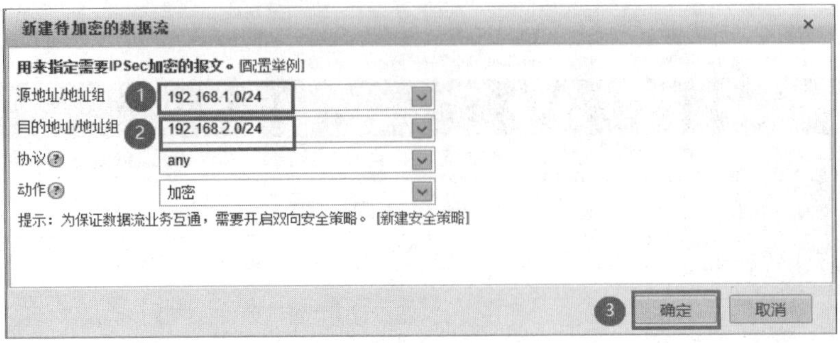

图 10.17　对企业总部访问分支机构的数据进行加密

4. 新建 SSL VPN 服务，满足企业出差用工访问企业总部内网的需求

步骤一：为企业出差员工添加访问企业总部内网的账号。步骤：点击对象→用户→ default →用户场景选择：SSL VPN 接入→新建用户 →输入新建用户登录名、密码→确定，如图 10.18 所示。

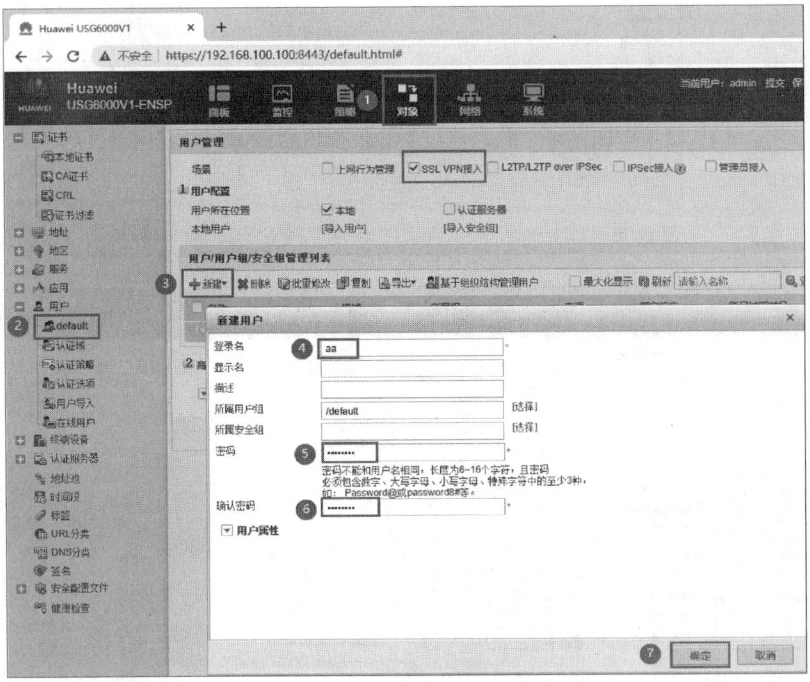

图 10.18　添加 SSL VPN 认证账号

步骤二：新建 SSL VPN 服务。步骤：点击网络→ SSL VPN →新建 SSL VPN →网关名称输入：SSL VPN →网关地址选择防火墙的外网接口 GE1/0/1 →下一步，如图 10.19 所示。

步骤三：启用网络扩展功能。如果新建 SSL VPN 出现错误，或者后期需要修改 SSL VPN 的配置，则点击 SSL VPN →编辑已创建的 SSL VPN 名称→启用网络扩展，在可分配

IP 地址范围内，输入分配给企业出差员工的 IP 地址：10.1.1.0/24 →路由模式：选择全路由模式，如图 10.20 所示。

图 10.19　新建 SSL VPN 服务

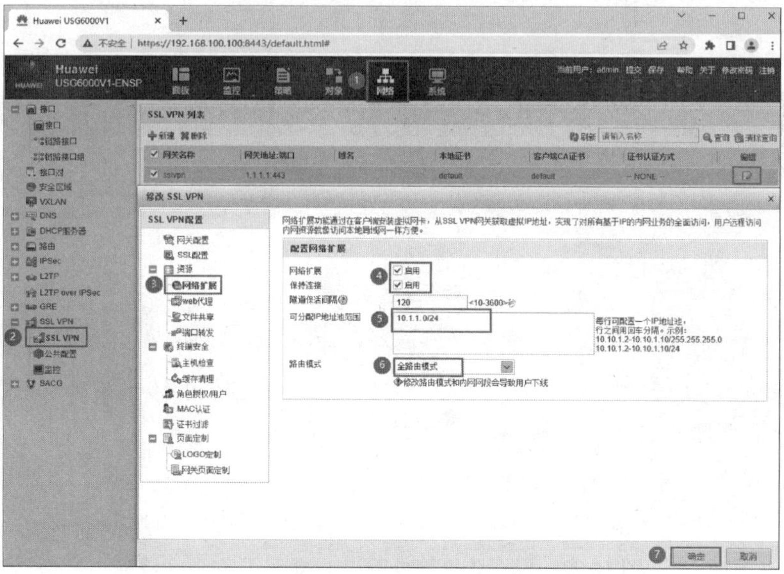

图 10.20　为 SSL VPN 启用网络扩展功能

步骤四：添加授权用户。在角色授权 / 用户窗口，添加能使用 SSL VPN 的用户角色。步骤：点击角色授权 / 用户→新建→角色名称输入企业出差用户→关联刚才创建的用户→业务启用：网络扩展→确定，如图 10.21 所示。

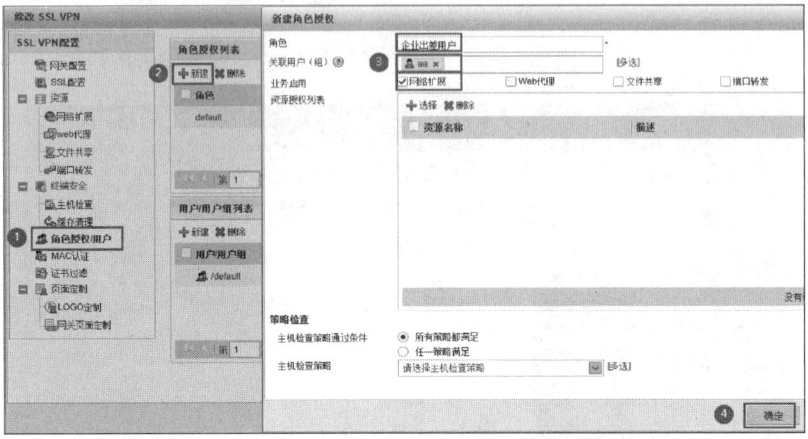

图 10.21　添加授权用户

10.3.6.4　分支机构防火墙配置

1. 配置分支机构防火墙各接口地址及所属的安全区域

步骤一：配置防火墙外网口 IP 地址和安全区域。步骤：输入用户名和密码登录防火墙后，点击网络→接口→ GE1/0/1 →选择外网口 GE1/0/1 归属的安全区域（untrust）→输入外网口 IP 地址→输入默认网关→确定，如图 10.22 所示。

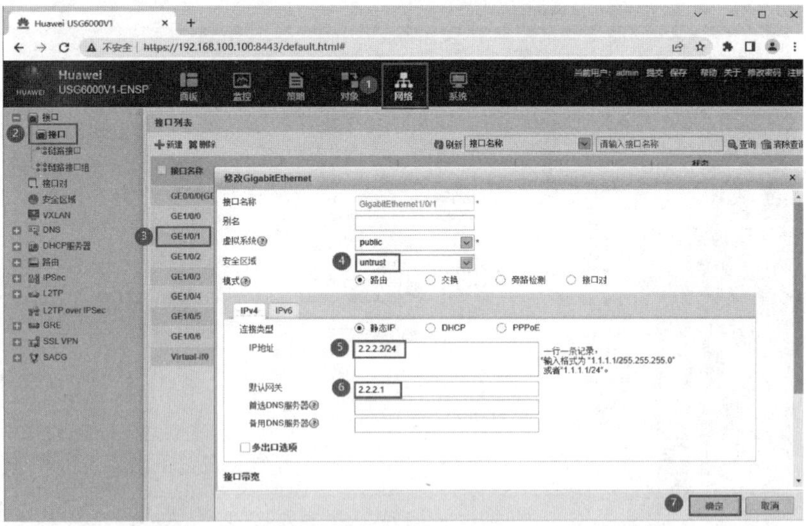

图 10.22　设置防火墙外网口 IP 地址及安全区域

步骤二：配置分支机构防火墙内网口 IP 地址和安全区域。步骤：点击网络→接口→ GE1/0/2 →选择内网口 GE1/0/2 归属的安全区域（trust）→输入内网口 IP 地址→确定，如图 10.23 所示。

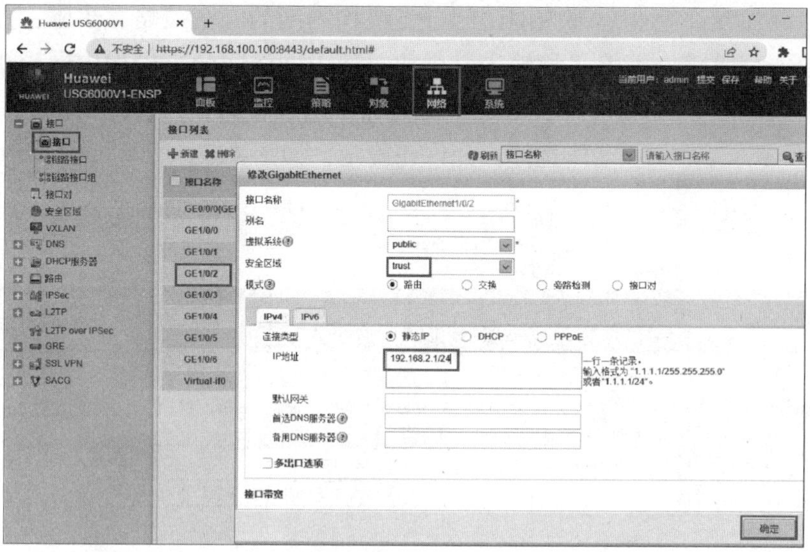

图 10.23　设置防火墙内网口 IP 地址及安全区域

2. 配置安全策略，满足数据流向要求

步骤一：添加允许内网访问外网的安全策略。步骤：点击策略→安全策略→新建安全策略→策略名称输入：允许内网访问外网→源安全区域选择：trust →目的安全区域选择：untrust →动作选择：允许→确定，如图 10.24 所示。

图 10.24　添加允许内网访问外网的安全策略

步骤二：添加允许企业总部内网访问分支机构内网的安全策略。步骤：点击策略→安全策略→新建安全策略→策略名称输入：企业总部内网访问分支机构内网→源安全区域选择：untrust→目的安全区域选择：trust→源地址/地区输入：192.168.1.0/24→目标地址/地区输入：192.168.2.0/24→动作选择：允许→确定，如图 10.25 所示。

图 10.25　添加允许企业总部访问分支机构内网的安全策略

步骤三：添加允许分支机构防火墙 FW2 自身访问企业总部防火墙 FW1 的公网 IP 地址。步骤：点击策略→安全策略→新建安全策略→策略名称输入：允许 FW2 访问 FW1→源安全区域选择：local→目的安全区域选择：untrust→源地址/地区输入：2.2.2.2→目标地址/地区输入：1.1.1.1→动作选择：允许→确定，如图 10.26 所示。

步骤四：添加允许企业总部防火墙 FW1 自身访问分支机构防火墙 FW2 的公网 IP 地址。步骤：点击策略→安全策略→新建安全策略→策略名称输入：允许 FW1 访问 FW2→源安全区域选择：untrust→目的安全区域选择：local→源地址/地区输入：1.1.1.1→目标地址/地区输入：2.2.2.2→动作选择：允许→确定，如图 10.27 所示。

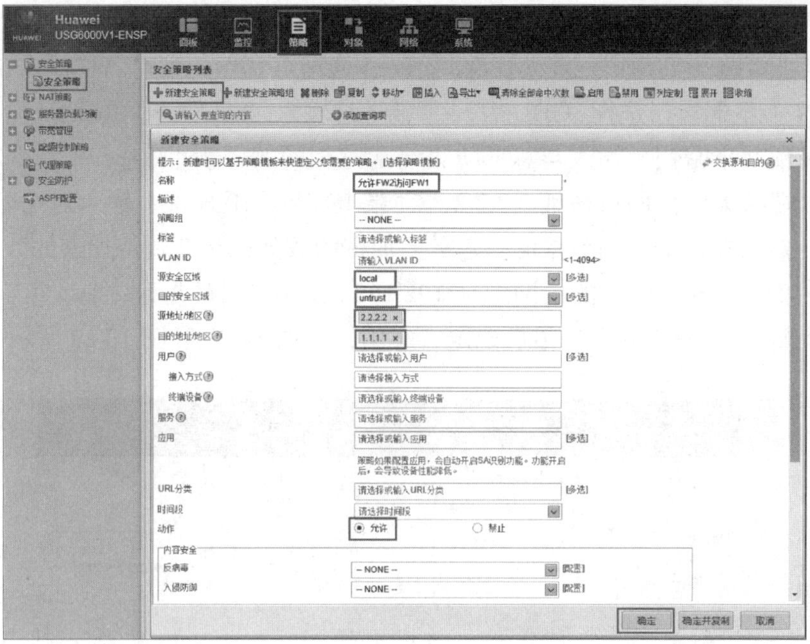

图 10.26　允许 FW2 访问 FW1

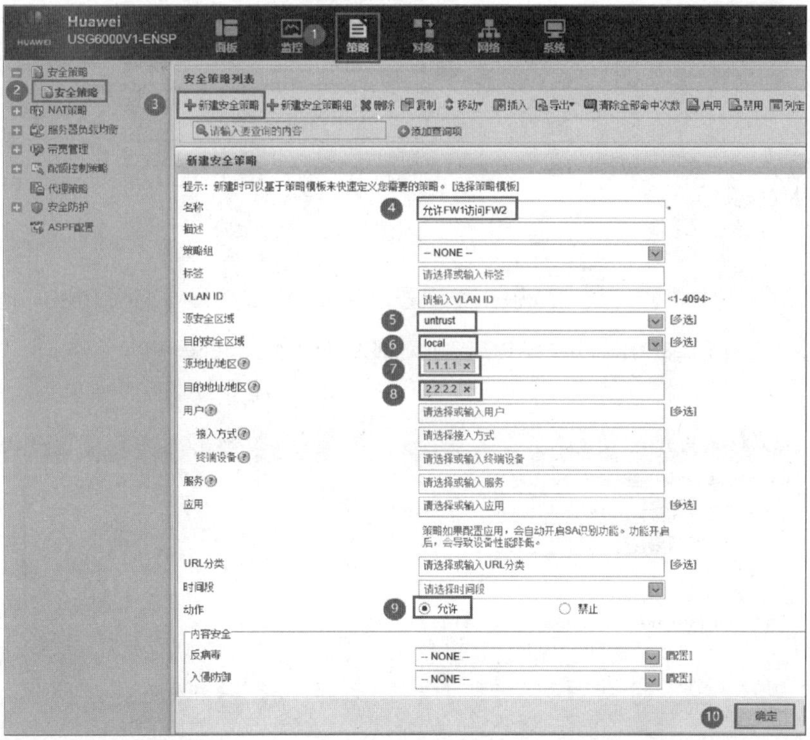

图 10.27　添加允许防火墙 FW1 自身访问防火墙 FW2 的公网 IP 地址安全策略

3. 配置 IPSec VPN，满足企业总部和分支机构互联互通的需求

步骤：点击策略→ IPSec →新建 IPSec 策略→选择点到点→策略名称输入：分支机构到企业总部 VPN →本端接口选择: GE1/0/1 →本端地址输入: 2.2.2.2 →对端地址输入: 1.1.1.1 →认证方式选择预共享密钥，输入与总部防火墙配置相同的密钥 12345678 →本端 ID 选择 IP 地址，输入本端防火墙外网口地址 2.2.2.2 →对端 ID 选择 IP 地址，输入对端防火墙外网口地址 1.1.1.1 →然后在待加密的数据流新建需要加密的数据→在新建待加密的数据流对话框，源地址 / 地址组输入分支机构的内网地址 192.168.2.0/24，目的地址 / 地址组输入企业总部机构的内网地址 192.168.1.0/24 →最后确定、应用，如图 10.28 和图 10.29 所示。

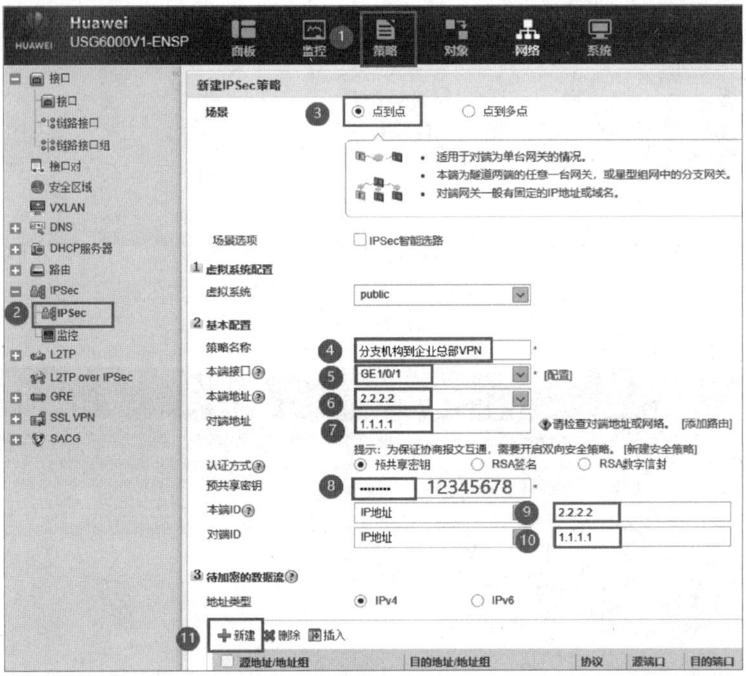

图 10.28　添加分支结构到企业总部 VPN 点到点 IPSec VPN 策略

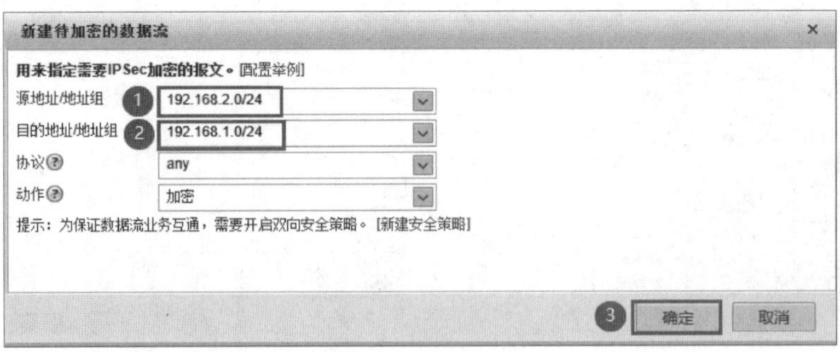

图 10.29　对分支机构访问企业总部的数据进行加密

10.3.6.5　给其他设备配置 IP 地址

（1）互联网路由器 IP 地址配置（在虚拟仿真实验中为了测试方便需要配置，在实践工程项目中，互联网区域不需要进行配置）。

```
<Huawei>system-view
Enter system view, return user view with Ctrl+Z
[Huawei]interface GigabitEthernet 0/0/0
[Huawei-GigabitEthernet0/0/0]ip address 1.1.1.2 24
[Huawei-GigabitEthernet0/0/0]quit
[Huawei]interface GigabitEthernet 0/0/1
[Huawei-GigabitEthernet0/0/1]interface GigabitEthernet 0/0/1
[Huawei-GigabitEthernet0/0/1]ip address 2.2.2.1 24
[Huawei-GigabitEthernet0/0/1]quit
```

（2）终端设备配置 IP 地址。步骤：双击终端设备→输入本机（IP）地址、子网掩码、网关等参数→保存，如图 10.30 所示。

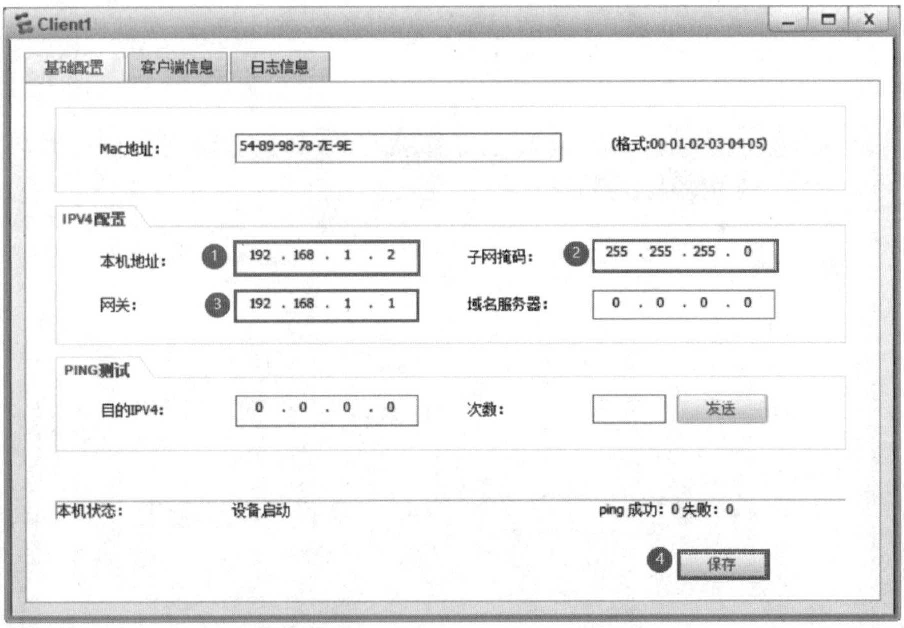

图 10.30　终端设备配置 IP 地址

10.3.6.6　测试网络连通性

1. 虚拟专用网建立成功标志

步骤：在防火墙的管理界面中，点击网络→ IPSec →会看到协商状态为成功：1，标志着 IPSec VPN 已经建立成功，如图 10.31 所示。

图 10.31　IPSec VPN 创建成功标志

2. 企业总部内网访问分支机构内网测试

步骤：在企业总部内网 PC 中，输入外网服务器 PC 的 IP 地址（目的 IPV4）→测试次数选择 5→发送→出现 ping 成功：5 的显示，表明总部 PC 访问分支机构 PC 成功，如图 10.32 所示。

图 10.32　企业总部内网访问分支机构内网测试

3. 分支机构内网访问企业总部内网服务器测试

步骤：在分支机构内网 PC 中，输入企业总部服务器的 IP 地址（目的 IPV4）→测试次数选择 5→发送→出现 ping 成功：5 的显示，表明分支机构 PC 访问总部服务器成功，如图 10.33 所示。

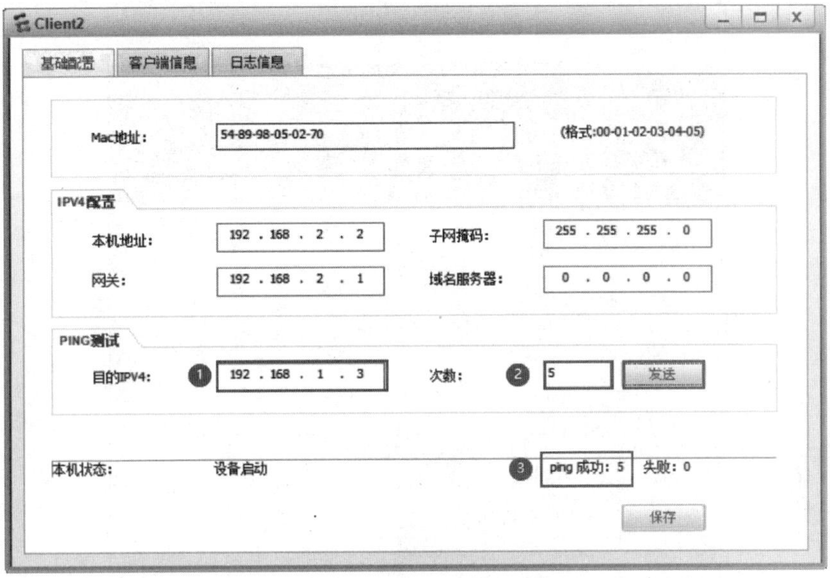

图 10.33　分支机构内网访问企业总部内网服务器测试

4. 企业出差员工访问企业总部内网服务器测试

方法一：企业出差员工在浏览器中输入 https:// 防火墙外网口地址，即可打开 SSL 登录界面，输入创建的 SSL VPN 账号（用户名）、密码，登录后即可访问内网资源，如图 10.34 所示。

图 10.34　企业出差员工访问企业总部内网服务器测试

方法二：安装 VPN 客户端软件 SecoClient，打开后输入企业总部防火墙外网口地址，登录也可访问企业总部内网资源，如图 10.35 所示。

图 10.35　通过 VPN 客户端软件登录

10.3.6.7　保存防火墙配置

步骤：在防火墙的右上角，点击保存→确定，保存防火墙的配置信息，如图 10.36 所示。

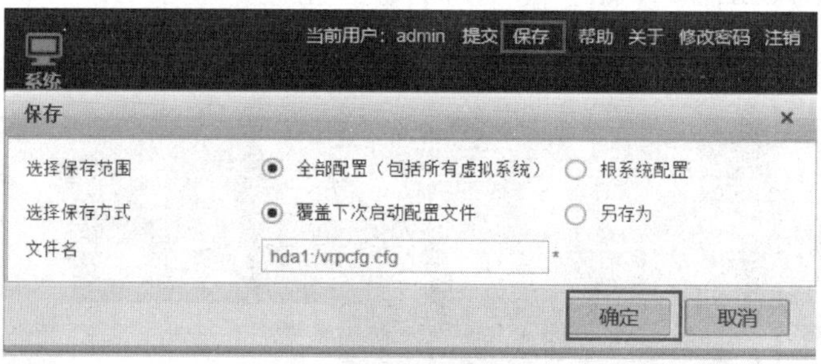

图 10.36　保存防火墙的配置信息

10.4　思政融入

居家办公的需求让人们开始加速思考新的办公状态。线上与线下结合的办公模式既能提高工作效率，也能减少沟通成本，因而开始被广为接受并应用。高效的办公模式依赖于我国网络建设的广度和深度。近年来，我国在移动支付、电子商务、超级计算、量子通信等众多领域已走在世界前列，我们网络人才必须在贡献社会进步和经济社会发展中具有坚定的“魂”。网络强国已成为当前我国网信事业发展的重要目标，是实现“两个一百年”奋斗目标和中华民族伟大复兴中国梦的必然选择。

10.5　思考与训练

1. 虚拟专用网的安全机制包含哪些?
2. 简述虚拟专用网的特点。
3. 根据组网方式的不同，虚拟专用网可分为哪两大类?
4. 什么是 SSL VPN？其跟 IP Sec VPN 有什么不同?
5. 企业员工经常出差在外，需要安全可靠地访问企业内网，请完成 SSL VPN 的配置，并写出配置步骤。

Project 10

Project 10　Deploy virtual private network to realize remote office

10.1　Project import

With the gradual expansion of the enterprise scale, remote users, remote office workers, branches and partners are also increasing, and the demand for key businesses is increasing. Employees on business trips need to connect to the enterprise intranet remotely for mobile office, headquarters and branches in different cities need to connect remotely, and networks between enterprises and business partners also need to be connected securely. The demand for remote network access is becoming increasingly urgent.

In the early solution, employees accessed the internal network through telephone dialing, which was not only costly but also insecure. If the network line is directly laid or the special line of the operator is leased, the cost is high and the implementation is difficult, which makes many small and medium-sized enterprises flinch.

With the development of information technology, there is a technology to establish its own private network through public networks (such as the Internet). This technology is called Virtual Private Network (VPN for short). Virtual private network establishes a trusted and secure connection through the establishment of a logical private network on the public network to ensure the security of data. The virtual private network solution can significantly reduce the enterprise's investment in network infrastructure, and also enable enterprises to focus on their own business, rather than the network, to improve work efficiency.

10.2　Relevant knowledge points

10.2.1　Overview of virtual private network

Virtual Private Network (VPN) technology refers to the use of Internet and other public network infrastructure, through tunnel technology, to provide users with a secure data channel with the same communication function as the private network, to achieve interconnection between different networks or between users and internal networks. Virtual refers to the end-to-end proprietary connection in physical terms. It only temporarily creates logical channels on the infrastructure of public networks (such as the Internet) for connection when users communicate; Dedicated is a logical dedicated channel, not a physical dedicated circuit. Virtual Private Network (VPN) is a temporary, secure network established by ISP (Internet Service Provider) and virtual private network technology for special users only.

At present, virtual private network technology is one of the important means to achieve secure transmission. It can be used to establish a trusted secure connection between remote users, company branches, business partners and the company's intranet, and protect the secure transmission of data. At the same time, by transferring the data flow to the low-cost IP network, the cost of remote network connection is greatly reduced. Because of its security, low cost and flexibility, VPN has become a hotspot in the field of network technology.

10.2.2 Characteristics of virtual private network

10.2.2.1 Virtual

Different from the traditional network, VPN does not use dedicated physical lines. It does not exist in the actual sense, but uses the existing public network to form a virtual network through configuration. It is a logical network. Its use of public networks for communication, On one hand, it can significantly reduce the cost of enterprise network connectivity; on the other hand, greatly improve the utilization of network resources.

10.2.2.2 Special-purpose

VPN is not available to any user connected to the public network, but only for specific groups or enterprises. From the perspective of VPN users, there is no essential difference between using virtual private networks and traditional private networks. As a private network, virtual private network, on one hand, maintains independent resources with the underlying bearer network, allowing only authorized users to use; on the other hand, VPN provides sufficient security to ensure that the internal information of the virtual private network is not invaded by the outside. Establish a reliable and secure connection between remote users, suppliers, partners, overseas agencies, etc. and the company headquarters to ensure the security of data transmission.

10.2.2.3 Convenient

For VPN authorized users, using VPN is as simple and convenient as using traditional networks. Although the VPN implementation is technically complex, it is transparent from the user's perspective.

10.2.3 Security mechanism of virtual private network

10.2.3.1 Tunnel mechanism

As one of the core technologies of virtual private network technology, tunnel technology can repackage the information data resource packets on the local area network to form data packets or compressed packets, so that they can be transmitted smoothly in the computer network, providing a layer of security for information data, reducing the risk of data loss during transmission, and increasing the security and reliability of data information during transmission.

10.2.3.2 Encryption protection

By using the encryption technology in the virtual private network technology, the plaintext in the network transmission is changed into the ciphertext to protect the security of the data information on the computer network, ensure the information security of users, and also provide security for the tunnel technology.

10.2.3.3 User identity authentication

Identity authentication is mainly used for remote access of mobile office users. By authenticating the user's identity, ensure that the user accessing the internal network is a legitimate user, not a malicious user. Legitimate users can successfully access enterprise resources, while illegal access will be prohibited.

10.2.4 Advantages of virtual private network and traditional data private network

10.2.4.1 Network construction is fast and convenient

The user only needs to connect each network node to the public network locally by means of dedicated lines and configure the network.

10.2.4.2 Reduce investment in network construction

Because it is a virtual private network built on the basis of public networks, it can avoid the high investment in software and hardware required for the construction of traditional private networks.

10.2.4.3 Save the use cost

Users use virtual private networks to form networks. They do not need to rent special lines like traditional private networks, nor do they need to build their own private network lines. This greatly saves link rental fees and network maintenance costs, and reduces the operating costs of enterprises.

10.2.4.4 The network is safe and reliable

The implementation of virtual private network mainly adopts the international standard network security technology. By establishing a logical tunnel on the public network and encrypting the network layer, the network data is prevented from being modified and stolen, and the security and integrity of user data are guaranteed.

10.2.4.5 Simplify users' maintenance and management of the network

A large amount of network management and maintenance work is completed by Internet service providers, and the basic network management work is completed by ISPs, simplifying the network management and maintenance work of enterprises.

10.2.5 Type of virtual private network

According to different networking methods, VPNs can be divided into two categories:

10.2.5.1 Remote Access VPN (Access VPN)

It is suitable for VPN dial in access of employees on business trips. Employees can access the enterprise intranet through remote dial-up at any place where they can access the public network, so as to access the intranet resources. As shown in Figure 10.1.

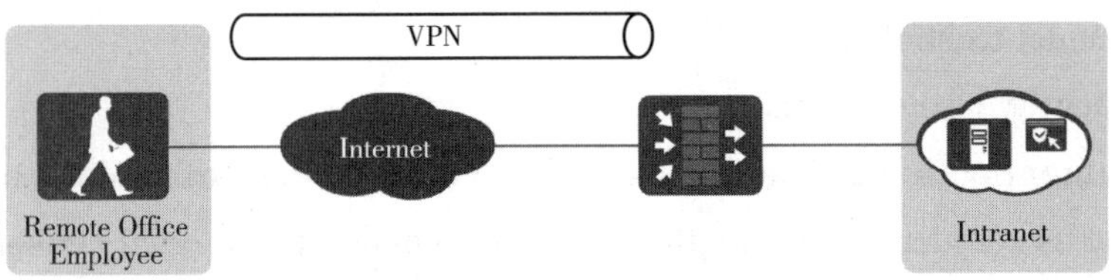

Figure 10.1 Remote access VPN

10.2.5.2 LAN to LAN VPN (also known as gateway to gateway VPN)

It is applicable to LAN interconnection between two remote organizations of the company. As shown in Figure 10.2.

Figure 10.2 LAN to LAN VPN

10.3 Project implementation

10.3.1 Project requirements

(1) Deploy virtual private network to realize interconnection between enterprise headquarters and branch networks.

(2) Deploy SSL VPN to enable enterprise employees to access the enterprise intranet when traveling on business.

10.3.2 Equipment list

This project mainly uses firewall, access layer switch and other equipment, and the specific equipment model and quantity are shown in Table 10.1.

Table 10.1 List of network equipment

Serial No.	Equipment model	Quantity	Function	Remarks
1	S5700	2	Access switch	
2	USG6000V	2	Firewall	
3	AR2220	1	Router	Operator equipment

10.3.3 Access switch project topology

The network topology of the project is shown in Figure 10.3。

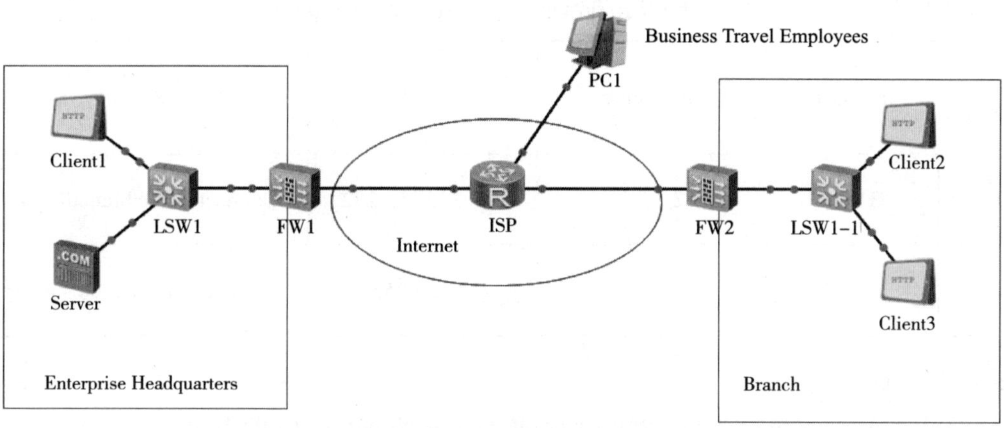

Figure 10.3 Network topology

10.3.4 Configuration ideas

The project adopts the following ideas:

(1) Initialize the firewall and allow the administrator to manage the firewall through the Web.

(2) Configure the interface addresses and security zones of the firewall.

(3) Configure security policies to meet data flow requirements.

(4) Configure IP sec VPN to meet the mutual access needs of enterprise headquarters and branches.

(5) SSL VPN is configured to enable business users to access the intranet resources of the headquarters normally.

10.3.5 Relevant information planning

10.3.5.1 Headquarters firewall

The address planning of the firewall is shown in Table 10.2.

Table 10.2 Enterprise headquarters firewall IP address planning table

Serial No.	Interface	IP address	Mask	Remarks
1	G1/0/1	1.1.1.1	255.255.255.252	Connect to the external network of the enterprise headquarters
2	G1/0/2	192.168.1.1	255.255.255.0	Connect the intranet of enterprise headquarters

10.3.5.2 Branch office firewall

The address planning of the firewall is shown in Table 10.3.

Table 10.3 Branch office firewall IP address planning table

Serial No.	Interface	IP address	Mask	Remarks
1	G1/0/1	2.2.2.2	255.255.255.252	Connect to the branch's internet
2	G1/0/2	192.168.2.1	255.255.255.0	Connect the branch's intranet

10.3.5.3 Router

The ISP address plan of the operator's equipment is shown in Table 10.4.

Table 10.4 Router address planning table

Serial No.	Interface	IP address	Mask	Remarks
1	G0/0/0	1.1.1.2	255.255.255.252	Connect the enterprise headquarters firewall
2	G0/0/1	2.2.2.1	255.255.255.252	Connect branch office firewall

10.3.5.4. Other equipment

The address planning of other equipment is shown in Table 10.5.

Table 10.5 Address planning of other equipment

Serial No.	Interface	IP address	Mask	Gateway	Remarks
1	Enterprise headquarters PC Client1	192.168.1.2	255.255.255.0	192.168.1.1	Intranet user PC of enterprise headquarters
2	Enterprise headquarters intranet server	192.168.1.3	255.255.255.0	192.168.1.1	Intranet server of enterprise headquarters
3	Branch office PC Client2	192.168.2.2	255.255.255.0	192.168.2.1	Branch office PC
4	Branch office PC Client3	192.168.2.3	255.255.255.0	192.168.2.1	Branch office PC
5	PC of employees on business trip	3.3.3.2	255.255.255.0	3.3.3.1	Employee on business trip

10.3.6 Operating steps

10.3.6.1 Initialize the firewall and allow the administrator to manage the firewall through the Web

Start the firewall, log in to the CLI interface of the firewall through the configuration line, and then initialize the firewall.

(1) Enter the default username admin and password Admin@123 , log in to the firewall.

(2) When prompted whether to change the password, select y.

(3) Enter old password Admin@123 , then enter a new password, and then enter the new password again to log in to the firewall. The new password must meet the password complexity requirements, that is, minimum of 8 characters with three elements of uppercase, lowercase, numeric and special characters. As shown in Figure 10.4.

(4) Enter the management port of the firewall, modify the IP address of the management port, and allow the firewall to be managed in the Web mode. The command is as follows:

```
<USG6000V1>system-view
[USG6000V1]interface GigabitEthernet 0/0/0
[USG6000V1-GigabitEthernet0/0/0]ip address 192.168.100.100 24    # The management address of the firewall is modified to 192.168.100.100, and the specific IP address needs to be determined according to the on-site management PC.
```

[USG6000V1-GigabitEthernet0/0/0]service-manage https permit # Allow administrators to manage firewalls in Web mode

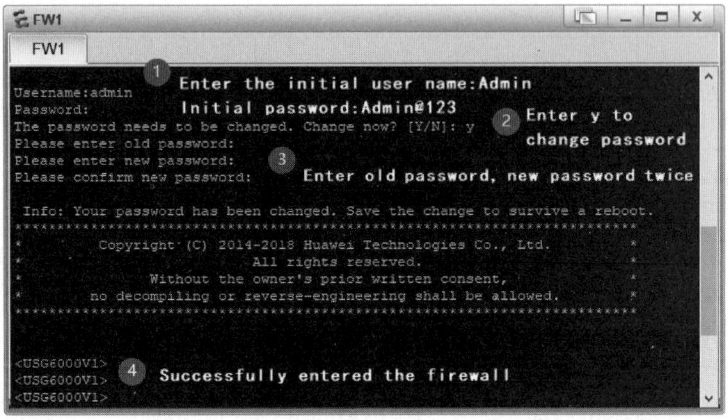

Figure 10.4 Modifying the password after logging in to the firewall through CLI

10.3.6.2 Connect to the G0/0/0 port of the firewall, open the browser, and manage the firewall through Web

(1) When deploying and installing firewalls on site, twisted pair cables are required to connect the management PC to the G0/0/0 port of the firewall. In order to learn in the experimental environment, we add a cloud to the firewall eNSP, and then modify the cloud parameter configuration to connect the cloud to the firewall G0/0/0 port, so that it can connect to the firewall. As shown in Figures 10.5 and 10.6.

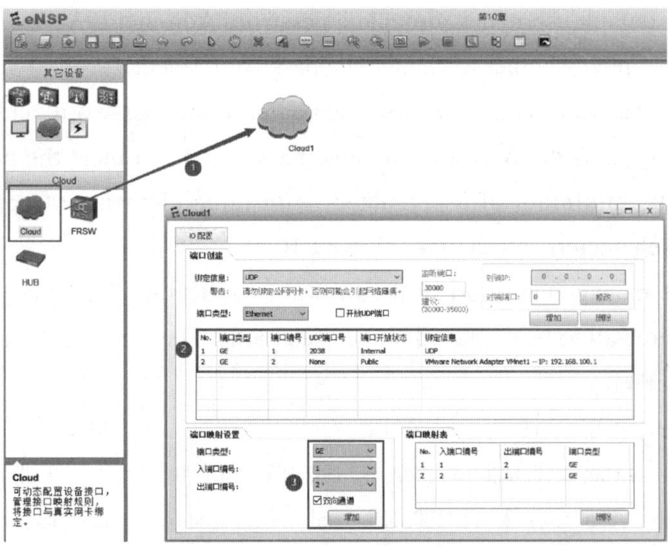

Figure 10.5 Add the network card of the management firewall in eNSP

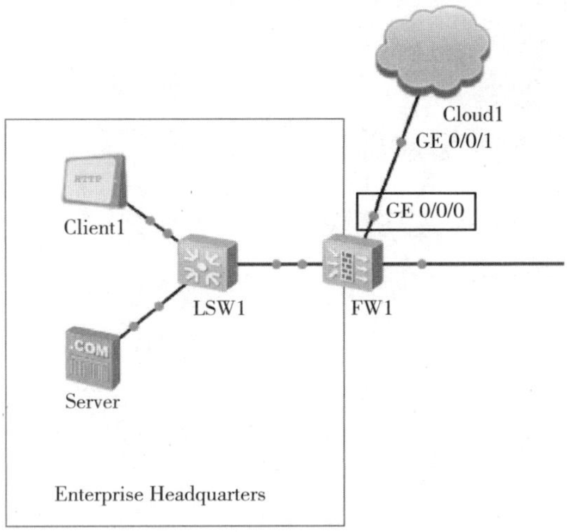

Figure 10.6 Management port for connecting cloud to firewall

(2) Enter https://192.168.100.100:8443 in browser, log in to the firewall in Web mode. As shown in Figure 10.7.

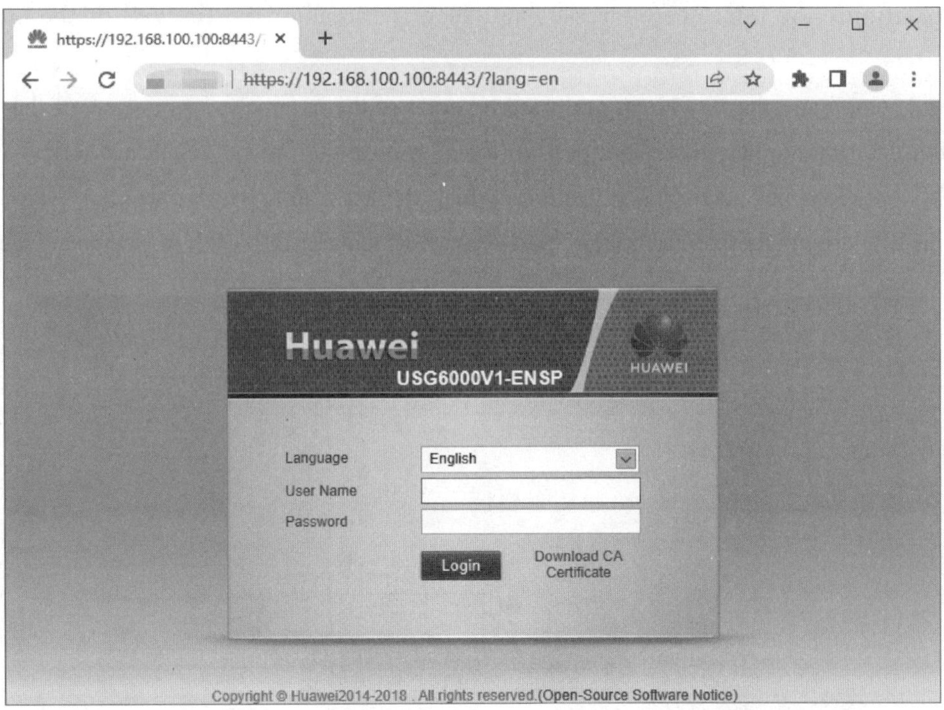

Figure 10.7 Log in to the firewall through the Webpage

(3) The branch firewall can log in by connecting to the management port of the firewall in the same way. As shown in Figure 10.8.

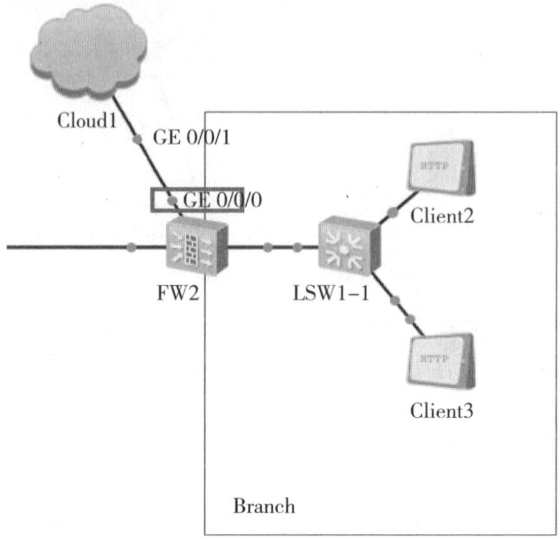

Figure 10.8 Management port for connecting cloud to branch office firewall

10.3.6.3 Firewall configuration of enterprise headquarters

1. Configure the interface addresses and security areas of the firewall of the enterprise headquarters

Step 1: Configure the IP address and security zone of the external network port of the firewall. Step: After entering the user name and password to log in to the firewall, click network →Interface →GE1/0 1 →Select the security zone (untrust) which the external network port GE1/0/1 belongs to →Enter the IP address of the external network port →Enter default gateway →OK, as shown in Figure 10.9.

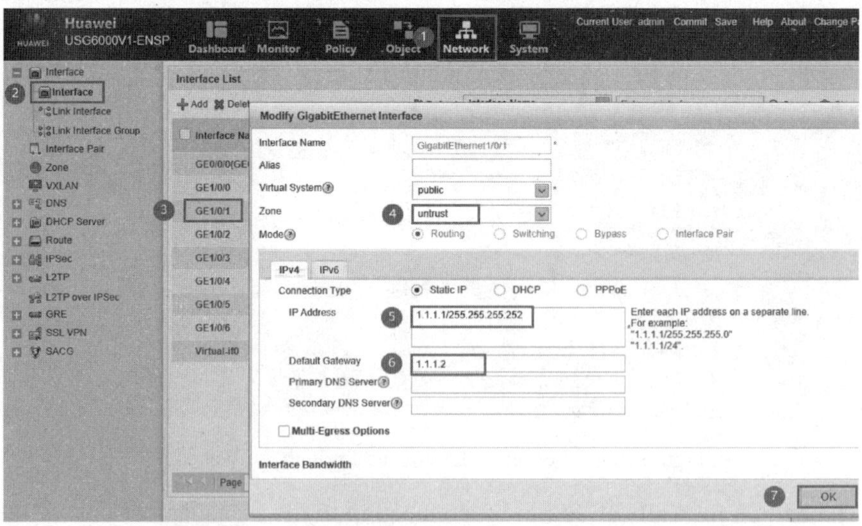

Figure 10.9 Setting the IP address and security area of the external network port of the firewall

Step 2: Configure the IP address and security area of the intranet port in the firewall of the enterprise headquarters. Step: Click network →Interface →GE1/0/2 →Select the security zone (trust) which the internal network port GE1/0/2 belongs to →Enter the IP address of the internal network port →OK, as shown in Figure 10.10.

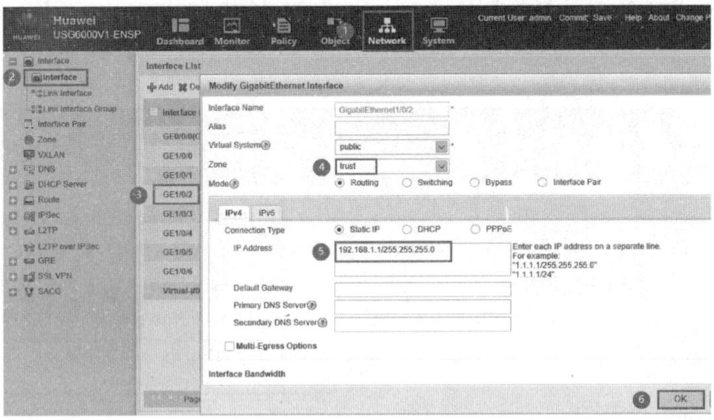

Figure 10.10 Setting the IP address and security area of the network port in the firewall

2. Configure security policies to meet data flow requirements

Step 1: Add a security policy that allows the intranet access to the internet.Step: Click policy → Security policy →Add security policy →Enter a policy name: Allow LAN access to WAN→Select a source zone: trust →Select a destination zone: untrust →Select an action: Permit →OK, as shown in Figure 10.11.

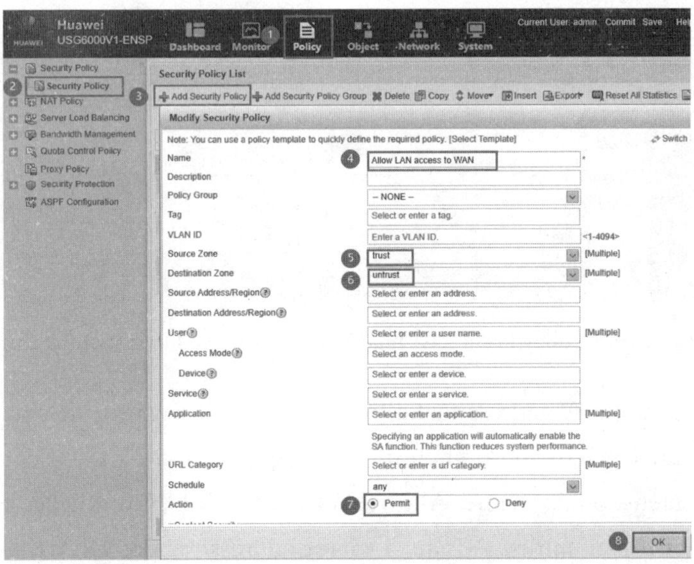

Figure 10.11 Add a security policy that allows intranet access to the extranet

Step 2: Add a security policy that allows branches access to the intranet of the enterprise headquarters. Step: Click policy →Security policy →Add security policy →Enter a policy name: Allow branch intranet access to the corporate headquarters intranet →Select a source zone: untrust →Select a destination zone: trust →Enter a source address/region: 192.168.2.0/24 →Enter a destination address/region: 192.168.1.0/24 →Select an action: Permit →OK, as shown in Figure 10.12.

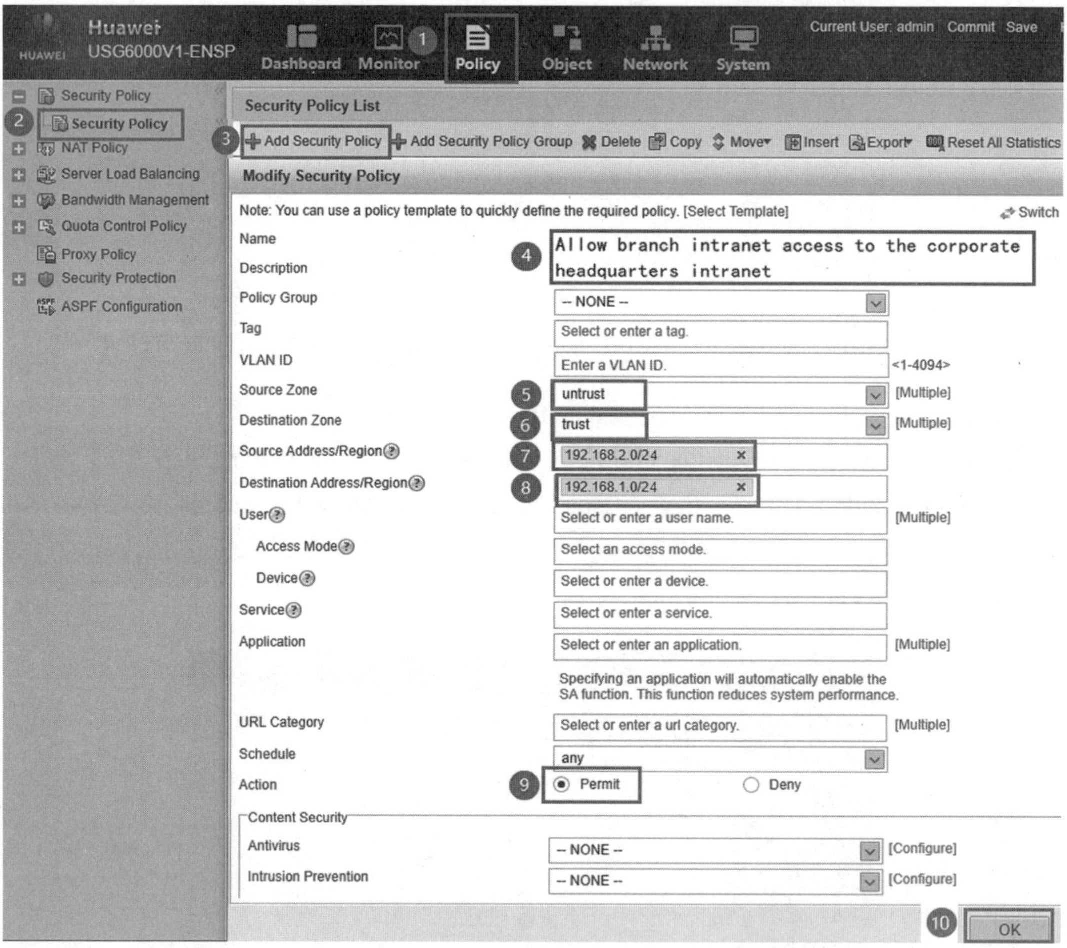

Figure 10.12 Add a security policy that allows branches access to the Intranet of the enterprise headquarters

Step 3: Add the public IP address that allows the firewall FW1 of the enterprise headquarters access to the firewall FW2 of the branch office. Step: Click policy →Security policy →Add security policy →Enter a policy name: Allow FW1 access to FW2 →Select a source zone: local →Select a destination zone: untrust →Enter a source address/region: 1.1.1.1 →Enter a destination address/region: 2.2.2.2 →Select an action: Permit →OK, as shown in Figure 10.13.

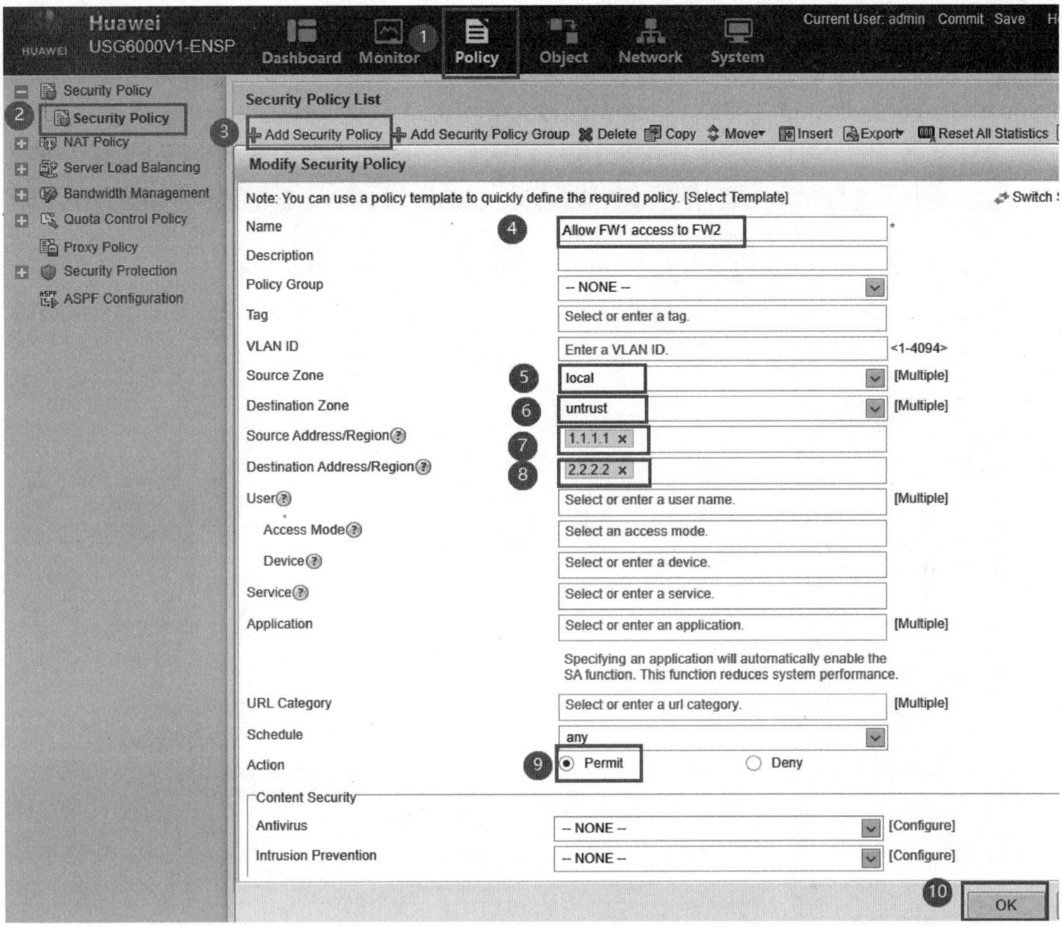

Figure 10.13 Allow FW1 access to FW2

Step 4: Add the public IP address that allows the branch firewall FW2 itself and the enterprise employees on business trips to access the firewall FW1 of the enterprise headquarters. Step: Click policy →Security policy →Add security policy →Enter the policy name: Allow access to FW1 →Select the source zone: untrust →Select the destination zone: local →Enter the source address/region: any (because in addition to the branch firewall that needs to access the public IP address of the headquarters firewall, enterprise employees on business trips also need to access the public IP address of the headquarters firewall, while enterprise employees on business trips have unfixed IP addresses) →Destination address/region input: 1.1.1.1 →Action selection: Permit →OK, as shown in Figure 10.14.

Step 5: Add the permission that employees on business trips can access enterprise intranet resources after being authorized. Step: Click policy →Security policy →Add security policy →Enter the policy name: Allow SSL VPN →Enter the source address/region: 10.1.1.0/24 (the

IP address assigned to the travel employee after the enterprise travel employee passes SSL VPN authentication and authorization) →Enter the destination address/region: 192.168.1.0/24 →destination action: Permit →OK, as shown in Figure 10.15.

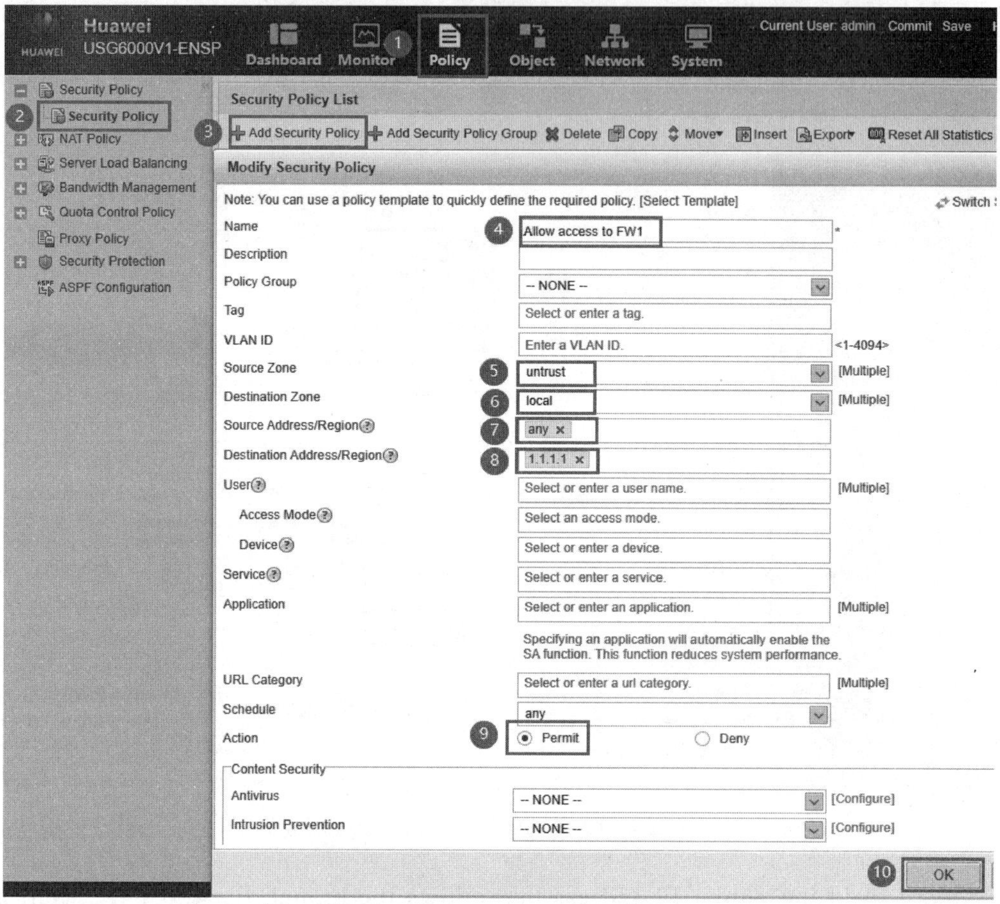

Figure 10.14 Allow extranet access to FW1

3. Configure IP Sec VPN to meet the needs of enterprise headquarters and branches for interconnection

Step: Click network →IPSec →Modify IPSec policy →Select site to site →Enter policy name: IPSec VPN →Local interface selection: GE1/0/1 →Local address input: 1.1.1.1 →Peer Address input: 2.2.2.2 →Select pre-shared key for authentication type →Select IP address for local ID and 1.1.1.1 →Select IP address for peer ID, enter the external network port address 2.2.2.2 of the peer firewall →and then create the data to be encrypted in the data stream to be encrypted →In the new cost secret data stream dialog box, enter the internal network address 192.168.1.0/24 of the enterprise headquarters in the source address or group, and 192.168.2.0/24 of the branch in the destination address or group →finally OK and apply, as shown in Figures 10.16 and 10.17.

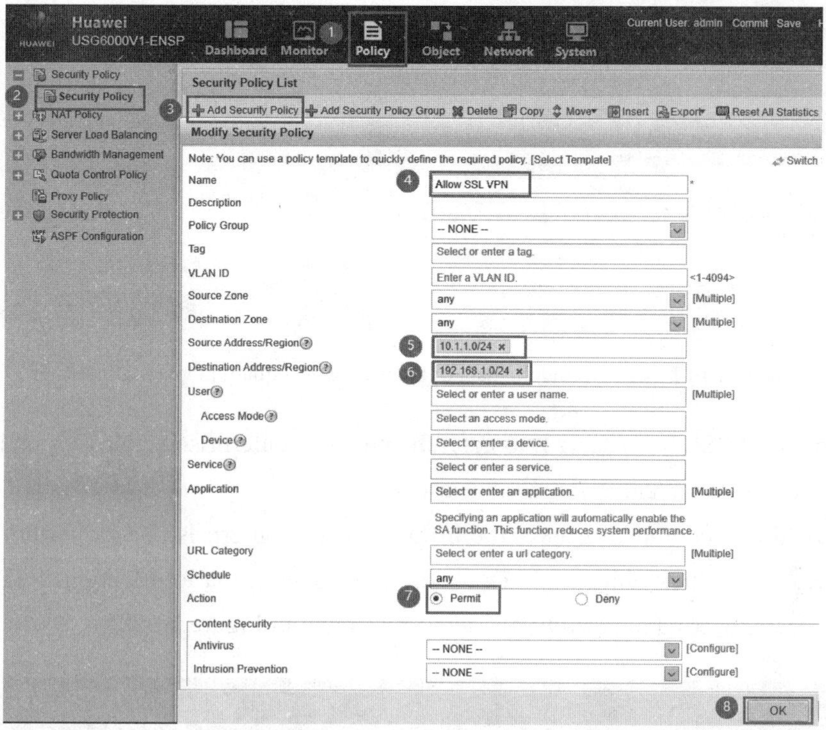

Figure 10.15 Add to allow employees on business trips to access enterprise Intranet resources after authorization

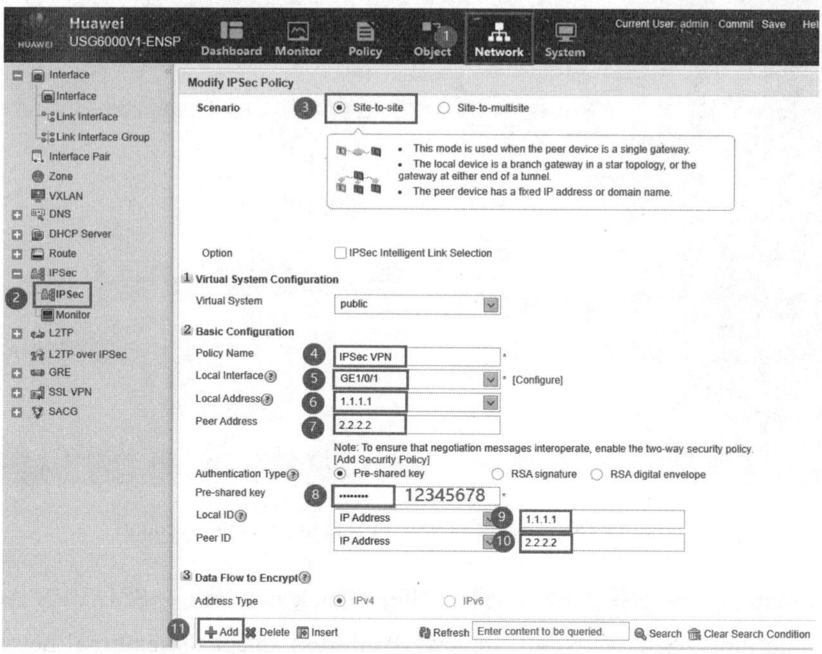

Figure 10.16 Adding a point to point IPSec VPN

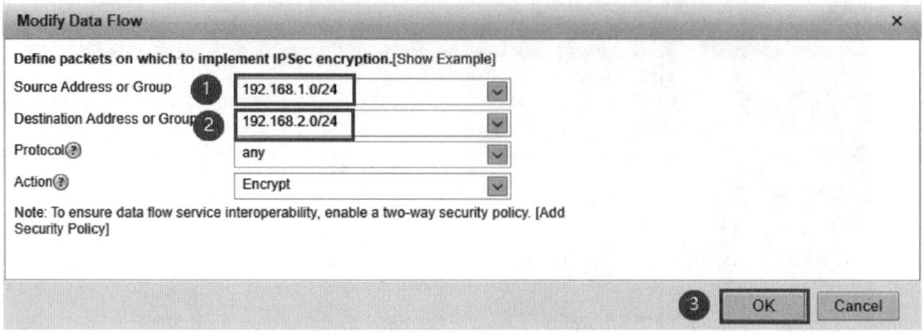

Figure 10.17 Encrypting the data of enterprise headquarters visiting branches

4. Create a new SSL VPN service to meet the needs of enterprise employees on business trips to access the intranet of the enterprise headquarters

Step 1: Add an account to access the intranet of the enterprise headquarters for business trip employees. Step: Click object →User →default →User scenario selection: SSL VPN access →Add user →Enter user name, password →OK. As shown in Figure 10.18.

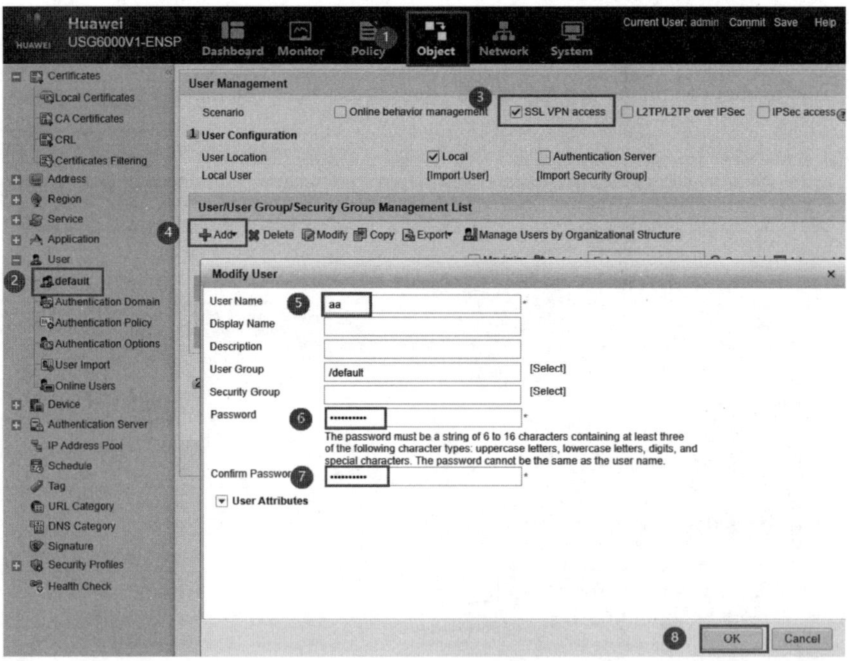

Figure 10.18 Adding an SSL VPN authentication account

Step 2: Create a new SSL VPN service. Step: Click network →SSL VPN Add SSL VPN →Gateway name input: SSL VPN →Gateway IP address select the external network interface GE1/0/1 of the firewall →Next. As shown in Figure 10.19.

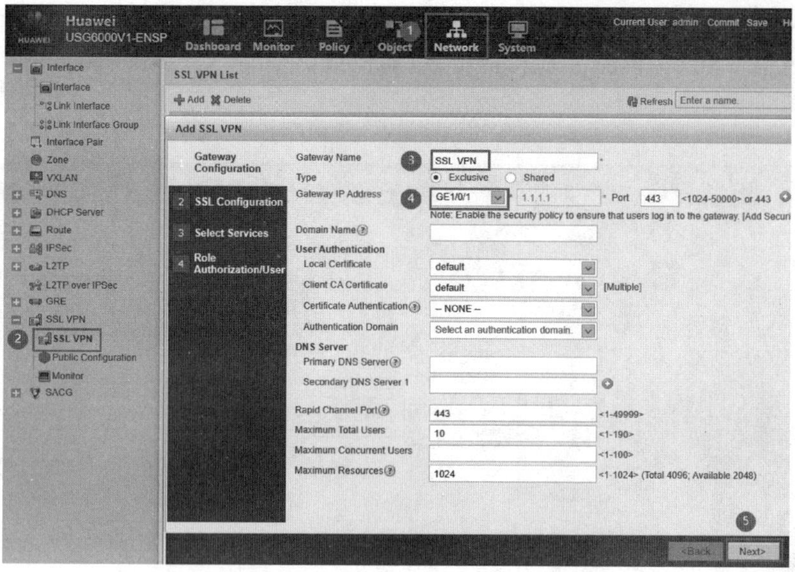

Figure 10.19 Creating an SSL VPN service

Step 3: Enable the network expansion function. If there is an error in creating a new SSL VPN, or the configuration of the SSL VPN needs to be modified later, click SSL VPN → Edit the created SSL VPN name → Enable network extension. Within the range of assignable IP addresses, enter the IP address assigned to employees on business trips: 10.1.1.0/24 → Routing mode: Select the full routing mode. As shown in Figure 10.20.

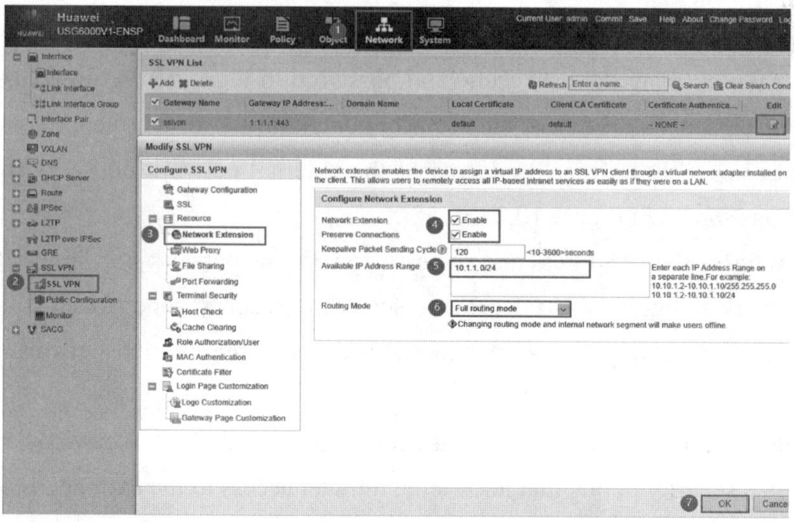

Figure 10.20 Enable network extension for SSL VPN

Step 4: Add an authorized user. In the role authorization/user window, add user roles that can use SSL VPN. Click role authorization/user Add → Enter the role name as a business travel users → Associate with the user just created → Enabled service: Network extension OK. As shown in Figure 10.21.

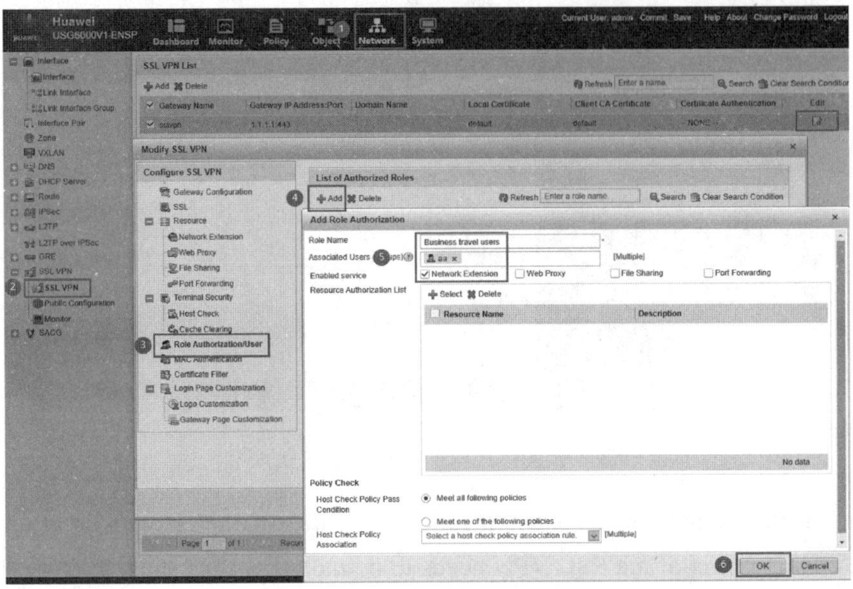

Figure 10.21 Adding an authorized user

10.3.6.4 Branch office firewall configuration

1. Configure each interface address and security zone of branch firewall

Step 1: Configure the IP address and security zone of the external network port of the firewall. Step: After entering the user name and password to log in to the firewall, click Network → Interface → GE1/0 1 → Select the security zone (untrust) which the external network port GE1/0/1 belongs to → Enter the IP address of the external network port → Enter the default gateway → OK, as shown in Figure 10.22.

Step 2: Configure the IP address and security zone of the network port in the branch firewall. Step: Click Network →Interface →GE1/0/2 →Select the security zone (trust) which the internal network port GE1/0/2 belongs to →Enter the IP address of the internal network port →OK, as shown in Figure 10.23.

2. Configure security policies to meet data flow requirements

Step 1: Add a security policy that allows the intranet access to the extranet. Step: Click policy →Security policy →Add security policy →Enter a policy name: Allow LAN access to WAN →Select a source zone: trust →Select a destination zone: untrust →Select an action: Permit →OK, as shown in Figure 10.24.

Step 2: Add a security policy that allows the intranet of the enterprise headquarters access to the intranet of branches. Step: Click policy →Security policy →Add security policy →Enter the policy name: Headquarters access to branch →Select the source zone: untrust →Select the destination zone: trust →Enter the source address/region: 192.168.1.0/24 →Enter the destination address/region: 192.168.2.0/24 →Select the action: Permit →OK, as shown in Figure 10.25.

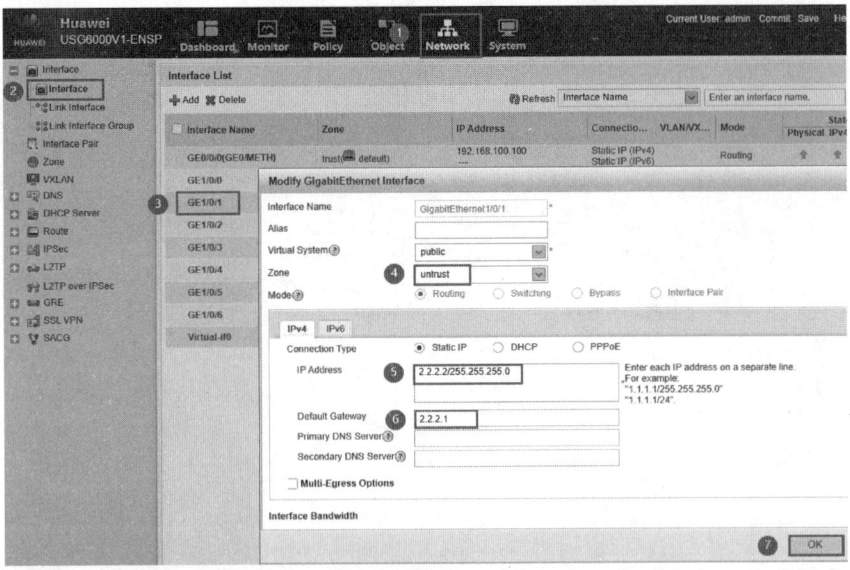

Figure 10.22 Setting the IP address and security area of the external network port of the firewall

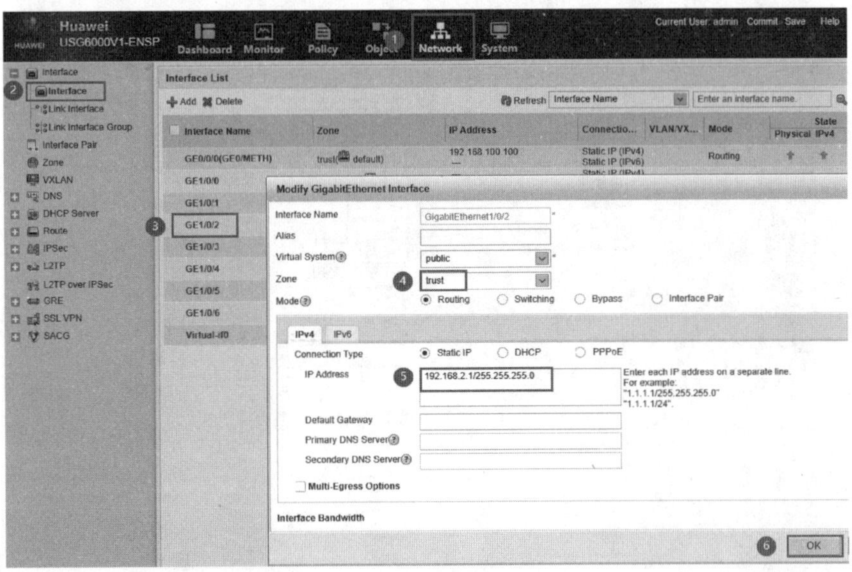

Figure 10.23 Setting the IP address and security area of the internal network port in the firewall

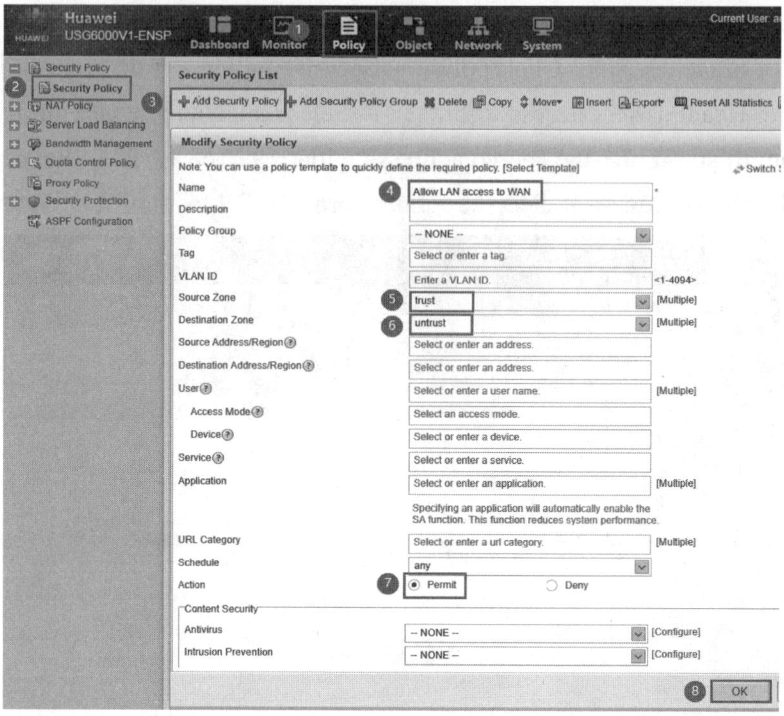

Figure 10.24 Add a security policy that allows Intranet access to the extranet

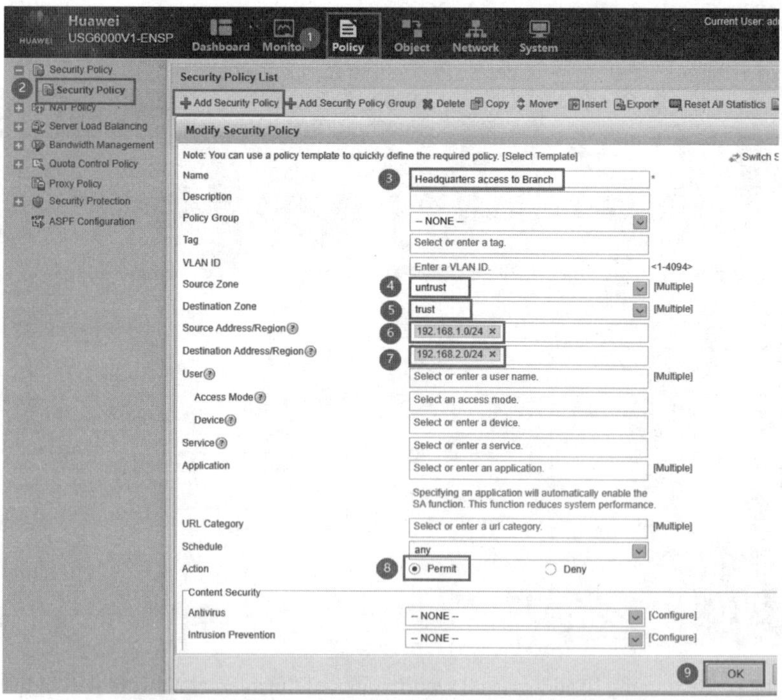

Figure 10.25 Add a security policy that allows enterprise headquarters access to the intranet of branches

Step 3: Add the public IP address that allows the branch firewall FW2 access to the firewall FW1 of the enterprise headquarters. Step: Click policy →Security policy →Add security policy →Enter a policy name: Allow FW2 access to FW1 →Select a source zone: local →Select a destination zone: untrust →Enter a source address/region: 2.2.2.2 →Enter a destination address/region: 1.1.1.1 →Select an action: Permit →OK, as shown in Figure 10.26.

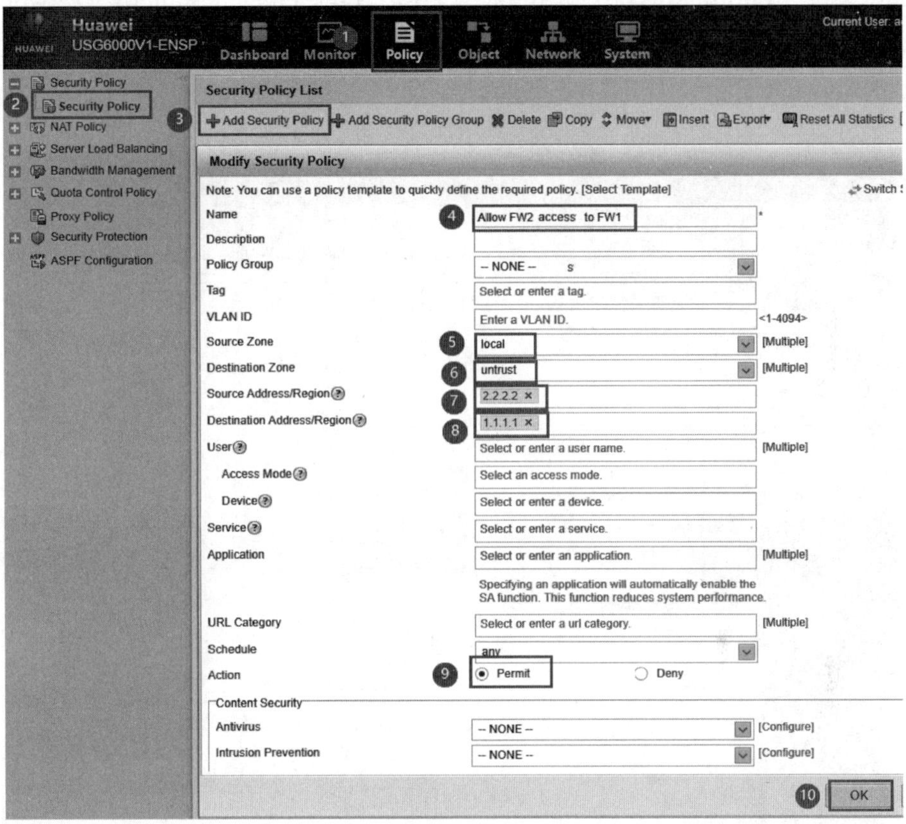

Figure 10.26 Allow FW2 access to FW1

Step 4: Add the public IP address that allows the firewall FW1 of the enterprise headquarters access to the firewall FW2 of the branch office. Step: Click policy →Security policy →Add security policy →Enter a policy name: Allow FW1 access to FW2 →Select a source zone: untrust →Select a destination zone: local →Enter a source address/region: 1.1.1.1 →Enter a destination address/region: 2.2.2.2 →Select an action: Permit →OK, as shown in Figure 10.27.

3. Configure IP sec VPN to meet the needs of enterprise headquarters and branches for interconnection

Step: Click network →IPSec Add IPSec policy →Select site to site →Enter policy name: VPN from branch to headquarters →Local interface selection: GE1/0/1 →Local address input:

2.2.2.2 →Peer address input: 1.1.1.1 →Select the pre-shared key as the authentication type, and enter the pre-shared key 12345678 →Local ID to select the IP address, enter the external network port address 2.2.2.2 of the local firewall →Select the IP address for the peer ID, enter the external network port address 1.1.1.1 of the peer firewall →Create the data to be encrypted in the data stream to be encrypted →Enter the internal network address 192.168.2.0/24 of the branch in the source address or group dialog box of the new cost secret, enter the internal network address 192.168.1.0/24 of the enterprise headquarters in the destination address or group →Finally OK and apply, as shown in Figure 10.28 and Figure 10.29.

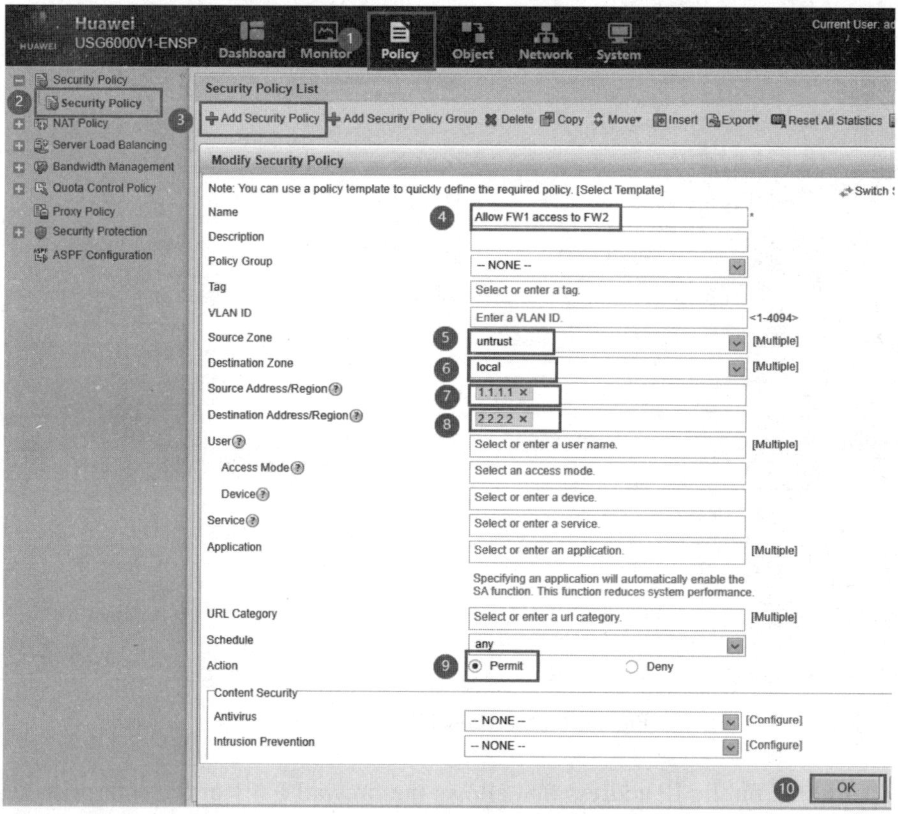

Figure 10.27 Allow FW1 access to FW2

10.3.6.5 Configure IP address for other devices

(1) IP address configuration of Internet router (configuration is required for testing convenience in virtual simulation experiment, but it is not required for Internet area in practical engineering project).

```
<Huawei>system-view
Enter system view, return user view with Ctrl+Z
```

```
[Huawei]interface GigabitEthernet 0/0/0
[Huawei-GigabitEthernet0/0/0]ip address 1.1.1.2 24
[Huawei-GigabitEthernet0/0/0]quit
[Huawei]interface GigabitEthernet 0/0/1
[Huawei-GigabitEthernet0/0/1]interface GigabitEthernet 0/0/1
[Huawei-GigabitEthernet0/0/1]ip address 2.2.2.1 24
[Huawei-GigabitEthernet0/0/1]quit
```

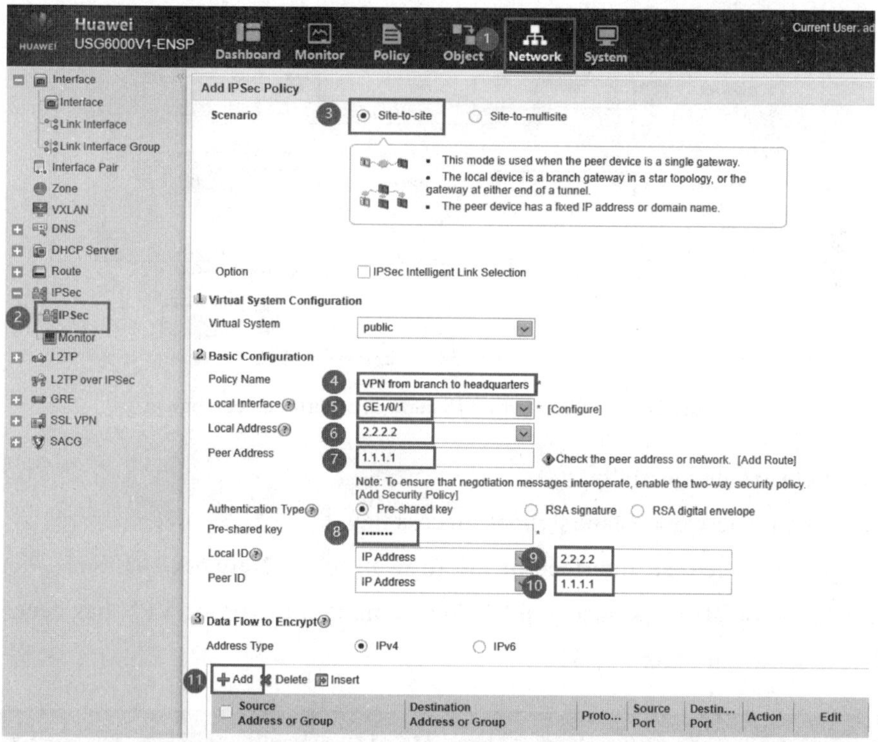

Figure 10.28 Adding a point to point IPSec VPN

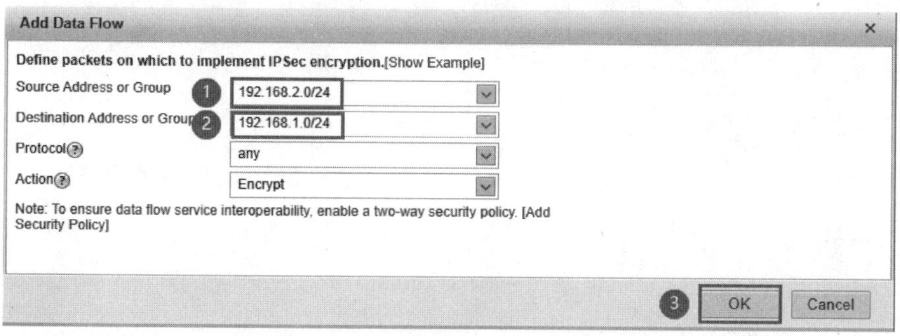

Figure 10.29 Encrypting the data accessed by branches to the enterprise headquarters

(2) Configure IP address of terminal equipment. Step: Double click the terminal device →Enter the local address, subnet mask, gateway and other parameters →Save, as shown in Figure 10.30.

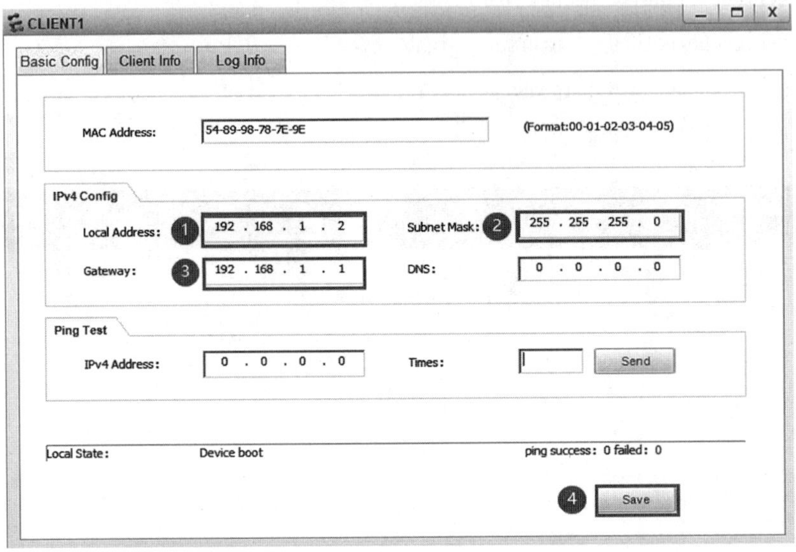

Figure 10.30 Configuring IP address of terminal equipment

10.3.6.6 Test network connectivity

1. Virtual private network establishment success flag

Step: In the management interface of the firewall, click Network →IPSec →and you will see that the negotiation status is Successful:1, indicating that the IPSec VPN has been established successfully. As shown in Figure 10.31.

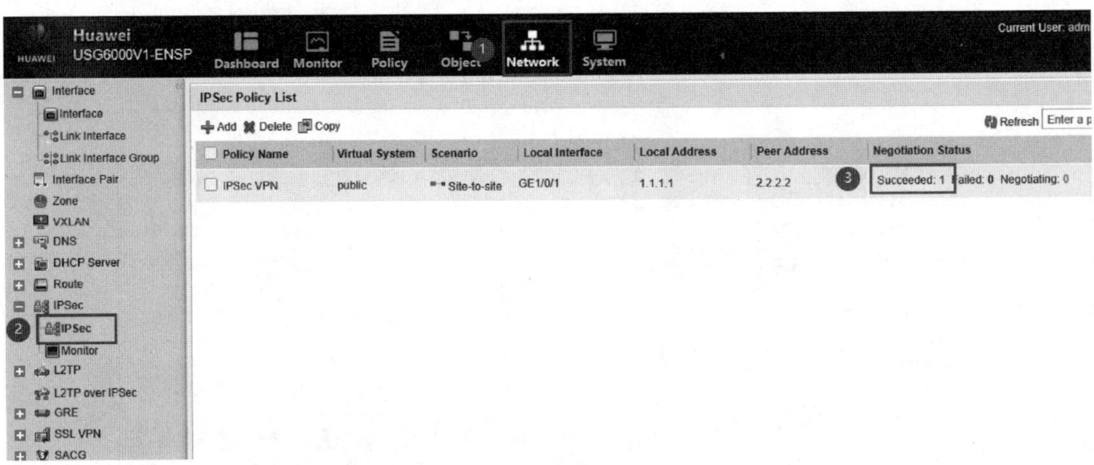

Figure 10.31 IPSec VPN creation success flag

2. Enterprise headquarters intranet access branch intranet test

Step: In the intranet PC of the corporate headquarters, enter the IPV4 address of the extranet server PC →Select 5 times for testing →Send →A successful ping message appears, indicating that the headquarters PC has successfully accessed the branch PC. As shown in Figure 10.32.

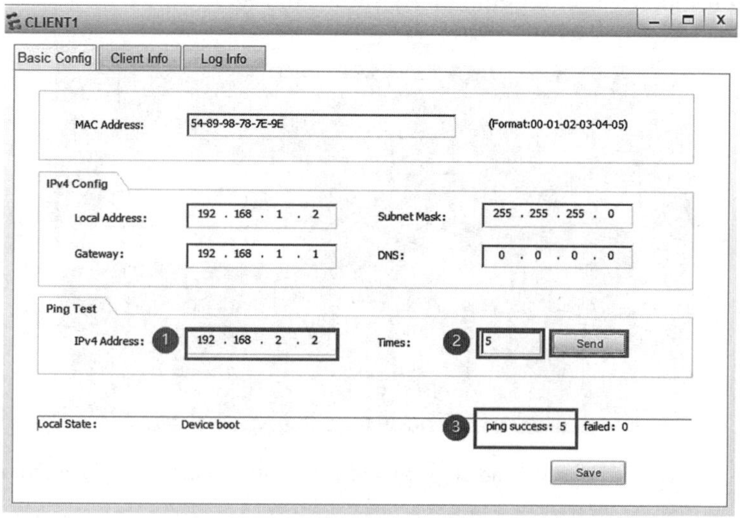

Figure 10.32 Intranet access test of corporate headquarters and branches

3. Branch intranet access enterprise headquarters intranet server test

Step: In the intranet PC of the branch office, enter the IPV4 address of the enterprise headquarters server →Select 5 tests →Send →A successful ping message appears, indicating that the branch office PC has successfully accessed the headquarters server. As shown in Figure 10.33.

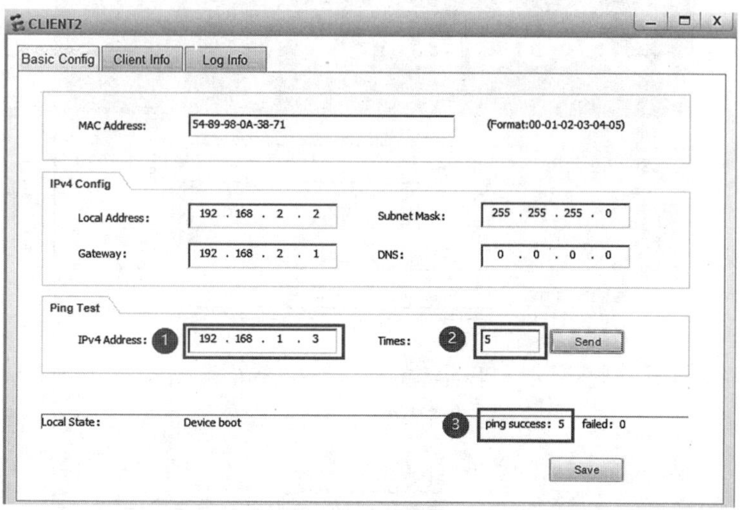

Figure 10.33 Branch intranet access to enterprise headquarters intranet server test

4. Enterprise business trip and employment access to enterprise headquarters intranet server test

Method 1: Enterprise travel workers can enter https://firewall external network port address in the browser to open the SSL login interface, enter the created SSL VPN account and password, and access the intranet resources after logging in. As shown in Figure 10.34.

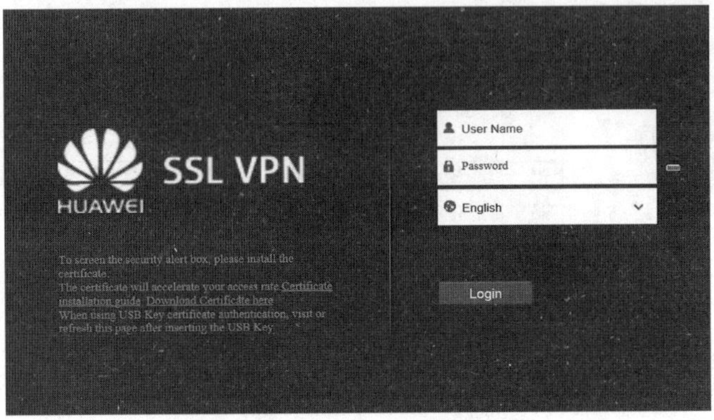

Figure 10.34 Test of enterprise travel and employment accessing intranet server of enterprise headquarters

Method 2: Install the VPN client software SecoClient, open it and enter the external network address of the corporate headquarters firewall, and log in to access the corporate headquarters intranet resources. As shown in Figure 10.35.

Figure 10.35 Log in through VPN client software

10.3.6.7 Save firewall configuration

Step: In the upper right corner of the firewall, click Save →OK to save the firewall configuration information. As shown in Figure 10.36.

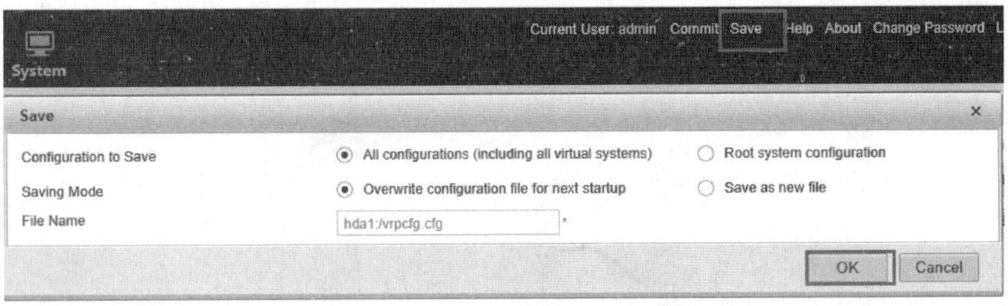

Figure 10.36 Saving firewall configuration information

10.4 Ideological and political integration

The need to work from home made people to think about new ways of working. The combination of online and offline office mode can not only improve work efficiency, but also reduce communication costs, so it is beginning to be widely accepted and applied. The efficient office mode depends on the breadth and depth of China's network construction. Although network technology originated in the United States, in recent years, China has been at the forefront of the world in many fields such as mobile payment, e-commerce, supercomputing, quantum communication, etc., and our network talents must have a firm "soul" in contributing to social progress and economic and social development. Network power has become an important goal of the development of China's network information industry, and is an inevitable choice for realizing the "two centenary goals" and the Chinese dream of the great rejuvenation of the Chinese nation.

10.5 Thinking and training

1. What are the security mechanisms for a virtual private network?

2. Briefly describe the characteristics of a virtual private network.

3. According to the different networking methods, what two types of virtual private networks can be divided into?

4. What is an SSL VPN and how is it different from an IP Sec VPN?

5. If enterprise employees often travel and need secure and reliable access to the intranet, please complete the SSL VPN configuration and write out the configuration steps.